Praise for *Overcoming Katrina*

A goldmine of real stories of Katrina survivors, this is the book everyone has been waiting for since Katrina. In their own words, New Orleans people tell how they survived Katrina, what they lost, and how they are enduring now. Stories of courage, racism, hope, abandonment, neighborhood, and struggle are all in this breathtaking book.

—Bill Quigley, Human Rights Lawyer, New Orleans, Louisiana

There is nothing more compelling than a collection of stories, straight from the hearts of people of all walks of life, to capture the essence and impact of a cataclysmic event like Hurricane Katrina and the immersion of New Orleans. This book is much like a portable museum installation, curated to illustrate the antecedents and the aftermath of a disaster and its impact on the souls of black folk and their community. We will be forever enriched by this splendid compilation with its rich variations on the themes of creativity and connectedness in response to adversity and exclusion, and in the wake of trauma and loss, hope and healing.

—Annelle Primm, Director of Minority and National Affairs for the American Psychiatric Association

These compelling personal narratives convey the rich African American family, community, and institutional life that have created the historic foundation of New Orleans. They are stories of hard work, dignity, survival, courage, and of heroic acts by ordinary people. And they are stories of an incompetent federal government, an indifferent president, and of citizens treated like an enemy in their own country. Read this book.

—Nan Woodruff, author of *American Congo* (Harvard, 2003)

Overcoming Katrina reminds us of a sad chapter in American history, but it also reminds us of the resilience of the human spirit. The authors are to be commended for bringing back the voices that are too easily forgotten about as we rush to deal with other crises. I have also seen both the pathos and the compassion they write about and offer testimony to the importance of their work.

—Ambassador James Joseph, Chairman, Louisiana Disaster Recovery Foundation

When one looks at the broad landscape of American history, it is often said that African Americans were conscripted to be keepers of the Dream, their task to remind a nation of its founding principle: all men are endowed by their creator with certain inalienable rights. *Overcoming Katrina* bears witness to the epic struggles that Black Americans have waged. The backdrop is a natural disaster; the real story is about a community determined to force a nation to keep its promises.

—Gary Puckrein, President of the National Minority Quality Forum

Palgrave Studies in Oral History

Series Editors: Linda Shopes and Bruce M. Stave

Overcoming Katrina

African American Voices from the Crescent City and Beyond

D'Ann R. Penner and
Keith C. Ferdinand

Foreword by Jimmy Carter

palgrave
macmillan

OVERCOMING KATRINA

Photo section credits: Page 1: Family photos courtesy Aline St. Julien, Narvalee Copelin, and Leatrice Joy Reeds Roberts. Page 2: Family photos courtesy Leatrice Joy Reeds Roberts and Irvin Porter. Page 3: Top right photo by Lloyd Dennis Photography; family photos courtesy Leonard Smith, Parnell Herbert, and Cynthia Banks. Page 4: Family photos courtesy Cynthia Banks, Denise Roubion-Johnson, and Keith C. Ferdinand. Page 5: Family photos courtesy Keith C. Ferdinand, Charles Duplessis, and Willie Pitford. Page 6: Family photos courtesy Willie Pitford, Mack Slan, and Rochelle Smith. Page 7: Top right photo by Riza Falk; bottom right photo by Gayle Dolliole; family photos courtesy Eleanor Thornton and Kevin Owens. Page 8: Top right photo by Nesossi Studios, Sugarland, Texas; bottom right photo by Vincent M. Bursey, photographer, Atlanta, Georgia; family photos courtesy Senta Eastern. Page 9: Top right photo ©2007 Jackson Hill, Southern Lights Photography, Inc.; family photos courtesy Yolanda Seals and Leslie Lawrence. Page 10: Bottom left photo by Valerie J. Love; family photos courtesy Velbert Stampley, Robert Willis Sr., and Toussaint Webster.

Cover Image: Aldon Cotton, pastor of Jerusalem Baptist Church, holding a folder with sketches for the new Jerusalem Baptist Church to be rebuilt across the street from the flood-destroyed, historic Jerusalem Baptist Church in Central City, New Orleans. Photo by Jackson Hill.

First published in 2009 by
PALGRAVE MACMILLAN®
in the United States—a division of St. Martin's Press LLC,
175 Fifth Avenue, New York, NY 10010.

Where this book is distributed in the UK, Europe and the rest of the world, this is by Palgrave Macmillan, a division of Macmillan Publishers Limited, registered in England, company number 785998, of Houndmills, Basingstoke, Hampshire RG21 6XS.

Palgrave Macmillan is the global academic imprint of the above companies and has companies and representatives throughout the world.

Palgrave® and Macmillan® are registered trademarks in the United States, the United Kingdom, Europe and other countries.

ISBN 978-0-230-60871-9   ISBN 978-0-230-61961-6 (eBook)
DOI 10.1057/9780230619616

Library of Congress Cataloging-in-Publication Data

Overcoming Katrina: African American voices from the Crescent City and beyond / edited by D'Ann R. Penner and Keith C. Ferdinand.
    p. cm.—(Palgrave studies in oral history)
    Includes bibliographical references and index.

    1. African Americans—Louisiana—New Orleans—Interviews. 2. Hurricane Katrina, 2005—Social aspects—Louisiana—New Orleans—Anecdotes. 3. African Americans—Louisiana—New Orleans—Social conditions—21st century—Anecdotes. 4. Disaster victims—Louisiana—New Orleans—Interviews. 5. New Orleans (La.)—Biography—Anecdotes. 6. New Orleans (La.)—Social conditions—21st century—Anecdotes. 7. Oral history—Louisiana—New Orleans. 8. Interviews—Louisiana—New Orleans. I. Penner, D'Ann R. II. Ferdinand, Keith C.

F379.N59N4 2009
976.3'35064092396073—dc22   2008035089

A catalogue record of the book is available from the British Library.

Design by Newgen Imaging Systems (P) Ltd., Chennai, India.

First edition: March 2009

10 9 8 7 6 5 4 3 2 1

At my father's funeral…I recalled a family trip out to California to see my Uncle Pat…. This was in the 50s, still very much a time of hard-line racial segregation. A White man was hitch hiking…. We stopped…and gave the man a ride to the next town. After the man got out, I asked my father why he had given the man a ride. He gave me a simple explanation, the simplicity of which still reverberates inside of me: "It was too late at night for anybody to be out there alone."

—Kalamu ya Salaam, *What Is Life?*
Reclaiming the Black Blues Self

Before this, everybody flocked to New Orleans from all over the world for all the games, Jazz Fest, and Mardi Gras. Millions of people. We embrace everybody with nothing but love. If we're selling food, but you don't have no money, we're still going to feed you, because that's who we are. But when we were in trouble, it was like we was the worst people in the world. They turned their backs in our time of need, even in that convention center where we done served thousands and thousands. I can't get the just of that.

—Eleanor Thornton, Waitress from Algiers

When all of this stuff happened, I was sitting in front of the television at the hotel in Greenville, Mississippi, and they're talking about we won't be able to get back in New Orleans for three years…. I was going back to New Orleans if I was the only preacher back. I'll pastor anything up in there. It was settled with me before I left.

—Aldon Cotton, Senior Pastor of Jerusalem
Baptist Church in Central City

Contents

SECTION ONE   RETIREES

SECTION TWO   AT THE HEIGHT OF THEIR CAREERS

SECTION THREE   THIRTY SOMETHING

SECTION FOUR   COMING OF AGE

Foreword

At the time of this writing, the 2007 Christmas season, there are at least twelve thousand homeless people in New Orleans. A recent study conducted by Columbia University's Mailman School of Public Health concludes that the traumatic aftermath of Katrina continues to plague even the youngest generation of Mississippians and Louisianans. By reaching back into pre-Katrina memories, the narrators of *Overcoming Katrina: African American Voices from the Crescent City and Beyond* allow us to understand the richness of pre-Katrina community life and the nonmaterial sources of trauma. Stories of survivors of Hurricane Katrina go beyond the mere retelling of the horrors of their tragedy and its impact on New Orleans and the Gulf Coast. They specifically speak to the concerns, dreams, hopes, and unfulfilled promises experienced by the African American community.

Three things moved me about the displaced Black New Orleanians whose stories comprise this book: their work ethic, faith, and patriotism. Having grown up in Plains, Georgia during the Depression era, I know how hard my black neighbors worked and struggled to make ends meet on a decidedly uneven playing field. These stories represent generations of African Americans, many of whom were born on or near plantations throughout Alabama, Mississippi, and Louisiana before a family member escaped to New Orleans in search of employment. As I have witnessed firsthand, Hurricane Katrina and its aftermath demolished the fruits of the labors of so many first-generation homeowners.

In the wake of intense tribulations unleashed by Katrina, spiritual faith may be the greatest unifying factor shared by these diverse voices. Several credit God with giving them the strength to endure inhumane circumstances and losses reminiscent of Job's trials. Faith generated the kind of love that reached out a compassionate hand, not only to other people of color but across the color line as well. Moreover, testimonies abound of a faith that renews the strength and clarifies the vision of the storytellers, thereby enabling them to rebuild their scattered communities after the storm.

Finally, I was struck by the intensity of pre-Katrina patriotism even in narratives equally suffused with an awareness of race-based discrimination. A striking number of these storytellers and their relatives risked death while serving our country. Even the younger generations showed pride in American displays of military prowess, were impressed by the speed and precision of American humanitarian aid after the 2004 tsunami, and wept with the survivors of 9/11. As a reader, you may be startled by the venom behind the assertions of betrayal as American citizens in the aftermath of Katrina. I interpret it as inversely proportional to the strength of their loyalty to the United States and identification as American citizens before the storm.

These stories provide an opportunity for Americans to reflect on how we want to be viewed internationally for our treatment of the most vulnerable in our midst. For all of the allegations of indifference and cruelty, I was heartened to read of the random acts of kindness by strangers and churches to displaced New Orleanians during their first month away from home. The power of a stranger's thoughtfulness after a week of indignities is credited by several narrators as an important element in the healing process. We, as compassionate Americans with an indomitable spirit, legendary resourcefulness, and zeal for equality, have a glorious opportunity to establish ourselves as leaders in the struggle for a more just world. Collectively, we can come together as a nation to rebuild not only homes, but trust and lives.

Jimmy Carter
Former U.S. President

Series Editors' Foreword

Like many readers of D'Ann R. Penner and Keith C. Ferdinand's *Overcoming Katrina: African American Voices from the Crescent City and Beyond*, we were riveted to our television sets during the last days of August 2005, as Hurricane Katrina tore through New Orleans and then, when the levees were unable to contain rising waters, as floods overtook the city. Hardly naïve about race and power in the United States, we—again like many readers—were nonetheless outraged at the images that repeatedly flashed before us: of mostly black New Orleanians, too stubborn or too poor to adhere to Mayor Ray Nagin's evacuation order, crowded for too many days in deplorable conditions without adequate food and water, implicated as criminals for taking necessary supplies from nearby businesses, victims of at best an emergency management system unable to cope with a disaster of such enormous proportions, at worst, of callous, racially inflected neglect.

Penner and Ferdinand were similarly outraged, as much by the ways the mainstream media presented what was happening to black New Orleanians as by actual events; and subsequently, by the demoralizing, nearly incomprehensible delays of federal efforts at rebuilding a city struggling to recover. And so they turned to oral history to create an alternative narrative, the results of which you hold in your hand: twenty-seven beautifully edited interviews, selected from a total of 275 conducted throughout the south over a three-year period between September 2005 and August 2008. We read of victims, certainly, but also of full and complicated human beings, as Penner and Ferdinand wisely chose to place "the Katrina story" within the context of a life story. We read of elderly men and women, with deep roots in segregated New Orleans, and of their children, the first generation to break free of the worst strictures of Jim Crow, who over generations built successful lives and dense communities, who lost everything, and who still retain the will—and the faith—to rebuild new lives, some in New Orleans, some elsewhere. We read of middle-class and even prosperous blacks, who had the familial and financial resources to get out before Katrina hit and who helped innumerable others; and of working-class blacks, including many women, with enough grit, smarts, and nerve to rise in the pre-Katrina world and who are again putting their lives together. We read too of how the most vulnerable—often elderly, sick, poor—lost the limited world in which they were nonetheless able to negotiate a satisfying life and who are now adrift, deeply disoriented in strange settings. And we read of young folks—and some older ones too—who find in the aftermath of Katrina an opportunity for a new start, new adventures, and new lives. We read also of real compassion extended to displaced New Orleanians by numerous volunteers and nonprofit groups, but not—it must be said—by the federal government, which rightly stands accused of unaccountable delays, unnecessary obfuscation, and intimidating amounts of paperwork.

One might read these narratives as exemplars of the uplifting "we survived" trope present in so many oral history interviews. Certainly these narrators did survive, and their survival undercuts shallow notions of victimhood. But their accounts, elicited by a skilled and passionate interviewer, give us something much more: insight into the profoundly painful—one might say tragic—losses suffered by black New Orleanians; an indictment of the city, state, and national governments charged with keeping all citizens safe; and, as Penner and Ferdinand have intended, an antidote to thin media stereotypes.

In doing so, they demonstrate oral history at its best: its capacity to offer thoughtful, dense, humane stories instead of soundbites, to render narrators as complex individuals, to represent a diversity of experiences. We are pleased to include *Overcoming Katrina* as the sixteenth book in Palgrave Macmillan's *Studies in Oral History* series, which is designed to bring oral history interviews out of the archives and into the hands of students, educators, scholars, and the reading public. Volumes in the series are deeply grounded in interviews and also present those interviews in ways that aid readers to more fully appreciate their historical significance and cultural meaning.

Linda Shopes
Carlisle, Pennsylvania

Bruce M. Stave
University of Connecticut

Preface

At the center of *Overcoming Katrina: African American Voices from the Crescent City and Beyond* are eyewitness accounts of abandonment and evacuation, heroism and terror, prejudice and generosity, and displacement and rebuilding, in the lengthy aftermath of Katrina. Narrators begin by talking about their pre-Katrina lives, providing some of the context missing from most of the Katrina stories currently in circulation. These details help to explain several aspects of the Katrina story, including why people were unable or unwilling to be evacuated from New Orleans before the storm, why the transition has been difficult for so many, and why many continue to miss New Orleans despite having successfully settled into new cities. The narratives conclude with descriptions of the new tactics used for overcoming the latest set of obstacles faced by New Orleanians of color both at home and in new cities.

The structure of the book reflects our desire to address two audiences: the general reader and the scholar. We have attempted to keep scholarly analysis, documentation, and the use of statistics confined to the introduction and the conclusion. Endnotes accompanying the narratives have been kept to a bare minimum and are used to direct the reader to additional sources on important points. Some readers may want to skip directly to the narratives without reading the introduction. Each chapter stands on its own and does not have to be placed in a larger context to have meaning.

The rich history of New Orleans is reflected in the voices of the narrators. No attempt has been made to phonetically reproduce dialects. Their spoken words, while sometimes considered incorrect if measured against the rules of traditional English grammar, uniquely express the concerns and hopes of members of this community. On occasion the narrators themselves standardized English or grammar. They had the last say on the style and content of their own chapters. They chose the pictures of themselves and one other person or place to include in this book. (Harold Toussaint exercised his right not to have his picture included in this book.) Some of the photographs were damaged by Katrina's floodwater or mold, making them all the more valuable for having survived at all. The narrations vary in length according to the talkativeness of the narrators and their locations during Katrina. We have chosen to group the narratives by generation, but we do not adhere to a strict age-based definition of a generation. Our generational categories depend on the individual's position in her work or life cycle.

D'Ann R. Penner transcribed the oral interviews as literally as possible. In the editing process, every attempt was made to retain the rhythm and flow of the conversations. Repetitive material and unnecessary words were deleted to maintain focus. Some chapters had the order of sentences modestly rearranged to cluster similar themes. Although taped interviews formed the basis for the individual chapters, the narrators and authors worked very closely to produce a final manuscript reflecting their most considered opinions. Notes from all follow-up conversations are in the possession of D'Ann R. Penner, and they are abbreviated in the footnotes as "PFN" (Penner field notes) with the accompanying dates noted. Following the example of David Farber (*Chicago '68*) and Richard Price (*Alabi's World*), the authors' words are distinguished from the narrators' words by the use of italics in the case of the former.

The authors have attempted to situate the experiences of the narrators in the context of the history of Katrina, New Orleans, and race relations in the South. There are many possible interpretations of each account. In the conclusion and epilogue, we offer some reflections based not only on these twenty-seven narratives but also on work the authors have previously done. Keith C. Ferdinand draws conclusions on the basis of his personal and professional relationships with thousands of displaced evacuees before and after the storm. D'Ann R. Penner interprets from the perspective of having conducted 275 (and counting) interviews since September 2005. We encourage you to engage in your own dialogues with the narrators, draw your own conclusions, and imagine your own scenarios for the future of New Orleans.

Above all, we wish to thank the narrators, who invested heavily in this project by sharing their difficult experiences. A number of people cared deeply enough about making these voices heard. Without a variety of supports, including creative, intellectual, monetary, moral, and tactical, from the following people, this book would still be a work-in-progress:

Barbara Andrews, Concetta Augustine, Elazar Barkan, Peter Bearman, Barbara Bekis, Jim Blythe, Gerald Bond, Karen Bradley, Thomas and Kathy Brady, Bob Brown, Shannon Burrell, Sadie Campbell, Mary Marshall Clark, Cheryl Cornish, Tim Cox, Ann Cvetkovich, Lloyd Dennis, Sylvia Denson, T.J. Desch-Obi, Guiomar Duenas-Vargas, Rachel Efron, Reggie Ellis, Ken Ferdinand, Jim Fickle, Susan Fountain, Bob Frankle, Kim Friedlander, Mike Fykes, John Hope Franklin, Dianna Freelon-Foster, Edward Friedman, Randy Gambel, Risa Gerson, Susan Glisson, Ken Goings, Lizabeth Grefrath, Ron Grele, Richard Greenberg, Margaret Greene, Vivian Gunn-Morris, Lee Hampton, Jean Handley, Helen Harrison, Gillian Hart, Flordia Henderson, Jane Henrici, Tamin Hill, Ivan and Bethany Illidge, Doug Imig, Ashira Israel, Penni Istre, Elaine Jackson, Nick James, Ardella Jeffries, Mugambi Jouet, Angela Keiser, Masoom Khan, Joe Kirchoff, Abraham and Reva Kriegel, Christine Kulke, Bill Lawson, Arline Lederman, Jonathan Lightfoot, Alan Litwa, Valerie Love, Paul Martin, Matt May, Matthew Mazur, Lucia McBee, Daphene McFerren, Alice Miller, Ada Mui, Marc Morial, Julia Morial, Gail and J. Herbert Nelson II, Sandile Ngcobo, Roxsana Patel, Joyce Parker, Sunsha Parker, Charles Payne, Kimberly Payne, Cynthia Pelak, Terrell Perry, Alessandro Portelli, Bill Quigley, Nelson Reynoso, Clyde Robertson, Rejane Roe, Mtumishi St. Julien, Asante Salaam, Susan Schwalb, Rachel Shankman, Desiree Shelling, Doris Smith, Kimberly Smith, Sara Smith, Brenda Square, Velbert Stampley, Amy Starcheski, Renae Stephens, Julia Taravella, Marilyn Taylor, Heidi van Es, Daniel Warshawsky, Dwight Webster, Sheila White, Charles Williams, Carolyn Wilson, Lakeitha Wilson, Mary Wilson, Abe Louise Young, Darius Young, Djamila Zaida, and Millie Zinck.

Special mention is reserved for Reverend Ben Hooks, former Executive Director of the NAACP, whose example in the anti-apartheid struggle for South Africa played an important role in this project's inception. The Benjamin Hooks Institute for Social Change at the University of Memphis subsidized some of the travel expenses incurred during the first year of research. The Amistad Research Center (ARC) at Tulane University and the Center for the Study of Human Rights at Columbia University both provided space for reflection and research.

The editors of the *Palgrave Studies in Oral History* series, Linda Shopes and Bruce Stave, worked quickly to bring these narratives into the public discourse as efficiently as possible and made numerous helpful suggestions. The anonymous reviewer's incisive critique improved the book's conceptual framework and widened the diversity of the chosen narrators. Chris Chappell, associate editor at Palgrave Macmillan, has been a peerless advisor.

All of the authors' royalties from the sale of this book have been signed over to Project HOPE (Health Outreach Prevention and Empowerment), a nonprofit under the auspices of the Association of Black Cardiologists (ABC), for disbursement to Katrina survivors still working to move beyond the long shadow of the hurricane. Keith C. Ferdinand's sections of this book are dedicated to the memory of Inola Copelin Ferdinand and Vallery Ferdinand Jr., and to his wife, his children, his aunt Narvalee Copelin, and his brothers. D'Ann R. Penner dedicates her sections of the book to: Tom and Kathy Brady, Bob Brown, Bob Frankle, Abe and Reva Kriegel, Anne Rainer, and to the memory of Reggie Zelnik. The book as a whole is dedicated to the people of New Orleans, victims and survivors of Hurricane Katrina's landfall and seemingly unending aftermath.

Introduction

Keith Copelin Ferdinand was raised in a corner of the Lower Ninth Ward in a small home, built by his father, Vallery Ferdinand Jr., a veteran of two wars. On September 9, 1965, Hurricane Betsy's ten-foot storm surge ruptured the levees near the Industrial Canal protecting the Lower Ninth Ward and flooded the Ferdinands, house to the roof level. Dozens of people, including Keith's paternal grandfather, were killed.[1] Two weeks later, Keith's mother, Inola Copelin Ferdinand, composed a handwritten, sixteen-page letter to her sister detailing the family's harrowing rooftop experiences and calmly tallying their material losses. "*We shall overcome* has more depth now than ever before," she concluded.[2] Her optimistic tone never faltered, despite a belief she shared with the overwhelming majority of her black Lower Ninth Ward neighbors that the levee had been blown up to protect downtown New Orleans and the French Quarter (a predominantly white neighborhood), as was done in 1927. This belief was reinforced by the observation of public buses sent to evacuate the white families living on Tennessee Street hours before the hurricane swept through.[3] An official investigation failed to uncover evidence that the levees had been blown in 1965.[4] The New Orleans Urban League estimated Hurricane Betsy touched more than forty thousand African Americans living in the Ninth Ward and wrought 1.6 billion dollars in damages.[5]

Forty years later, her youngest son Keith became a victim of the floodwaters that inundated 80 percent of New Orleans, a crescent-shaped bend in the Mississippi River, in the aftermath of Hurricane Katrina. Keith lost his cardiology clinic with its cutting-edge medical equipment, his medical practice with 7,600 patients, and his elegant home in New Orleans East, all achieved after a lifetime of eighty-hour work weeks overcoming the barriers faced by the generation of young blacks born in the segregated South of the 1950s. Post-Katrina, Keith and family fled from Jackson, Mississippi, where they observed the storm, to Atlanta, where he founded Project HOPE. *Overcoming Katrina: African American Voices from the Crescent City and Beyond* chronicles this and other histories of overcoming from the era of segregation to the present day in the words of black New Orleanians, twenty-six residents and one expatriate.

Hurricane Katrina: Landfall and Aftermath

At 10:00 a.m. on August 28, 2005, New Orleans Mayor Ray Nagin declared a mandatory evacuation for Orleans Parish.[6] Nearly 20 percent of Orleans Parish residents (100,000 of 484,000) decided to "ride out the storm," as locals refer to braving hurricanes. Before nightfall on day one of the storm, approximately nine thousand people took shelter in the Louisiana Superdome, an enormous multipurpose facility with a fixed-dome structure designed to house sporting events and exhibitions. It sits on thirteen acres in the heart of the Central Business District (CBD). By daybreak, another eleven thousand people would join them. Most remaining New Orleanians, including ten of our narrators, waited for the storm at home or gathered together with family members at the home of the person with the largest, most structurally solid dwelling.

In the Superdome, called the "refuge of last resort," the electricity failed before Hurricane Katrina first made landfall in Louisiana. Katrina moved onshore at 6:10 a.m. on August 29, 2005, near Buras-Triumph, as a Category Four hurricane. Television viewers witnessed the winds of Katrina ripping at the metal shell of the Louisiana Superdome and the façade of the Hyatt Hotel. Several of the major network newscasters were sequestered in the downtown area. They reported to the nation at 9:00 a.m. that the city appeared to suffer only a glancing blow.

As devastating as the 120 mph winds were, most of the catastrophic impact resulted from the rupture of ill-designed and poorly maintained canals that laced through the city and became a conduit for floodwaters. At 6:50 a.m. the Industrial Canal that separates the Lower Ninth Ward from the Upper Ninth Ward began to crack from the pressure of a sixteen-foot storm surge. Less than an hour later nine hundred feet of floodwall collapsed. Eight feet of water overwhelmed the Lower Nine and the Parish more than an hour before the newscasters' prognostication. The breach of the 17th Street Canal contributed to the flooding of Gentilly, an integrated neighborhood on the east side of City Park, and Lakeview, an upper-middle class, predominantly white neighborhood on the west side of City Park. Before the day was over the London Avenue Canal was breached in two places, submerging the Seventh Ward, the Eighth Ward, and the Upper Ninth Ward. New Orleans East became a reservoir of storm surge waters that followed the pathway of the Intracoastal Waterway known as MR GO (Mississippi River-Gulf Outlet).

Almost all of the narrators in this book were more traumatized by the events that transpired in the immediate aftermath of the hurricane than by the landfall, beginning with the flooding and ending with mass evacuations to forty-six different states. We distinguish this from the natural disaster by some variant of the phrase "the aftermath of Katrina."

By day three the Superdome was rumored to house thirty thousand newly homeless people. On day four, August 31, 2005, Mayor Nagin, who rose from the Seventh Ward to become a millionaire, changed the mission of the New Orleans Police Department (NOPD) from "search and rescue" to "law and order." The National Guard, primarily tasked with protecting private property, also began to arrive on day four. Blackhawk and Chinook helicopters filled the sky. The Louis Armstrong New Orleans International Airport (Armstrong Airport) accepted its first group of survivors. Dazed people, including Eleanor Thornton and Rochelle Smith, began filling up the Ernest N. Morial Convention Center, a sprawling riverfront exhibition hall.

It was not until the evening of day five, Thursday, September 1 that Michael D. Brown, director of FEMA (Federal Emergency Management Agency) acknowledged that thousands were stranded without resources in and around the ten-and-one-half block convention center. In response, a military helicopter dropped the first batch of food and water: enough MREs ("Meal, Ready to Eat") for approximately twenty-five people. The Gretna Police Department sealed off the Mississippi River Bridge to foot traffic, even though the first exit led to Algiers, which had running water, empty schools and parks, and hospitable New Orleanians.[7] Rubber bullets were fired by police officers at foot passengers trying to walk to food, water, and medical assistance. Attempts to evacuate the most medically critical from the Superdome were complicated by shots fired at a Chinook helicopter attempting to assist with the evacuation. A National Guardsman was wounded inside the Dome. By day five, an estimated three thousand people were crowded under the I-10 overpass on Causeway Boulevard in Metairie, Louisiana, which became a makeshift staging ground and campsite. Every day additional people joined the throngs in front of the Superdome and convention center waiting for sustenance and transportation. Lieutenant General Russel Honore, a Louisiana native and Joint Task Force commander, entered the city and attempted to set a different tone: "Imagine being rescued and having a fellow American point a gun at you," he instructs. "These are Americans. This is not Iraq." He then attempted to reorient the troops to begin viewing their work within the context of a humanitarian-relief mission. In the meantime, Governor Kathleen Blanco took to the airwaves to assure Louisianans that 250 military police would soon be assisting in New Orleans with rescues and "urban warfare."

At noon on September 2, a thousand soldiers and police in full battle gear stormed the convention center.[8] Not meeting the anticipated resistance from the men, women, and children huddled inside, they assumed their positions throughout the 1.1 million square foot center within twenty minutes, and began, on day six, to distribute food and water to the storm victims. The second military airdrops of food and water were made in downtown New Orleans. Fifty trucks with National Guard military police arrived in New Orleans. Governor Blanco continued to contradict General Honore's new tone by warning that this installment of troops had recently returned from active duty in Iraq: not only were they "locked and loaded," she hinted, "they also know how to shoot to kill." CNN reported that hospitals across the city, most of which lost power on day two of the storm, remained filled with patients and personnel, including Cynthia Banks, Denise Johnson, and Jermol Stinson.

On Saturday, September 3, day seven, federal troops received authorization to shoot-to-kill at their discretion. At Armstrong Airport the crowd had grown to four thousand people waiting to be evacuated. By Sunday, two hundred thousand displaced people had already arrived in Texas, the state that ultimately attracted the largest number of Katrina survivors. In addition, thirty thousand people had been evacuated from the Superdome, twenty thousand from the convention center, four thousand from the Armstrong Airport, as well as all of the patients in University, Methodist, and Kindred Hospitals, and most of the people on Causeway Boulevard. *Overcoming*'s narrators who chose to stay in New Orleans were finally released from the post-Katrina drama in the Crescent City.

Eyewitnesses

"Knowledge of suffering cannot be conveyed in pure facts and figures, reportings that objectify the suffering of countless persons. The horror of suffering," writes Rebecca Chopp, the dean of the Yale University Divinity School, "is not only its immensity but the faces of the anonymous victims who have little voice, let alone rights, in history."[9] This book tells the stories of the people behind the numbers in their own words. One strength of the oral histories is the chance they offer to learn the firsthand experiences of the survivors: to be on the roof in Eastern New Orleans for three days with Leonard Smith, a retired career Navy veteran; or at a Superdome loading dock with

Kevin Owens, a maintenance man, after he was separated from his wife, Elise, after complying with military evacuation orders; or with deacon Harold Toussaint from the Ninth Ward, as he sought to negotiate expedited evacuation from the Armstrong Airport for the forty-one elderly people, predominantly white, in his care.

This book's twenty-seven narrators are very diverse. Politically, they range from Mack Slan, a conservative businessman who disparages the younger generation for not sharing his ability to make "good, rational decisions," to Kalamu ya Salaam, who was followed by the New Orleans Police Department for several years as a militant defender of Black Power in the late 1960s and '70s. Eight of the narrators, including one woman (Narvalee Copelin), served in or supported the armed forces. Seven narrators actively participated in the civil rights movement. There are four clusters based around traditional New Orleans kinship bonds: two cousins, two brothers, one mother and her son, and one aunt and her nephews. Spirituality matters to the majority of the narrators, who identify themselves as Catholic, Muslim, or Protestant. *Overcoming* also includes the stories of one severely disabled young man (Jermol Stinson) and a retired auto mechanic on disability (Pete Stevenson).

What they all share in common is that they are African Americans and New Orleanians. We chose African American narrators not because we think that black voices are more important than other voices, nor are we trying to suggest a hierarchy of suffering. For months after Katrina in Lakeview, former million-dollar mansions remained uninhabitable, and thousands of residents were displaced. Working-class whites in St. Bernard Parish struggled valiantly with levels of ruin rivaling that of the Lower Ninth Ward. Ideally, this book would be one volume in a series of first-person works on Katrina that included Asian, Hispanic, and white perspectives as well. Moreover, we acknowledge that voices representing government officials, rescue workers, and white property owners in New Orleans are important and would provide different perspectives on our subject. Government officials, however, have had no shortage of venues to defend their actions and amplify on their experiences.[10] By the same token, very moving accounts of storm and rescue experiences from Lakeview, integrated New Orleans neighborhoods, and the Gulf Coast are readily accessible to the interested public.[11]

What ultimately compelled us to focus on the voices of African Americans were several things that set their experiences apart from most non-blacks, including Hispanics and Asians. First, the numbers warrant our focus. Specialists believe that as many as 80 percent of the post-landfall survivors in New Orleans were African American.[12] It was clear to news analysts at the time that blacks comprised the overwhelming majority of the crowds at the Superdome, the convention center, the Louis Armstrong Airport, and the staging ground on Causeway Boulevard underneath the I-10 overpass in Jefferson Parish. These arenas and transfer stations were the scenes of unique conditions that fostered a different dialogue about the aftermath of Katrina than the more traditional natural disaster discourse of howling wind, fallen trees, and physical devastation. According to 2000 census data, 73 percent of people living in the flooded neighborhoods of New Orleans were black or 80 percent of the households.[13] In the enduring aftermath of Katrina, the percentage of African Americans in New Orleans has dropped from 70 to 55 percent of the city's total population.[14]

Second, since Katrina, there have been several impassioned, urgently written books to explain *Why New Orleans Matters*, to borrow the title of Tom Piazza's 2005 book.[15] This nostalgic literature, however, has been dominated by white voices, and often written by literary authors who adopted New Orleans for a spell.[16] Many of the narrators in this book believe that the Crescent City had a unique, deeply historical significance separate from white experiences. Black New Orleans was viewed as a miracle, "our own little kingdom," in the words of Harold Toussaint,[17] a narrator who traces his roots back to maroons.

In the immediate aftermath of the storm, the desire for a black-centered work on the meaning of New Orleans was discussed on e-drum, an African-centered political and cultural listserv moderated daily by Kalamu ya Salaam. Asante Salaam, Kalamu's oldest daughter, titled the developing project: "How ya' mama'nem? Love Letters from New Orleans."[18] The three books that do focus most closely on the African American experience of the storm ask questions of intense interest to academics and activists.[19]

Very little literature exists about what New Orleans means to blacks, especially to those who live outside of the tourists' arc of awareness. One argument of this book is that black New Orleans has always been about more than just support for a sugar- or tourism-based economy. Most black New Orleanians didn't need the artificiality of Bourbon Street, when they had their own neighborhood bars, corner stores, city parks, and open markets. *Overcoming* approaches the question of why New Orleans matters from perspectives of twenty-seven African Americans who lived, loved, worked, and celebrated life and death there before the storm scattered them across the country. It is filled with examples of narrators recalling the neighborhoods of their youth or young adulthood, and what New Orleans meant to them before they were forced to leave. These narratives are memorials to the corner stores, the Baptist churches, the community health clinics, and those streets where the aunties stood on the corner, and whose physical traces have now all been washed away.

The third reason we chose to limit the scope of this book to black voices was the tenor, focus, and content of the media attention African Americans received in contrast to non-blacks. Once the news media became aware that New Orleans had flooded, three complementary storylines emerged that fit common stereotypes. The first took a "social disintegration" view of the aftermath; it was symbolized by the photograph of young black man leaving a store with a flat-screen television set in his hands. The peak of public outrage occurred when shots were fired at one of the first Chinook helicopters engaged in airlifting out Superdome survivors three days after landfall. As Katrina's aftermath unfolded before a watching world, black survivors were called, among other things, "scumbags," "looters," "thugs," "hoodlums," and "refugees."[20] On the basis of media coverage, Sajeewa Chinthaka, a Lakimba photojournalist, noted her "absolute disgust" at the televised behavior of (black) New Orleanians in the aftermath of Katrina, which she compared unfavorably to the reaction of Sri Lankans who lost everything after the tsunami in 2004.[21] Approximately three weeks after the last New Orleanians had been evacuated, the members of the news media admitted that these reports of rampant crime and violence were either taken out of context or grossly exaggerated. They drew their conclusions on the basis of military and police witnesses, who were in New Orleans after Katrina's landfall, not on the testimonies of displaced black New Orleanians.[22]

The media's second storyline was reserved for brave and resourceful Katrina heroes, who appeared to be predominantly white.[23] Initially highlighted was the "Cajun Navy," Douglass Brinkley's apt nickname for the white boaters of the hinterlands of New

Orleans, who directly rescued countless victims from their houses across the city. Praise was heaped on three young white Duke students who "borrowed" a press identification badge to get past the military perimeter to see how difficult rescuing someone from the convention center actually was. What was missing, an omission this book seeks to correct, were stories of African American men and women who weathered the storm, fed their neighbors, and saved lives.

Destitute black New Orleanians, or so they appeared after surviving the hurricane, living without air conditioning for days, and wading through threateningly deep water with their salvageable belongings in a plastic bag, were the focus of the third genre of media stories after Katrina. Some commentators were moved by their plight, others spun their tragedy as proof that the "Great Society" had failed once and for all.[24] These misperceptions were deepened in the popular consciousness with reporting from shelters in Alabama, Georgia, Louisiana, Tennessee, and Texas, where displaced individuals of color were housed. A conflation of the working poor and lower-middle-class blacks with unemployed individuals on welfare ensued. This is illustrated by the experiences of disaster response volunteers. For example, a white nurse working at a military base in San Antonio, a processing center for evacuees transported from the Superdome and convention center, described in a letter home the families she assumed were from the Ninth Ward as people who "probably hadn't worked, didn't have a job, [and] were caught in a cycle of poverty."[25] Similarly, another volunteer working in Austin estimated that "seventy percent of New Orleans was below poverty level."[26] There are several reasons for this mistaken equation of black New Orleans with overwhelming poverty, in addition to prevalent stereotypes that have been well-documented elsewhere.

Decade after decade, many of the poor in New Orleans worked excruciatingly hard, as the narrators of this book make abundantly clear. A study conducted by the nonprofit Initiative for a Competitive Inner City (ICIC) concluded that New Orleans residents living in economically distressed urban areas were more than 87 percent employed.[27] Being "nickel and dimed in America," as Barbara Ehrenreich so memorably popularized the phenomenon,[28] is often the situation of men and women in low-income jobs of the variety that dominated the service industry in New Orleans before the storm. Disentangling the assumption of slothfulness from the reality of constant financial hardship is necessary to accurately view the impact of Katrina's aftermath on displaced African Americans.

At the time of the last census in 2000, only 3 percent of households in New Orleans were receiving cash assistance and just 8 percent in the Lower Ninth Ward.[29] Even in the Lower Ninth Ward, frequently hailed by outsiders as the symbol of black poverty, the poverty rate was only 33 percent.[30] However, 60 percent of all Lower Ninth Ward households were occupied by homeowners, and most of the homes were already paid for.[31] Since Katrina, much has been written on the imagined lives of the poor drawing heavily on statistical analyses of quantitative data or theoretical literature about what it means to be black and poor in America.[32] One feature of this book is that it provides a record of men and women considered poor speaking directly about their work, struggles, and accomplishments.

Taking a lesson from two sociologists, Mary Pattillo-McCoy and Mitchell Duneier, we have been careful to select as many representatives from the working poor, the lower- and upper-middle-classes, as we do from the underemployed. Well before

Katrina, working class black men (and women) had been rendered practically invisible, not only by the media but also, Duneier asserts, by white and black sociologists themselves. These working-class men, he discovered, traditionally believed in "hard work, family life, and the church."[33] As Pattillo-McCoy points out, on a good day, "poor urban ghettos" receive most of the media attention, even though only one in four African Americans actually live below the poverty line in the United States.[34] Sixteen of these narrators achieved lower-middle or upper-middle class status by early adulthood, even though only two could be described as having experienced a middle-class childhood. Of those not yet retired or on disability, eight individuals in this book were classifiable as working class at the time of their interviews.

Several narrators volunteered to participate in the project to set the record straight and honor the memory of the people who raised them. In light of the loss of so much written documentation and material culture to the floodwaters and subsequent black mold, the voices and experiences of eyewitnesses should not be allowed to be washed into the bayou like so much debris.

Neighborhoods

Crescent City neighborhoods play a consistent role in forming these narrators' core identity: the neighborhood of birth, of childhood, and of adult home. Oral history most effectively explores questions of subjectivity,[35] which go to the heart of questions of identity and power.

In the aftermath of Katrina, most media attention focused on a handful of black neighborhoods—the Lower Ninth Ward, Pontchartrain Park, and the housing developments—in ways that offended many former residents. The black experience in New Orleans is broad and layered. Before Katrina, *Overcoming's* narrators had spent a significant portion of their life in one of the following neighborhoods: Algiers, Bywater, Carrollton, Central City, the Garden District, Gentilly, Mid-City, New Orleans East, Tremé, Pontchartrain Park, the Upper or Lower Ninth Ward, and the Lafitte and B.W. Cooper Housing Developments. Two were living just outside of Orleans Parish, one in the suburb of Metairie (Le Ella Lee) and the other in the hamlet of Violet (Yolanda Seals). Metairie is on the East Bank of Jefferson Parish and Violet is in St. Bernard Parish. Both interviewees spent considerable time in Orleans Parish on an almost daily basis and thought of themselves as New Orleanians. Six of the narrators were born outside of New Orleans, and six grew up in other states.

As Richard Campanella, the foremost geographer of New Orleans, points out, the residential patterns of blacks and whites in New Orleans on the eve of Katrina had evolved up to that point deeply rooted in the histories of slavery, segregation, and housing discrimination, but *not* in the oversimplified way outsiders tend to assume.[36] Whites were (and remain) firmly entrenched on the highest land in the city in Uptown and along Lake Pontchartrain's southern shore, but they also occupied some of the lowest land in the city in Lakeview. Conversely, some of New Orleans' high ground in Uptown was occupied by blacks as well. The uptown African American enclaves date back to an era when wealthy white families on the broad, oak-lined streets nearest the Mississippi River preferred to have the men and women of color who worked for them live within walking distance. The land hugging the Mississippi River along the industrial wharves in Uptown was originally less desirable to wealthy whites because of what Campanella calls "nuisance" factors, including noise and pollution, and so it was settled by the blacks who worked there.

Under Jim Crow laws, African Americans were forbidden to buy or rent property in Lakeview, Gentilly, and New Orleans East, until lawsuits brought by the NAACP made the restrictive deed covenants and the steering practices of realtors illegal.[37] The tragic irony of these battles for integration is that when African Americans began to exercise their rights to buy property in Gentilly and New Orleans East, property that was originally homesteaded by middle-class whites, they inadvertently bought some of the lowest lying land in the city. The reason that New Orleans East was predominantly African American in July of 2005 was not because the land was undesirable or dangerous, or because it came about as an indirect legacy of slavery, but because whites began moving out to both banks of Jefferson Parish as soon they saw that they were in danger of being outnumbered. The men who developed Lakeview and later New Orleans East were confident that new drainage systems and ever stronger levees would overcome the risks of building on low-lying land in a city that often flooded after a hard rain. This belief in technology's ability to constrain nature combined with a vision of a more prosperous city based on increasing trade also led to the building of the Industrial Canal and MR GO. The breach of these structures during Katrina led to the inundation of the entire Ninth Ward voting district, which encompasses not only the Lower Ninth Ward but also New Orleans East. By contrast, New Orleans' West Bank (Algiers) was considered economically underdeveloped, lacking canals or direct access to the Gulf of Mexico, but it was spared flooding after Katrina's landfall. In 2005, Algiers had been predominantly black for more than thirty years. The swath of the city originally known as "back of town" marked the habitable perimeter of the city before modern drainage systems opened up Lakeview and now lies sandwiched between the high ground along the Mississippi River and the low-lying Lakeview–Gentilly–New Orleans East corridor.

For those unfamiliar with the geography of New Orleans, there are a few more things worth describing. New Orleans has an east and a west bank, separated by the Mississippi River. Orleans Parish borders St. Bernard and Jefferson Parishes. The children of Italian, German, and Irish immigrants moved to St. Bernard Parish from the Lower Ninth Ward after school integration and Hurricane Betsy. Whites relocating from New Orleans East a decade or two after Betsy chose Jefferson Parish as their new home. Before Katrina, Gentilly and Mid-City were among the more integrated middle-class neighborhoods in New Orleans.

You can tell something about a New Orleans resident (probably including their age as well) by where they place the dividing line between Uptown and Downtown. All but the youngest narrators in this book use the traditional Canal Street divide. No one in the group uses the post-Katrina version of the boundaries of Uptown, as being essentially the same as those of the Garden District.[38]

The Ninth Ward originated as a voting district, and technically includes half of Pontchartrain Park, all of New Orleans East, the Upper Ninth Ward (including Bywater and Desire), and the Lower Ninth Ward (including Holy Cross). However, almost no one from Pontchartrain Park and especially New Orleans East, identifies themselves as from the Ninth Ward, unless they were raised in the Upper or Lower Ninth. Although New Orleans East now has a reputation for having once been a black middle-class powerhouse, it should be remembered that there were zones of poverty and numbers of shoddily constructed apartment complexes that ran along the service roads of I-10 and down some major thoroughfares like Crowder Boulevard. Narrators experienced the storm from both these extremes: from a lakefront home on Bullard Avenue (Willie Pitford), and from a flimsy apartment near I-10 (Eleanor Thornton).

These cultural and familial memories are not without frequent reminders that for most New Orleanians on the eve of Katrina, the city was not an idyllic environment. The New Orleans economy was based on service industry jobs and low pay.[39] The city was burdened with failing schools, a corrupt criminal justice system, substandard housing, and a large number of unemployed or underemployed black youth who engaged in a life of drugs and criminality.[40] In several disadvantaged neighborhoods, including housing developments and older uptown enclaves, areas with high rates of homicide, robbery, and rape were adjacent to areas where thousands of tourists came to eat rich food, drink to the point of inebriation, and dance to roots music.

Race and Rescue

In the long aftermath of Katrina, politicians, activists, and media pundits argued whether racism played a role in what were embarrassingly slow rescue efforts for a nation that takes pride in its wealth, might, and can-do attitude. Some speculated that if New Orleans had been a majority white city, every effort would have been made to execute a total evacuation before landfall. Even without a successful evacuation of a predominantly white city, this line of argument goes, food, water, and sanitation facilities would not have been begrudged to the stranded, whose food, medicine, and cars had been submerged in the flood water. Others argued that bureaucratic incompetence caused delays that hurt Gulf Coast residents of all colors equally. Some Katrina survivors asserted the conviction that the flooding of New Orleans in predominantly African American sections of town was deliberate; a type of ethnic cleansing at worst, brutal urban renewal at best. Those with more passion explained the phenomenon by suggesting that the levees had been deliberately dynamited. Dyan French Cole, better known as "Mama D," a well-respected activist in New Orleans, testified to her beliefs before a televised session of the Congressional Hearing called to investigate the Katrina fiasco.[41] A less widely reported version of the nonnatural breaking of the levees thesis was that the levees were underdeveloped and deliberately undermaintained in predominantly poor, black neighborhoods so that they would naturally give way under the pressure of a storm surge, thereby protecting the wealthier, white areas of town without leaving evidence of criminal tampering with federally maintained levees.

Interactions between blacks and whites occur not only in and around New Orleans but also throughout small towns and large metropolitan areas in six states: Alabama, Arkansas, Georgia, Louisiana, Tennessee, and Texas. These eyewitnesses spent time in predominantly white areas ranging from rural towns to exurbs with expensive gated communities. These areas represent the second wave of white flight coming after some blacks managed to integrate the suburbs that resulted from the first wave, which had in turn, arisen from the integration of city schools.[42] These narratives allow you to draw your own conclusions on the state of race relations in the South at the turn of the twenty-first century.

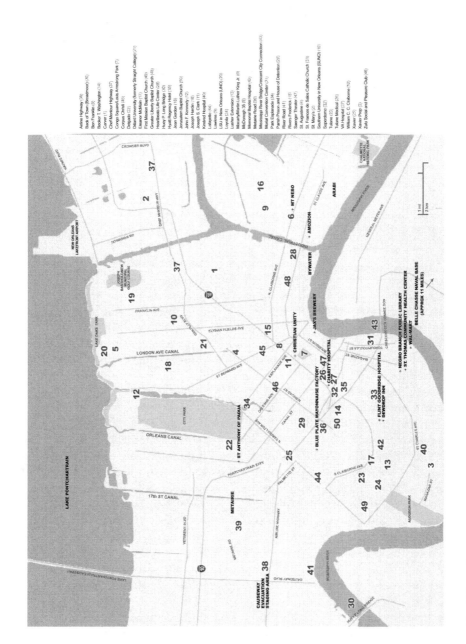

Airline Highway (36)
Back of Town (Broadmoor) (36)
Ben Franklin (3)
Booker T. Washington (34)
Carver (1)
Chef Menteur Highway (37)
Congo Square/Louis Armstrong Park (7)
Corpus Christi (46)
Delgado (22)
Dillard University (formerly Straight College) (21)
Eleanor McMain (17)
First Mission Baptist Church (45)
Greater Liberty Baptist Church (48)
Heartbeats Life Center (28)
Huey P. Long Bridge (30)
Hyatt Regency Hotel (30)
Jean Gordon (18)
Jerusalem Baptist Church (50)
John F. Kennedy (12)
Joseph Hardin (16)
Joseph S. Clark (11)
Kindred Hospital (40)
Lafayette (44)
Lawless (9)
LSU in New Orleans (LSUNO) (26)
Loyola (24)
Luther Emerson (13)
McCarty/Martin Luther King Jr. (8)
McDonogh 35 (6)
Memorial Baptist Hospital (4C)
Melanie Road (30)
Mississippi River Bridge/Crescent City Connection (43)
Morial Convention Center (31)
Park Esplanade (34)
Parish Prison and House of Detention (43)
River Road (41)
Rivers Frederick (15)
Saenger Theater (47)
St. Augustine (4)
St. Francois de Sales Catholic Church (31)
St. Mary's (2)
Southern University in New Orleans (SUNO) (10)
Superdome (32)
Tulane (5)
Tulane Medical (26)
VA Hospital (27)
William C. C. Claiborne (10)
Xavier (25)
Xavier Prep (3)
Zulu Social and Pleasure Club (46)

New Orleans, Louisiana.

Source: Map by Masoom Khan, masoomkhan@gmail.com

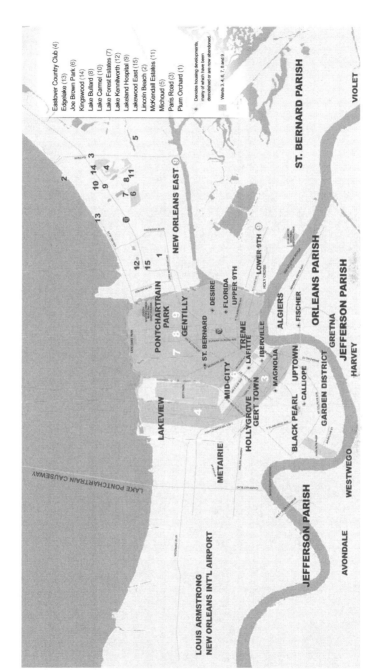

New Orleans Wards 3, 4, 6, 7, 8, and 9.

Abbreviations

ABC	Association of Black Cardiologists
ADD	Attention Deficit Disorder
ARC	Amistad Research Center
AALP	African American Leadership Project
CARE	Community Awareness Revitalization and Enhancement Corporation
CBD	Central Business District
CDC	Center for Disease Control
CPR	Cardiopulmonary resuscitation
CSC	Churches Supporting Churches
EEOC	Equal Employment Opportunity Commission
FEMA	Federal Emergency Management Agency
FMSS	Field Medical Service School
GTU	Graduate Theological Union
HOPE	Health Outreach Prevention and Empowerment
ICIC	Initiative for a Competitive Inner City
KIPP	Knowledge Is Power Program
KKK	Ku Klux Klan
LSU	Louisiana State University
MCAT	Medical College Admission Test
MSW	Master's of Social Work
MRE	Meal, Ready to Eat
NOPD	New Orleans Police Department
NOW	New Orleans West
OPP	Orleans Parish Prison
POW	Prisoner of War
PTSD	Post Traumatic Stress Disorder
ROTC	Reserve Officers' Training Corps
SBA	Small Business Association
SUNO	Southern University of New Orleans
UCLA	University of California at Los Angeles
WPA	Works Progress Administration
WWF	World Wrestling Federation
YUP	Young Urban Professionals

RETIREES

ONE

Aline St. Julien

Aline St. Julien, the youngest daughter of a Creole mother from Lockport and a black father with roots in Santo Domingo, was born in 1926. She spent her childhood in the Sixth Ward, raised a family of seven children with her husband of thirty-six years in the Seventh Ward, and, after 1982, lived independently in Gentilly until Katrina. The youngest of five children, Aline was the only child in her devoutly Catholic family to attend parochial school. Her attendance at Xavier Preparatory School was made possible by the largesse of her brothers, who were then serving in the military. From an early age, Aline embraced her blackest roots. During the struggle for integration in the '50s and early '60s, she served on national and local boards, including the National Board of Black Catholics and the New Orleans Urban League.

This interview occurred on February 15, 2008, one week before her eighty-second birthday, at the dining room table of her one-bedroom efficiency apartment in a complex owned by the Catholic Church. An ornately framed, abstract painting by a Haitian artist with crescendos of blue-green waves and the hint of a flamboyant dancer dominated the wall behind the table and set the mood for the interview. Aline was wearing a ribbed pink sweater with a precisely tied, rayon, peach scarf. Her elegant jewelry, including a silver bracelet doubling as a key chain, matched perfectly.

Her history embodies the complex interactions between peoples, arts, music, and food that is New Orleans. Her deep knowledge and variegated experiences reflect the vibrant

*culture of Tremé and the Seventh Ward. Her narrative is rich with descriptions of North
Claiborne Avenue before cement pillars replaced towering oak trees, of Creole family tradi-
tions, and of integration struggles. She witnessed Katrina from a hotel room in Memphis,
Tennessee, where she had driven from New Orleans with her children. After an exile in
Maryland, Aline chose to return to Louisiana. Even her post-Katrina stroke has not suc-
ceeded in repressing her passion for dancing. Aline continues to teach yoga and line dancing
to nuns and grandchildren alike.*

I was born in 1926 and raised in the Tremé area of the Sixth Ward. My first recall was
when I must have been four years old. We were sitting in front of an open furnace at
1230 N. Johnson Street where we lived. We didn't have air conditioning and all of that,
no electricity. We had lamp light and a little potbelly stove. The pipe from the stove had
to run like an arrow across the ceiling and then it would go out the side of the house.
Our family rented a two-story, shotgun double house for twenty-five dollars a month.
We sublet the upstairs to help pay the cost.

It was nice growing up in Tremé because we had a culture. Every Saturday you'd
walk N. Claiborne Avenue. That was our strip. The oak trees were still there, and that's
where most of the culture and black people were in that area. We went to the market
there, and we had a lot of black-owned businesses and stores. On N. Claiborne was a
coffee company where we could watch them roast the beans and brown it up right in
front of us. We got fresh coffee and fresh chicken. The further down on N. Claiborne
you went, you had the chickens running around in cages, and the vendors would
butcher it for you. They'd pluck it and hang it up. When I went to Paris, I saw chickens
hanging naked on a hook just like I used to see when I was a young girl.

My maternal grandfather was the product of a Spanish colonizer. We were closer
to whites than to blacks in culture because of my mother. My mother was a country
woman from Lockport. People used to tease them because they didn't want to say
Bayou Lafourche. They would say, "Y'all come from Bayou Lafourche." "No, we don't!
We come from Lockport." My mother was proud of her life in Lockport. They lived
where everyone was farmers and their neighbors were white and they used to birth each
other's babies and everything. My father and mother migrated here to New Orleans for
better quality of jobs.

My father's people came from enslaved Africans in Santo Domingo. While my
mother never acknowledged the slave background, she was proud that he was from
Santo Domingo. To her he was "Mr. Duminy." I challenged her one day, and I said,
"Well, ma. If you feel the way you talk, why did you marry daddy and daddy's dark?"
She said that he was a gentleman. So if you were black and you were a gentleman, then
you were accepted.

My daddy had been in the service in World War I but was discharged with a disability. My daddy was cigar maker at the El Trellis Cigar Company. I was fortunate
enough to have my daddy and my mama home. My daddy was there all the time, helping my mama wash dishes and everything. They'd wake up at 5:00 in the morning
after having put clothes in a big galvanized tub in the middle of the yard on a char-
coal furnace, and that would boil all night. Then they'd wake up and they'd get their
clothes washed with a washboard. Once a week we washed. We had a big yard and
clothesline poles. The breeze would make the sheets dance in the air.

My grandfather was in the hog market business. I remember them sticking a pig in his neck and the blood would come out in the bucket. During that time we had blood sausage. Mama would make hog head cheese from the head of the hog. She would have all those little cups on the table—the collagen from the bones would be like jelly when they cooled. I would go on my bicycle on a Saturday and sell them. They were two for twenty-five cents or fifteen cents a piece. So that's the kind of chores I did, since my daddy didn't work. My mother was so proud of my efforts. Daddy had a disability pension, which was a pittance. Just thirteen dollars a month, but a long time ago that was good. We could buy a loaf of bread and sausage maybe for a nickel or a dime or three loaves of day-old bread for ten cents. Mama knew how to sprinkle water or whatever she used to freshen it up before putting the bread in the oven.

My mama was a really good seamstress. She'd make different kinds of dresses and suits. She made me a suit out of a man's pants. Then she worked in white women's houses sometimes. White men had two families during slavery time, so you had the white Duplessises and the black Duplessises. So when mama'd go work for white people, they kind of knew her as Mrs. Duminy because she light-skinned enough to pass for white.

"We Were Brought up as Creoles"

I remember when I was a young girl, and I went riding in City Park. A white guy in City Park said, "Nigger get out of here." I got frightened and turned around and went back. We used to be on the Bayou St. John riding our bicycles, but anybody could call you a name, and anybody could chastise you, like, "You don't belong here." In the church, we sat in the last pew or we had to go upstairs. When I went to the Xavier Preparatory School, the sisters would sit in front of the screen, and we'd sit in the back.

I used to go to confession and tell all my sins, even that I stole the country sausage out of my mama's beans. I believed everything that was taught at the Catholic Church. We also believed that Catholic people were the only people that were going to be saved.

Mama didn't want us to listen to the real blues and jazz records. She wanted us to listen to classical music. The people next door would have the "low-down blues" and I'd be singing them, and she'd say, "No, you don't." So we didn't have any real music to listen to, but Wayne King on the radio when we got our radio, and we didn't have a radio for a long time. Most of our leisure time was spent sitting on the front steps singing, playing concerts, playing guitars, or playing coon-can outside and going to the movies.

We didn't call ourselves "Black" during that time. That was considered a bad word. We'd call ourselves "colored." So one day in school, they sent a letter home asking mama when was the first time she told us we were Negroes and mama got hurt. She said, "We're not Negroes. We're Creoles." When I got old enough to open up the dictionary and start reading things, I discovered that Creoles could be almost anything.[1] They had Creole tomatoes and Creole horses. When I got up in the classroom and said I was Creole, my teacher told me, "Sit down. You're a Negro." I felt hurt. We were Creoles. We believed that we were better than Negroes. This Creole business, it was nothing but the slave master taking everything his money had bought, including sex.[2] So that's why I tell them I don't know why I should be so proud of my history.

My husband, Harold St. Julien, and I almost grew up together. My husband grew up on Toulouse Street near the water pumping station. The families were thrown together because his uncle married my aunt. So the two families intertwined and that's how Creoles used to be. Every time we would have a picnic, my mom would go with her sister. My future husband's mother and my mama's sister were close, so they were often at the picnics. My husband was one of eight children, we had five on our side, and my mama's sister had seven. So you could see that when we had a picnic, we didn't need any outsiders, even though sometimes they had another family that would come along too. The only place we could go away from New Orleans was Pass Christian. We could go to Pass Christian because they had a black minister that had a church on the Gulf of Mexico in Pass Christian. It was beautiful to us because we would go once a year.

Family Life

My husband was four years older than me and I thought he was so much older. His father was an alcoholic, and he was the oldest son. Harold, therefore, took the role of breadwinner for his family. His mother respected him so. He did not fight in World War II. The reason they didn't take him was that he had a hernia. He wanted to go; he tried every branch of service.

In my day you had the romance, the courtship, the cedar chest, the engagement ring, and then the wedding. Your boyfriend gave you the cedar chest and you prepared for your married life. The whole family bought things or made things. Mama would make me aprons and gowns, and people would give me sheets. Or you'd go shopping and you'd buy something for your cedar chest. Mama was making me flannel gowns to keep me warm. She made my husband pajamas. Your cedar chest would be almost filled by the time you got married.

Then you were set to begin a life that you never knew before. Two people getting together, they had to have things. In those days, you weren't supposed to go out till nine days. I didn't believe in any of that. We got married on March the fourth in 1946 right after the war. I had just graduated from school. I taught my husband to dance. After he learned, we won two waltz contests.

We had our first baby in '47. I had seven living children, but I had three that I lost. Two miscarriages and one baby was full-term but he didn't have a cranium, so he lasted about nine hours. So I should have ten children. When I was having all those babies, every time I turned around I was pregnant. My father-in-law is one of nineteen children. So that's why I say about the St. Juliens, you don't marry a St. Julien, because they're baby factories.

If you were Catholic, you were made to believe that you were better than somebody else. Whites would help you. They'd call you "nice colored people." Where my husband worked, he was a skeleton key, or a jack of all trades. He started in a packaging room wrapping packages. He learned how to fix ranges, then refrigerators, and sewing machines, and before you know it, he was all over the place. Mr. Lee Lehleitner, who owned the place was a very fair-thinking man.

I was one of the first in New Orleans to let my hair go natural. My mama wanted to die. I said, "Yes, I have kinky hair, and it's good. There's no such thing as good hair and bad hair." "Good hair" was supposed to be straight hair. So when I got to be on my own and thinking for myself, I said, "This kind of thinking will never do."

We'd go once a week with the children to Lincoln Beach when they finally built the beach that my people could go to. Way out in the sticks, on the outskirts of eastern New Orleans. I used to get in that pool and tell the children: "Watch, mother's going to be Este Williams." Later, when they passed by Pontchartrain Beach Amusement Park where the white beach was, they'd ask, "Why can't we go there Ma?" I said, "because we're not invited." We were always trying to protect our children from the real truth. One day they were looking at television. "Ma, Jackie Robinson is black." I said, "Yeah, he's black, and daddy's brown, or you yellow, or you this."

I have so many proudest accomplishments as a mother. The first proudest accomplishment was that my son went to the seminary. Every Catholic family dreams of that. And they not only got Mtumishi, named Michael at birth, then they got Eric, my second oldest son. Mtumishi was in the seminary ten years. They put him out along with eight seminarians. They wanted to join the civil rights movement, and the Catholic Church wouldn't allow them.

Leadership

I was raising my children, and I was on the Board of the National Office of Black Catholics. The bishop used to pay my fare to go and come. I had my seven children, and I was going to these meetings in Washington, DC, Chicago, sometimes in Texas. And I was on the Community Relations Council of Greater New Orleans. This was a mixed race and religious group of people in the 1950s and '60s. They had these people like Senator Michael O'Keefe, Moon Landrieu, and Helen Mervis, a Jewish woman. They formed a panel of women. For six years, I coordinated that panel and we would have to give talks on our experiences with discrimination. Naturally most of us blacks were Creole, and I had two or three Jewish women. We went all over to churches, the synagogues, the universities, schools, everywhere. We wanted to change the hearts and minds of white people. That effort lasted for six years.

I was in the Catholic Church and the Catholic Church was integrating these big organizations. I was on the Bishop's Human Relations Commission. I was on Mayor Schiro's Human Relations Committee. Most whites at that time were all segregationists, but we were going to try to get together. They had people on the Human Relations Committee like a professor from Xavier University. While Warren McKenna was teaching at Xavier, a twenty-five-year-old cop roughed him up. We couldn't get it redressed. He was going to sue the cop or the city, but before you know it, it was hush-hush and nothing was done.

The church started leaving the city as whites moved to the suburbs. In 1954, you had to integrate the schools. The whites went slowly. And the Catholic Church went even slower. So what the Catholic Church did was start moving the churches out in the suburbs. They started running with the bigots. I used to say to myself, "Are they teaching us the same things? 'Turn the other cheek.' 'You were going to get what you deserved when you got to heaven,' and all of that stuff." I asked, "Why do they act so ugly?"

I don't belong to none of them anymore. I got so frustrated because I could see where we weren't getting anywhere. They'd come up after the talk, and they'd tell me,

"You are nice colored people. Where could I meet other people like you?" They really didn't get the message. I just got myself away from that because I said, "All I am is a guinea pig."

I decided I wanted to go to a church where I could feel my culture. I started going to St. Francis de Sales Roman Catholic Church. They had a white priest there, Father Joseph Putnam, but he was blacker than most of our black priests. I remember the first time I went to the church and heard our type music. The tears started coming out of my eyes. I felt like I had come home.

Our media is so biased. The beautiful things and accomplishments that happened to us, they don't tell us. Like the Black Indians make their own costumes.[3] They make it a family affair. Everybody is working toward making these beautiful costumes, and they are so beautiful! You know when I saw the beauty of the costumes? They were on display at the Delgado Museum of Art in City Park. Again I tell you, I got emotional. The tears came down because I saw the beautiful work, the intricate work, and then I walked in the back of them. I never paid attention to how much work is in the back of them, and the crown.

Grieving and Independence

The day my husband died a picture was taken of him with his golf clubs. We had just come from a cruise. We'd gone to St. Thomas and Puerto Rico. When we came home, he decided to go to a golf tournament to raise money for sickle cell anemia research. The last thing I told him when he walked off was, "Beat them up baby." I said that because before I used to always say, "You go golfing all day, and you get angry when you come home and I'm in Dillard University's library. You have nerve."

My husband was killed by lightening in 1982, and I thought I could not go on. And then I started traveling. I heard a lecture on Egypt by Asa Hillard, an Egyptologist, and I came home, and I'm sitting in the bed reading the paraphernalia, and all of a sudden I just said to myself, "I'm going to Egypt!" I couldn't get people to do the things that I wanted to do. So I went on this tour with students and other people, and we had the best of times. After the tour guide would tell us what he said about Egypt, then Asa Hillard would get us straight on it, and he would clarify some of the things and tell us the real truth. I also went to Senegal, Gambia, and Goree Island. I saw through the Door of No Return. It was an eerie feeling I felt when I was over there. I just went traveling all over, and then when I came back home, it was like starting all over again. The grieving was just so heavy, and I thought I wasn't going to make it.

So when I was 56, I graduated from SUNO (Southern University of New Orleans in Pontchartrain Park) with a B.A. degree. I went to Howard University to do some graduate work but was still grieving and homesick and all of that weighed on me. I did an essay on female circumcision, not the kind of research I should have been doing in the condition I was in, because it just made me feel worse. I stayed there for about one semester and then came home.

When I came back, my sister-in-law took me to a spa, and that's where this physical fitness focus started. I got back into living and the most important thing to me was the independence. I started my yoga classes three times a week, I did swimming twice a week, and I rode my bicycle all around the Dillard campus.

I decided I would have a class reunion to see all my old classmates, and I did. The guy I asked to take me said yes, but then he had a death in the family and had to leave town. So I said, "I'll find somebody else." And this was a blind date. Somebody told me this guy could dance. So that was the beginning of a thirteen-year relationship with James Aaron, dancing.

I lived in Gentilly Garden with the Dillard students. The students were there, and then they had a few tenants. That's a very convenient location because I had easy access to the drugstore, the library, the bank, and the post office. Everything was right in that section of Gentilly at Broad Avenue and Elysian Fields Avenue. So it was easy. I had things just the way I wanted them. I was independent. I went where I wanted to go.

I danced the weekends. I'd go to mass on Sunday, and after mass, I'd go to the nursing homes and dance, because I'm quite a dancer. The old people would be eating and they'd start tapping their feet. I'd waltz for them especially, and the organist knew I like "Mr. Bojangles," which is a waltz that people don't know. But I danced it so beautifully by myself, and I'd go to the extreme, and the old people would be just looking at me. I'm looking in their eyes too, and I'm saying to myself, "You're not seeing me, you're seeing you, baby. You're waltzing." I took them back into past memories.

"It Took Forever to Get Me back to New Orleans"

So when Katrina came, I went with Mtumishi. I had never left for a storm before. Mtumishi called me, "Ma, you're going to come with us." I prefer going with Mtumishi because, well, the ride is nicer. And he knows where he's going, and he's got all these mechanisms that I don't have access to. They call it progress. He pulls out the cell phone, the car talks, and I feel comfortable.

Mtumishi had his family and their children. My sister who is three or four years older than me was with us. We were too many people, so they, for lack of a better word, "dumped" the elders. They sent us to my sister's son's house in Maryland. It was a farm. It was nice, but when you leave your house thinking you're going overnight, you're miserable. You want to be home.

Mtumishi and his family came to Baton Rouge from Memphis. His two daughters were living in Baton Rouge, so he bought a house big enough for them there. Nilima lived there, and it was a place for me to go. I knew my children were there, so I came to Mtumishi's house in Baton Rouge. I think that was my eightieth birthday. Everybody in the family was there. All I could do was cry because it was such an ordeal to be away from home, when it happened like that.

I was able to rent in Bluebonnet Apartments in Baton Rouge. That's Jimmy Swaggart's school. It's in the student's dormitory. So I didn't have a real apartment like this one. I had to create a makeshift kitchen. I got a little portable stove and a little portable refrigerator. And my kitchen was in a corner on one side, and the sink that you usually just wash your face and brush your teeth in? That's where I washed my dishes and did everything. I bought a little pan that fit snug in, so I made do.

But it was hard for my children. They had to travel from New Orleans to Baton Rouge. All of them live here in the city now. Everybody had a day. I said, "That's why it's good to have seven children."

I was an active person, and now I'm a different person. I've got to accept it, and I really don't, because by the time I got to Baton Rouge, I said, "I'm going to start my yoga again," and then I joined the YMCA, and did my swimming twice a week. I moved back to New Orleans in February, 2007.

This wall over here has pictures of all my children, and I'm so proud of that. I sit on my rocker and look at them. These pictures in my bookcase are of all my grand-children and great-grandchildren. Mtume and Asante are in San Diego. This is Kinii, she's an author. She goes to Brazil. There's some feeling she gets over there. The youngest, Tiaji, is in Washington. She worked for the State Department for a while, and the United Nations. That's Tutashinda.

I did yoga for twenty-five years, and I tell my yoga people, "Listen to your body and use the props." That's all I can do now, use the props. I had perfect balance. Now I could fall by looking up. To get down is not very hard, but I'm having difficulty getting up. But I will not stop. My grandchildren learn so much from me. They're in awe of me, some of them. Just the other day, Mtume, Asante's brother comes here, and he says, "Ma, I want to do yoga with you." So he comes here one day and brings another cousin of his and we move the rocker and all three of us are in here doing yoga. And I teach them the line dance, so all of them think I'm just a fun grandma, and they invite me to their classrooms.

What I like best about being back in the city is just being able to see what Katrina really did to us. It's amazing when you go around. I mean this is my territory. I used to be going around every day. When I saw the devastation in the Lower Ninth Ward, I said, "They wiped those people out." And some houses that were not destroyed by Katrina have fallen apart, but they're still there. It's just sad. But I like being back home. I'd rather be here than in a less devastated city because you never really feel satisfied until you're home. What I missed is the culture.

Conclusion: "We've Got Beautiful Memories"

What I dream for this city would be going back in time. If we could have things the way we used to have them, and I think it's age that makes me feel that way. I can't get used to the cell phone.

I just felt when Katrina came the rug was pulled from under me. I used to save things but between Katrina and the moving from one place to the other, I don't know what I have and I don't know how to look for things. Where are my books on Egypt? So Katrina messed all that up: my dancing and my independence. My children didn't want me to drive again. They're all my parents now because of the stroke. I know most of the things they tell me are true, but I want to be rebellious.

This morning when I was taking my coffee, I realized that there was so many things I could remember from just taking a cup of coffee. When I was a girl, my younger sister and me would sleep in the bed. We were in the Lafitte Project.[4] Now they want to tear them down, but they had good people. We prayed to get in the projects, and we prayed to get out and get our own home. I heard my daddy coming up the steps with those big Brogan's shoes they used to wear. He had two cups of coffee for me and my sister. "Wake up, wake up." "Oh Daddy, let us sleep, daddy." "No, you got to have something warm in your body!" So we had to sit up and drink that coffee. When my husband found that out, he always brought my coffee to the bed after we were married. And for my mother-in-law, serving a cup of coffee to a priest was like a work of art. She'd have a kettle of hot water on, and she'd pour the hot water in the cup, and let it sit there a little while and pour it out. She'd have to wipe the cup good before she poured the coffee in, and then she'd serve it to you. She put the sugar in a dish and cream in a creamer. After I returned to New Orleans, I brought my sister a cup of coffee for her. She would just lavish a cup of coffee. It wasn't this instant coffee and all of that; it was drip coffee.

I've had a good life. I was always a proud woman. I carried myself with dignity, that's just my way.

Narvalee Audrey Copelin*

Born in 1926, Narvalee Copelin grew up in the Lower Ninth Ward at a time when it abounded with green space, where struggling first-time homeowners could stake their claim to the American dream. The Lower Ninth Ward is legendary for producing activists who challenged segregation directly.[1] Narvalee inherited her organizing spirit from her father, Reverend Noah Copelin, and passed it on to her nephews, Kenneth and Keith Ferdinand and Kalamu ya Salaam (see chapters eleven and twelve). Drawn to water from a young age, Narvalee served in the Navy before becoming a high school business teacher. A world traveler, a dancer, and a devoted sports fan, Narvalee speaks as an expatriate who left New Orleans many years before Katrina.

This interview took place at Narvalee's ranch-style home in Los Angeles on June 19, 2006.[2] Her immaculate house was furnished with custom-made furniture purchased in the late 1960s. Evidence of her recent eightieth-birthday celebration filled the living room. The wood-paneled den was lined with sketches from Brazil and a modernist print of Kenneth Ferdinand playing the trumpet. During the interview conducted at the dining room table, Narvalee was surrounded by family scrapbooks and photo albums with rare photos dating back to the 1800s. Narvalee, a stylish woman whose favorite color is purple, was radiant and animated when she discussed her family and the Lower Ninth Ward of her past. By contrast, post-Katrina memories were spoken slowly and suffused with a heavy sense of loss.

Narvalee's memories were of growing up a pastor's daughter in the depression-era Lower Ninth Ward before the period of white flight, the epidemic of cocaine, heroin, and other drugs, and the deterioration of the neighborhood's infrastructure. The portrait she draws of her father, mentor to several generations of young people from the Lower Ninth Ward, provides a glimpse of the unique spirit of the neighborhood in earlier days. Even prominent Republican Baptist preachers possessed a dancing spirit, dressed to kill, and appreciated the power of music. Emotionally, Narvalee never left New Orleans, where she maintains close ties with her friends and family to this day. She reminds us of the extent to which Katrina-related loss and death among the elderly reverberate among communities of the heart irrespective of the physical miles separating them.

I was born in '26. I can remember a little about the Depression time, because my parents, Noah and Theresa Copelin, used to put the money in an envelope, and I had to get on the bicycle and take the house payment to the person who owned the house. She lived on Rampart Street in the Lower Ninth Ward. She would write in this little black book whatever amount of money I gave her. Sometimes she would say, "Tell your daddy that wasn't enough." One day she wrote "paid in full" in the book and gave me a large envelope. Upon returning home, daddy told me I would not have to go anymore. The land and house were for the family. Just about everybody in the Ninth Ward owned his own property. I never heard of anybody paying rent until I grew up and started going out as a young adult in Uptown.

The only house I remember is 1311 Lizardi Street, and I think that when my parents came to the Lower Ninth Ward, they bought that house. My boys told me that that was the only house standing in the block after Katrina. It was built high off the ground. My daddy owned three adjacent lots: in the middle was our house; on one side of our house there was a lot where my father used to grow the corn, string beans, potatoes, cabbage, lettuce, tomatoes, and bell peppers; on the other side was a lot for playing. In playing house at lunch time, we didn't worry about anything. We'd go out there and grab a tomato and bell pepper, rinse them off, put on some salt and pepper, and go under the house—it was our play house.

Now when I was coming up, we knew everybody. Just like family. They had the Copelins, the Watsons, the Jolicoeurs, the Lukes, the Lees, the Dixons, the Amoses, and the Cooks. Most families past our block would get their mail at our house. Where we lived at 1311 Lizardi was the stopping point for the mailman, the boundary between established homes and reclaimed swampland.

At first we were the only blacks in the block from St. Claude Avenue to Villere Street. All my playmates coming up were the little white girls in the two blocks ahead of me. They had this big open playground, and they could all go there and play. But in front of my house was another big open space, where Thomas Alva Edison Elementary School's at now. The men started cleaning off the lot, and they would come there and play ball. But I would go to the white playground and play because that's where my friends were.

The most fun thing we did was to play ball and roll around in tires and see who could come down the levee without getting killed. You would fit yourself inside a big open tire, hold on, and just roll. Or ride a bicycle and see who could come down that levee without getting hurt. I used to love going over the Industrial Canal and St. Claude.

We didn't even know what drugs were. Sometimes we had fights with each other, just with your fists or with a cabbage. We used to use veggies and throw them around at each other. We'd break a tomato on somebody in a minute.

I can remember, I had a fight at the grocery store, and before I got home, my mama knew about it. I was so upset. I said, "I know who told you." They used to call her Miss T. She tells everybody's business. There was a Miss T on every corner. There wasn't any sense in doing something wrong because somebody was going to tell on you.

We didn't lock doors. All those houses the two blocks before our house didn't have any fences. Those white people didn't have any fences between their properties either.

The whites knew all of the blacks and the blacks knew all of the whites. It would be about nine or ten o'clock, and I would be on the streetcar coming home by myself.

The conductor knew me, "Little Copelin, you are running mighty late today." I would get off the streetcar and walk home, and there'd be one or two street lights.

My girlfriends who lived in what they called uptown New Orleans had never even been below the Industrial Canal. And when they came down there, they were just shocked. They'd never seen all of this luscious green grass and everything where you could run and play. It was St. Augustine grass, a thick, dark, wide-bladed grass that grows almost like a vine on the ground.

After they went down there to visit, a lot of my friends started buying property down there. Right after World War II, the Italians started moving out. I can remember the last family to move. After the war, I had a little white friend of mine who came to visit me. He told the cab driver the address and the cab driver said, "You sure this is where you want to go?" When he got to my house, he said, "Why did he ask me if I was sure?" I said, "Boy, you are from Florida. You know what they were talking about." But in the time before that, he would have had no problem whatsoever. Life changed to that extent in a few short years.

I went to Southern in Baton Rouge, which was really in Scotlandville at that time. When I'd get $2.02, the price of a ticket, I was on the train coming to New Orleans.

In the '50s and early '60s, you know, several teachers who taught at Joseph A. Hardin Elementary School, the school down there named after a pioneering Black doctor and Liberian Consul, purposefully moved down to the Lower Nine so they would be closer to it. We just had ministers, doctors, lawyers, everybody had moved down there because it was such an open space, and you could breathe all this good fresh air. A lot of people had moved down there.

I moved out to Los Angeles in the '60s. When I first moved out here, I would go to visit constantly, because I wanted to see my parents and whatnot. When I left it was still more or less where you knew everybody, and when I returned for a visit and saw all these new houses and strange people, I was shocked. It had grown just that fast.[3]

Reverend Noah Copelin

Noah Copelin was my father. My father's father used to live on Rampart Street. He was a little man, neat as a pin. That's the only grandparent I can remember. When he died, I can remember going there. There were all these people in this big yard frying fish and boiling crawfish. Everybody was sitting down, eating and having fun.

My father's nickname was Dude. He smoked cigars. My father's church, Greater Liberty Baptist Church, was at 1230 Desire Street in the Upper Ninth Ward. My father was a very, very exciting preacher. He used to cry a lot. My daddy used to work selling life insurance also. He was energetic.

I can't remember when my father brought my mother to New Orleans. I think she was working in somebody's house. He took her out of that and brought her to New Orleans. My mother was a healthy, strong woman. My mother wasn't slow, but she didn't rush. I never ate peanut butter. We had three hot meals, because my mother never worked. At our house, my mother would make ice cream and all the kids would come in the yard and play. She lived to be ninety-six years old.

I've never known my daddy without a car. He had this old car, and they used to have to push it and put water in the radiator. If the car broke down, he would take it to some man who was in the neighborhood. He was the only one for a while in the neighborhood who had a car. Daddy always had a car full of people.

My father was a Republican. He went to President Dwight D. Eisenhower's inauguration. The one thing I learned from my father was never sell out. His church was a little church. The man who owned a lumber company, another white man from the city, and three other white men came to the house to see my father. They wanted to build him a brand new church, if he would get everybody in the Ninth Ward to vote a certain way, Republican. I was sitting on the floor in the living room, because if daddy was home, I didn't care who came, I was going to be right under him. They were talking and talking, and he said, "no." I can remember this man from the lumber company said, "Do you realize we're going to give you a brand new church, and you don't want that?" My father said, "I do not want it. A person votes according to the way they want to vote. The only thing I tell my people to do is vote. I don't even tell them to follow me." When they left, I can remember my daddy said, "I want you to learn one lesson. Never sell yourself to anybody."

I used to get so upset with him when he would send me out there to pick up litter. I'd stop right by the line between our house and the neighbor's. And he'd say, "I still see litter. You didn't pick up Miss Shamina's." I'd have to go pick up her litter. That whole darn block had to be straight. He used to get out there and cut that grass. When he got to the front of the house, he'd go on down the whole block. He said he did it just so our block could look nice. We were the only Blacks in the block. The whites didn't pay him a dime.

As long as you didn't curse or anything, it was fine with my father. We were all just raised as free spirits. There was nothing we weren't allowed to do, as long as we weren't harming anyone else. "Do it, remember you have to pay the consequences but don't get anybody else involved in anything you can't stand the consequences for," my father used to say.

They told me he used to like to dance when he was young. I think that's why he understood me so well. I would go out and not get home till real late. I used to get on the table to show my daddy what I had seen in the night clubs. Because my father was a minister, I did not hang out in the Ninth Ward. I'd always go uptown, out of respect for my parents.

Noah Copelin was a dresser. They called me, and told me daddy had gone to the hospital. I say, "Oh well, that's the end. He's not going to lie there and wait on anybody." When I got there, my sister Inola had decided they were going to bury him in his robe that he used to preach in. I said, "no way! My daddy needs to be in one of his blue suits."

Segregation

My father, my brother Sherman, and I were in the car, and they used to have ferries down in Violet, a small, predominantly Black community in St. Bernard Parish on the east bank of the Mississippi River. My father was preaching at this old church. On our way coming back, I wanted to see the water in the Mississippi River. Daddy told Sherman, "Take Narvalee out there to see the water." We were standing just looking over and a white woman accused my brother of saying something to her. We didn't even see the woman. When we looked up, all of these men were approaching us. The head man on the ferry knew my father, came over and was told, "This so-and-so said something to my wife." We started going back to the car. The head man said, "No, that's the preacher's son. I know he wouldn't do that." My father was in the car. When we got to the car, and we were in there, they started throwing rocks at the car. I told my daddy, "I will never go in the country again." I have never forgotten that ferry ride and those rocks. My father just said, "Those people are ignorant and they don't have God in their lives." I just looked at them as a little girl. From that day on I became very skeptical of white individuals.

One man bought property on the corner of our block, and he tried to be more segregationist than the other white residents, and I told other white people in the block that he had looked at me funny and had made some remark. They went down there and told the family, "We've been living here and we don't have any problems." The family didn't stay very long.

My father used to drive me to school every morning, when I went to McDonogh 35, the only high school they had in New Orleans for Blacks at that time. I finally got him to let me come home in the evening on the street car with a group of friends. They would have these signs, "for colored only," behind which we were supposed to sit. We'd pick up that sign and just scoot it all up and down the street car. The conductor would say, "Where's the sign?" Nobody would answer. Now he's looking for the sign, and we'd start dropping books on the floor. We did it just to irritate the white people who were on the street car. We used to have such fun.

In high school, all my little friends were Creoles, and half of them you didn't know if they were white or Black, so they would go to the movies after school to the Saenger Theater on Canal Street. Blacks weren't allowed in the Saenger Theater, so I'd wrap something around my head. I'd be showing my little black face, but I'd do this funny talking, and they thought I was from some country somewhere. Several of my friends could pass for white, so they would buy the tickets.

The people in the neighborhood helped build St. David Catholic Church in the '30s. The fathers were white. The parishioners were all Black. The whites had St. Maurice Catholic Church built in 1857, and that was located on St. Maurice Avenue in what is now called the Holy Cross neighborhood of the Lower Ninth Ward. They helped build St. David because they didn't want the Blacks coming to their church.

The sheriff of St. Bernard Parish, old Bill Stanley, lived in one of the four houses in our block. He was mean as hell. He had a little daughter, Alice Mae. Alice Mae could play with me as long as her daddy wasn't home, but when it was time for her daddy to come home, she disappeared. There was one family who had a daughter. The tale goes, one of the young Black boys around my brother's age started going with her. Then he was dead one day. And they swear that Bill Stanley had him killed. I really don't know it to be true, but that's the reputation he had.

When I grew up, I was the one who used to take Keith and them to all the little devilish things. They would go along with me to the ball games. We used to go down to the French Quarter and we'd see what was going on. You know there was so much segregation during that time, you couldn't go to a lot of things.[4] But the things that we could partake of, we did.

Conclusion: After Katrina

They tell me they've got a lot of cars up under the freeway. I don't want to go down there. The only way I'd go is if it's an emergency that I would have to go to my nieces, nephews, and friends. I have got a lot of friends who are survivors of the hurricane in New Orleans. I have some pictures of Katrina stuff. I used to live next door to Myra in New Orleans. Now she has a trailer next to her girlfriend's house on the east side. She said she had to fix it up in order to tolerate living in a trailer. She's a teacher down there and she's just trying to wait and see.

We were supposed to have another high school reunion. They didn't have it before Katrina, so I know it's over now. The people who were in my class are scattered. I kept in close touch with some of them. I've lost three girlfriends. Just couldn't take it. One of them had a heart attack. The other was sort of ill before and just gave up on it all. One is in a nursing home. Her husband told me that he had put her there. They had a lot of suicides that weren't even reported.

I have one friend, Audrey Giddings, who moved to Virginia. She's not going back. Another one who built her house from scratch is now in some little town in Texas. She's not coming back. Two are in Georgia; they're not going back. They say they just don't have the will to go through that again. I think it's because of their age. Only one lived through Betsy. She wouldn't even go back to look at her house. She sent a nephew to get her personal things and put her house up for sale. There are some assisted living places that are open in New Orleans, and she was telling me, she had gone to visit those, so she's waiting for one of those to come up, because she's been living in a trailer in Kenner on the west bank in Jefferson Parish.

I saw where they were going to build at least one thousand new places for the underprivileged to return to New Orleans. I said, "I'll believe it when I see it." They had a big crime there around the Lafitte Housing Development last week. They are all just frustrated.

You really need faith and a solid spiritual background to endure hardships. That's what the younger generation doesn't have, and that's what I kept trying to instill in my kin. If you don't have that faith to fall back on, you are subjected to all those other things.

Leatrice Joy Reed Roberts*

Born in 1929 on Gordon Street in the Lower Ninth Ward, Leatrice Roberts' memories of growing up in the segregated South during the Depression are lightened by her recollections of community sharing and intergenerational connectedness that made her feel invincible. Leatrice transcended color barriers as a graduate student at Louisiana State University (LSU) in Baton Rouge, and as an educator of white children in the first decade of integrated public schools in New Orleans. As a teacher, she taught a generation of students to strive to fulfill Martin Luther King Jr.'s dream. Before the storm, Leatrice and her husband of forty-eight years left their home in Pontchartrain Park and drove to Paris, Texas, ninety miles northeast of Dallas. For a half-century, Pontchartrain Park was a middle-class haven for New Orleans blacks, including teachers, transit operators, city workers, attorneys, and physicians. At the center of Pontchartrain Park, towering oak trees edged the fairways of the eighteen-hole Joseph M. Bartholomew Sr. Municipal Golf Course.

This interview of Leatrice, the granddaughter of the influential Ninth Ward preacher, Bazile Jolicoeur, took place on August 16, 2007, in the study of Leatrice's spacious, new suburban home in Duncanville, Texas, fourteen miles from Dallas.[1] Leatrice was dressed in the one pantsuit she retains from her life in New Orleans.

Leatrice's narrative illustrates the strength of character of the Lower Ninth Ward pioneers. These pioneers took advantage of any openings inadvertently created by virulent racism to gain a better life for their children. She exemplifies the upbeat, resilient spirit of post-war African Americans. Even the experience of losing everything she owned at the age of seventy-eight was, in her rendition of the ordeal, redeemed by God's grace and her principles. The account ends with reflections about the rituals of Crescent City sharing and accountability, and how they bear fruit and ease worry, even in Duncanville.

My grandfather Jolicoeur died on April 1, 1949. If he'd made one more year, my grandparents would have been married for fifty years. I was only nineteen. "Don't you let these professors turn your head from what you know is thus sayeth the Lord," he told me right before I entered college.

My grandfather was a very large man. He never went out in the street, if he was not fully dressed in a suit and tie. Reverend Jolicoeur's thoughts were before his time.

He started a scholarship program in the '40s. Each church would submit a list of the names of deserving kids to whom would be given fifty dollars each.

My grandfather pastored Amozion Baptist Church in the Lower Ninth Ward at 907 Deslonde Street. He was a thus-said-the-Lord kind of preacher, directly from the Book. He'd give you a text, and you could go and read it for yourself to see that it was what the Lord had said. He was generous with me. All I needed to do was bury my head in a book, and I could do no wrong. The word "bored" was not in my vocabulary. When he saw I wasn't reading, he would ask, "Did you read the dictionary yet?"

My grandfather was the president of a group of ministers throughout New Orleans. It was a Baptist organization. These churches joined together for worship and charitable deeds. So I was the Jolicoeurs' granddaughter. They just demanded more of people they thought were "succeedable" material. They were not going to let me fail. Everybody was your support group.

My grandfather was a counselor. People then used ministers as spiritual advisors, because they knew they could say what needed to be said and get advice. Reverend Copelin and Reverend Jolicoeur were outstanding men of God. They were respected, as were teachers and nurses.

"Nurse of the Ninth Ward"

When my grandpa gave my mother the money to go to school to go to register, he thought she was going to Straight College to be a teacher. There was a doctor from St. Bernard Parish, who my grandfather had worked for at one time. He knew my family, he encouraged my mother to be a nurse, and he helped her. Before my grandfather knew anything, my mother had enrolled herself into the Flint-Goodridge School of Nursing and she was living in that school. He did not like the connotation that he had heard about nurses and doctors that the former were there to service the doctors. I think my mother and Dr. Rivers Fredericks convinced him that was not the case. She graduated in 1924. There were very few jobs for black registered nurses. She worked with Dr. Fredericks.

My mother often imitated what she imagined was the rationale Huey P. Long, the governor of Louisiana from 1928 to 1932, used to convince legislators to train black nurses: "We got all our girls in that Charity Hospital.[2] And we got our girls washing these Negroes. Why don't we put some of these Negro girls in there to wash those Negro men?" That's how my mother got a job. She was one of the first to go into nursing in Charity Hospital because she was qualified as a registered nurse. Nursing was her passion.

My mother worked at Charity Hospital. When my mother got the job at Charity, she was making fifty dollars a month, and she got a little extra because she worked at night. She was working twelve hours a day.

Most of the people went to Charity Hospital, which accepted uninsured patients; we were all uninsured until Albert Dent came up with the penny-a-day hospitalization insurance plan for patients who wanted to use Flint-Goodridge Hospital.[3] Evidently a lot of these people must have had diabetes, because Charity Hospital would give her the medicine, and on the weekend, when my mother was home, we would walk around the neighborhood and she would give them their medicine. It was difficult for them to get to Charity Hospital in the CBD. They called her the Nurse of the Ninth Ward. My mother loved people, and people loved her.

There was one branch library in New Orleans for black people.[4] It was at Philip and Dryades way uptown. When she was off, my mother would take me to the library. We'd get on the street car, ride to Canal. Get on a bus to go up to Dryades. Ms. Guichard, the librarian, did everything. The stacks were closed and there was no card catalog. My mother would say, "Get me some books for Leatrice," and she'd go get a stack. I read a lot of biographies. We'd take the books, have lunch, and then we'd come back home. That was an all day trip.

The Depression and the War

All of us were with my grandmother and grandfather. My mother didn't have the responsibility of cooking and taking care of me, because my grandmother stayed home and took care of all of us. The peddler would come around in the morning, and she would get her fresh vegetables grown in the gardens in the Lower Ninth Ward and down in the St. Bernard Parish.

At that time, people helped each other, and that's what made us rich. When the WPA (Works Progress Administration) was established, people got supplies. They shared those with the preacher and with us. And my grandmother was an extremely good baker, and she'd take that bran flour and she'd fix it with raisins and honey, and did all kinds of things. I would go and take it to the people. "Hey, how you doing? I got some sweet potato bread for you." There was a man, Mr. Cephas, who worked for somebody down at Shell Beach in Plaquemines Parish, facing the Gulf of Mexico. In the morning when he was coming up from Shell Beach, we would have soft-shell crab and grits for breakfast.

I had a girl friend Eunice that lived all the way at the end at Florida Walk, the last street in the Lower Ninth Ward that ran parallel to the cypress swamp on the northern border. We went there by way of a beaten dirt-and-shell path through the woods. Eunice always would come up to my house, and I never could go to Eunice's house. So one time I decided I was going to go to Eunice's house. I got on my bike (I rode my bicycle everywhere), and I road that beaten path into the woods. Then I returned back to Gordon and Urquhart Streets from Florida Walk, and before I got back home, somebody had called my grandparents and said, "Why is Leatrice on that bike way back here?" And they were waiting for me. You see the type of community it was. In elementary school as a child, I went back where I knew I should not have gone, and the word got home before I did. Look at the cohesiveness of that community, the taking care of each other, protecting each other. Look at the love.

Now in my situation, we were the only ones on Gordon Street with a telephone. So my grandmother would get the messages and write them down. In the afternoon, I had to deliver messages to people. There was one lady, whose husband's name was George. We used to call her Miss George. Miss George worked at Grant's, a ten-cent store. She worked in the candy-and-cookie department on Canal Street. Every time I'd go there, she'd offer me candy.

World War II impacted me. I had to have World War Problems I and World War Problems II. I was the one in my eighth grade class that collected the money for the war bonds. You paid twenty-five cents for a stamp. When you get the book filled, you turn it in and get a bond that at maturity could be cashed in for twenty-five dollars.

After World War II that's when the influx of people to the Lower Ninth Ward began. They were coming from the center of New Orleans, and they bought the plots for two hundred dollars. They were getting out of historical New Orleans because that's where they just had shanties and bad houses. They just wanted to get in something that was better than where they were. A lot of them were first generation homeowners.

Education in the Segregated South

Coming up, I had a lot of walking to do. That made me athletic. For elementary school, I had to walk all the way to Macarty School in the 1600 block of Caffin between Claiborne and North Roman, when McDonogh 19 Elementary School was only two blocks from my house. I always questioned, "Why couldn't I go there?" "Because you are a Negro and Negroes did not go there" was the inevitable reply. At the end of our garage, there was a piece of fence that was broken. Through the crack between the fences, I played with my neighbors who attended McDonogh 19. I knew more than they knew because I was always the teacher. One day I asked them, "What is biology?" They couldn't answer. Our teachers told us, "You have to be the best." It was drilled in us. "Good, better, best! Never let it rest until your good is better, and your better is best." "Excellence is the key to success," and, "You're going to have to do better than others." They didn't seem to get that at the white school.

In eighth grade, I went to Booker Washington Junior High. It was a fine new building with a cafeteria and an auditorium. It was so pretty. I was just so proud to be in that school. Ms. Hill played "Little Star" on the piano. Our teachers were demanding.

I went to McDonogh 35 (35) for high school. It was a school where the music department was superb. We gave operas. Look at what a force it had on me that I remember it today, and that was way back in the '40s. Some of those teachers that taught me were at McDonogh 35 in 1921, when my mother graduated. So I could do no wrong. "Oh, you the little Jolicoeurs' daughter," they would say. I got out of high school in '45.

I also went to business school as I was going to 35. Same month as I graduated from high school, I got a secretarial certificate. When I went to college, I used my secretarial skills. The first year I took all my notes in shorthand. I typed all of my papers.

I went to Dillard University. That was my first degree. My mother sent me. I didn't have any spending money but what my people gave me: lunch money. But I was never unhappy, because I never wanted those things I knew I couldn't have.

Integration: "A Struggle Every Step along the Way"

For my master's degree, I went to LSU in Baton Rouge from 1954 to 1956. I was already teaching kindergarten in New Orleans, so I went to summer school until my last semester. Then for one semester, on Wednesdays and Thursdays, I'd work during the day, drive to Baton Rouge, have class from 7:00 to 10:00, and drive back to New Orleans. I was a young person then. I was driven.

I went under protest from my people. They didn't want me to go up there because they were doing ugly things to us at that time. Whites shot some of our black boys attending LSU while I was there.[5] One night, I was coming home, and there was a panty raid. Thank God I knew how to drive at Carnival time.

In deep segregation, we'd have to get off the sidewalk for whites to pass. But it was different at LSU. I came along, and they must have thought that I was poison because they'd get off the sidewalk whenever they saw me. I'd walk into the women's lounge and everybody walked out. Oh, if I wasn't laughing at them in my heart. These poor people! After I got my Master's, I had another layer of confidence. I knew who I was since I was five years old, when I started school and I was already ahead of the others in my class because I could read already, but going to LSU was an experience.

Family Life

I started LSU in June 1954 and got my M.Ed. degree in June 1956. I got my "Mrs." in December of 1956, when I married Edward Roberts of Uptown. The "Mrs." is still standing tall. We lived in an apartment by Dillard. I typed papers for my husband when he went to grad school, because he didn't go to grad school until after we were married. I do all the cooking, but my husband cleans up.

My son, Edward Reed, was born in 1960. We moved into a house in Pontchartrain Park when my son was five. It was on Pauline Drive. It was a little bitty house. We had three bedrooms, a living room, and a dining room. It was quaint. We only had one bathroom. My mother lived across the street.

The first few years of desegregation were very stressful. So after working all day, Edward and I would come home, but we would keep the babysitter upstairs an extra thirty minutes, so that we could enjoy a martini together before cooking dinner, spending time with our son, and grading papers.

Oh God, that was a lovely neighborhood. The golf course in the center made it so nice—I'd walk around it every day, approximately two-and-a-half miles.

My son started playing in grade school with the Southern college band. He went to LSU but didn't like it, so he came home and finished at SUNO. For graduate school, he went to Howard University. He finished a two-year program in social work in one year, so he's got an MSW (Master's of Social Work) degree. Before the storm he was working as a social worker, but playing saxophone is his passion.

What I loved best about New Orleans: the people, the food, the customs, being able to go places, the festivity of the city, Mardi Gras, and going out to parties in long, beautiful ball gowns. I used to have a whole closet full. In my younger days, B.B. King and Ray Charles would come and give a show at the Dew Drop Inn. My husband is an extraordinary dancer. We would dance all night long.

Teaching Career

When I went to McDonogh 40 School in 1949, it was way down in Gentilly Woods, which borders Pontchartrain Park. The people were good, dear people, but they were very limited, and very poor. They loved the teachers dearly. When I was in New Orleans, I would see some of them and it was just like meeting family. I'd go to their churches. I went to every parent's house. Sometimes they wouldn't let me go in the house, and I just sat on the steps. That's how I could learn my child and teach the child. I would do income taxes, and read and write letters for them. I was at that area for seventeen years.

Then in 1972, I went to William C.C. Claiborne Elementary School located at Louisa and Mirabeau Streets in Gentilly Woods. That was an integrated school. It was 80 percent white, and 20 percent black at first. I was in a fishbowl. There were other black teachers, but I was teaching these people's little darling five-year-olds. They wanted to know, "What is this black person doing?" I had more volunteers conducting surveillance. Every time we would go out in the yard, the man across the street would come out to his yard. "Yes, sir, I'm not going to do anything to your child." Within ten years it was 85 percent black. Gentilly Woods started selling out to black people.

I was from the old school. When you saw me, you knew this was the teacher. I stayed there for ten years, and then I retired. I rode my bicycle to school some days. The children loved it. I tried to instill excellence in the children.

Katrina

We'd never left for a hurricane. I let my husband make the decision. I never thought that I would leave New Orleans. Edward and I packed a few things and we drove to Paris, Texas. We got a hotel room thinking we would return home in a couple of days. When it became clear that we would need to be away from New Orleans for at least a few weeks, one of our nieces advised us not to stay in Paris, because of its notorious turn-of-the-century fairground lynchings and present-day struggles.[6]

So we moved to Duncanville and rented a small one-bedroom apartment, while we sorted things out. My husband drove back to New Orleans to handle the gutting of our house and other business. The closest hotel he could get was in Mississippi. On his back to the hotel one night, he got pulled over by a police officer for going ninety-five miles per hour. Edward was so exhausted he just wanted to get to a bed.

When we were trying to decide whether to return to New Orleans permanently, my godson, a physician, said, "I know you love Amozion Baptist Church and New Orleans, but I don't think you should come back for health reasons."

Early studies, like the one by the National Resources Defense Council in November 2005, warned of the dangers of airborne mold, especially for the elderly.[7]

So we decided to buy a house here. I'm pleased that my husband and I could come here and start a whole new life at seventy-eight years old. We're stocking a whole new house after fifty years of marriage. I'm supposed to be throwing things out, not buying furniture and sheets. When I was buying sheets, they were percale sheets. I go to buy sheets now at three hundred thread counts. What is that? None of the sheets is worth a penny with a hole in it because it wrinkles when you wash it. I'm not going to iron all those sheets. I sent some to the cleaners.

When we stood up for our fiftieth anniversary, the pastor at our new church of about three thousand people said, "I want you to know something about these people. These people came up here with nothing. They'd lost everything after Katrina. These people now have a house, these people are active members of the church, have positions in this church, and they've never stopped looking up." What are we to do? I don't see that as such a great accomplishment.

We've never been people on the dole. I'm not looking for anybody to do anything. But the people in Dallas have given us so much, including the clothes I wear. These pants I'm wearing today are what I left New Orleans in. There's a jacket that goes with it. And these shoes I left with. Everything else I lost. I don't have any ball gowns in this house. Don't even have really a cocktail dress. My sorority sisters all pitched in and sent me several new outfits. For the first time in my life, I own a pair of blue jeans.

Right now there's a man from Louisiana that raised those tomatoes we're going to have with our salad. And he says, "I've got to help the aged." He brings a little bag of home-grown vegetables, and he passes it out in church on Sunday. Sometimes he brings okra. So one day, because of my giving nature, I went in my pocketbook. He said, "I can't take that." I said, "I'm not paying you for the vegetables. I'm paying this for you to get some more seeds."

I don't like going back to New Orleans now. I want to be back in New Orleans, but I don't feel good when I get back to New Orleans. It's not what I saw before. The trees don't look the same. I go to look where I used to go to walk. It's all shambles. The golf course is gone. It now resembles a subtropical jungle. Wesley Barrow Stadium, a former neighborhood landmark, isn't there. The two schools are not there. Most of my friends will be scattered until they die and come back in a casket.

Leatrice and Edward had their family home demolished. Only a cement slab now remains. They have hired a local contractor to restore the house across the street originally owned by her mother and occupied by their son before the storm.

The contractor is saying to me, "You shouldn't leave all of these decisions to me." I said, "Baby, if you do it, it's going to be alright. You just go ahead and fix the house livable." The contractor is a person who got his apprentice training from his father. His father went to Amozion Baptist Church. His father and I went to Macarty together. And my mother used to go and take medicine to his mother. In that young man, I have every confidence.

Conclusion

I learned early on in life to make myself content with what I have. It prepared me for Katrina. I'm not in sackcloth and ashes. In that house I had things of my grandmother, things of my mother, my own things. Three generations. There was a chifforobe of my grandmother and my grandfather that I had paid five hundred dollars to get refinished. That was an antique for me. I had a big old rocking chair from the living room. And then I had it no more. So I'm not sorrowful about my things. I am sorry about the people and the hardships they're enduring.

My proudest accomplishments are that I have taught many, many people. I have touched their lives. They still remember me and will talk to me, when I see them. They're proud. I've always been a lady. I have high moral standards. I do not think highly of myself, but I'm pleased with what I'm doing. There are people that recognize that I've done the work of the Master. I have a feeling within myself that I have contributed, that my Christian principles are showing.

Irvin Porter

Born in 1924 in Iberville Parish, Louisiana, Irvin Porter recalls his sheltered life as the youngest child, before he was plunged into active duty in the Pacific theater during World War II. Irvin's military experiences fueled his determination to no longer accept the second-class citizenship he had endured living in rural Louisiana. After registering to vote, Irvin attended college in Baton Rouge, where he met his future wife, Alice Morrison. Alice's family owned one of the first black taxi companies in New Orleans. The couple, both teachers, raised four daughters and pioneered Pontchartrain Park where they sought the middle-class life depicted in television sitcoms of the 1950s, with a brick ranch home, substantial yards, and access to quality schools. In 2006, Irvin was a winner of Tom Joyner's contest, "Real Fathers, Real Men." Listeners to the show write letters nominating exemplary African American fathers. Each Tuesday the winning letter is read on the air and one thousand dollars is awarded to the honored father. Irvin used his award to purchase a riding lawn mower.[1]

Irvin gave his account on February 14, 2006, from the kitchen table of his sparsely furnished rental house in Baton Rouge. He was not keen on being interviewed at first, yet he eventually told his story with the same quiet flair and aptness with which he has lived. Irvin's smile is riveting, especially when his eyes light on his wife of fifty-five years. His hair is white, his eyes sharp.

Irvin's short narrative uses the terse language typical of members of his generation, accustomed to a lifetime of struggle under challenging circumstances. Irvin watched the aftermath of Katrina on television with his wife and oldest daughter's family in Germantown. Irvin's account shows how confusing the reconstruction process has been for his generation, even for someone like Irvin with a master's degree and well-educated children. Irvin returned after Katrina to rural acreage near his birthplace.

In 1903, my father, a farmer who raised sugarcane and garden vegetables, bought a small piece of land in Sunshine, which is near Plaquemines Point, Louisiana, approximately sixty-five miles northwest of New Orleans. I was born in 1924. Economically, we made it. Other children considered us rich. The property out in the country—that's been clear for umpteen years.

Most of the work I did coming up was as a carpenter, but I did other jobs. I did manual labor and worked in a lumber job. By being the youngest, I never was subjected to too much, because all my brothers would all look out for me.

During World War II, I fought in the Pacific theater two-and-a-half years. After I came from the service, I guess I had a little attitude, being a little bit different from now. If I wasn't militant, at least I had a little attitude of what I wasn't going to take. So I wanted to do two things after I returned. I wanted to get a legal driver's license and register as a voter.

After federal troops withdrew from the South in 1877, multiple voting tests and regulations were passed into law during the 1898 Louisiana Constitutional Convention that were designed, in the words of one politician to "let the white man vote" and to "stop the Negro from voting." Before 1957, for a black man in southern Louisiana to exercise his right to vote often meant risking his life.[2]

I went to register to vote in the little registrar's office in Plaquemines Point. One of my brothers went with me and stood outside, while I went inside to attempt to register. I failed at registration the first three times. So after the third time, the registrar took me to an old faded newspaper clipping he had on this little bulletin board and pointed out to me that it said you had to be able to write. Now to keep from having a problem about my penmanship, instead of using cursive, I had used manuscript. I did my signature in cursive. But he didn't understand that was writing. That was one of his reasons for rejecting it. Also he asked me if the leading white people knew that I had come to register. I said, "I didn't ask nobody if I could come to register." About the fourth time, I was registered.

There was a World War II veteran who also was able to register. As it is you had about two or three country stores. People go out there and talk. So this fellow came one night and told me that a certain fellow had made threats and advised us not to come to the voting booth. And he expressed such a desire to vote. I told him I was going to vote. If he wanted to go with me, fine. But I needed him to do one thing. "Don't tell nobody what time we were going." And I'd pick him up with my oldest brother: everybody in the area knew him, because he was one of these people folks could get along with. I picked him up and we went down there. I got the surprise of my life. I walked in there prepared to go down if necessary. No telling what was going down that day. I walked in that voting place and they called me Mr. Porter. Not an unpleasant thing was said, no gestures or anything. After we walked out of there and came back unscathed, that was the end of it. That probably was in '49, '50, or '51.

The GI bill helped with some things, like college tuition and a down payment on some property.[3] I got my B.A. degree in Education from Southern University at Baton Rouge, and an M.A. from Louisiana State University in New Orleans. I got married in 1951.

When it came time to find a job in Baton Rouge, however, I felt at a disadvantage because the positions were sewn up through connections, either relatives or pastors, and I had no one in the city of Baton Rouge. By contrast, New Orleans was eager for new teachers, and they sent someone to Baton Rouge to interview me.

I was a Special Ed teacher for the emotionally disturbed, a carpenter, a master auto mechanic, and an adjunct at a technical college. My teaching philosophy was if I could help a child have the best hours of his day, then I was happy. I wanted to give the students relief from what they were going through away from school.

Opening up Pontchartrain Park

In 1959, we bought our family home on St. Ferdinand Street in Pontchartrain Park. At that time we bought the house for $18,200, a lot less than a new car now. (In the late '50s, house prices in The Park ranged from $14,300 to $25,000.[4]) I had the GI bill. And the down payment was not large, but it did take a little bit out of what you did have. My wife was a junior high art teacher at Carver High School in the Desire Project at the time.

If you had the means and bought a lot and developed your plans with your architect and contractor, you could have a two-story house, like a few doctors and lawyers. The developers had about five or six floor plans. The houses were not prefabricated completely, but there were sections. The plans they had, you couldn't get them to change. Alice wanted a window over the sink to face the street, but they said they couldn't do it because they already had plans made out. The only concession I could get them to make was to line my house perpendicular to the other plots. The lot was wider at the back, than at the street, so the angle they had laid out for the house, I didn't want that. I wanted it parallel with the street. That's all I could get them to do. There were choices some times of what kind of floor you wanted, or that type of thing.

The kicks part of it was when we were asking the salesman for certain things, and he was frank about it. He had gone to Southern with us, so he was somebody we knew, and I was asking him about having them fill the yard up and grade it off, and put appliances in and stuff like that. He told me, "Uh uh." I said, "They're advertising that's what they're doing in such and such an area."

He said, "Yeah, but we don't have to do that here. The people are buying the houses. So we don't have to do any of that to get the people to buy." From their contractors to the salesmen, everybody got as much as they could, and gave you as little as they could get by with. They didn't even put a light bulb in the place! No Venetian blinds or anything like that came with the house. It was completely bare.

We had three bedrooms and a bath and a half. Ours had hardwood floors and tile in the bathroom because I thought it was worth what they were charging for that. The tile in the bathroom wasn't level. When I insisted that they straighten that out, they took the grout buckets and slid them across the hardwood floor. The contractor wouldn't repair that damage.

To me New Orleans was just a place to work, because the job was available there. On the weekends mostly, I would come out to Baton Rouge or out in the country. I recall one time my neighbor, I believe he was an accountant, moved to New York, but he came back because New Orleans was a much better place to raise his children, and he missed the golf course. Well, the golf course was not much further from me than it was from him (I'm on one street and he's on the other) but I never worried about the golf course. The other things in New Orleans, I could get there if I was living out in the country. Because the few times we go to the museum or whatever, if I'd be in the country, I could drive down there and do the same thing. So really there wasn't much to New Orleans that appealed to me.

Family

Alice and I raised four children, and we have four grandchildren. I'm satisfied with them. I never worried much with friends. At church I have acquaintances, but not too much of visitation or bumming outside the church.

Like anything else I do, we made a mistake and we spaced our kids out. What the thought in my mind was that four years apart would keep us from being in a pinch financially from this one graduating and that one. The people who had them all close together and got rid of it worked out better.

A memory I treasure was when Gail was in the Catholic girls' school, St. Mary's, and there was what you called a Ring Ceremony and she asked me to be there. So after school, I went. I got there and the program was about to start. Here these girls come running, asking Gail if they could borrow her daddy to walk down the aisle with them. They didn't have a daddy or granddaddy or uncle or nobody to walk with them. Her response when they asked to borrow her daddy was "No indeed!" I walked with the "orphaned" girls anyway. By the time Gail walked down there, you'd have thought she was the one who didn't have a daddy. I really didn't know it meant that much to those girls. So since that time I've told other men, if by chance they had a daughter in a Catholic school and Ring Day came, put everything else aside and be sure to be there.

A regret I have is when Gail was moving to Greensboro, and she had this rental truck and towed the car, and I wasn't there to help her with that. That's another time I was working and wondering how she was making out. I worked hard. Papa showed us how to work, and he said, "It wasn't work that killed a man; it's not knowing how to work that killed a man."

I'm not too much on traveling. Alice expresses sometimes a desire to travel, to go and see where she hasn't been, but I'm content not to worry about going anywhere, just to go somewhere to see something.

Katrina

The weekend before Katrina, Irvin and Alice were in Arkansas attending a family reunion with their children. They rented a van and took along their photo albums. After the storm, Alice and Irvin intently watched television looking for missing family members at their eldest daughter's home in Germantown, Tennessee, where they regrouped after the family reunion.[5]

Before the storm, I had a brother, who lived a block away in Pontchartrain Park. He and his wife were stranded five or so days. She was immobile. They had to be rescued out of their house by boat. They had to stay at the funeral parlor several days, and there was no way to contact them. Communication in the 504 area was out. And there was those days we didn't know what had become of them.

After the storm, my brother's wife took down with pneumonia. Even after they were rescued and several hospital stays, days, weeks, didn't help anything, and his wife passed. So I guess in their minds, her death was hastened by the ordeal. I don't know about my brother because he may not have a desire to keep on. He's in Alexandria, Louisiana with one of their sons because he has limited mobility and his daughter-in-law is a nurse. If she doesn't take care of him physically, she can hire those people that would, and she could monitor and see that it's done properly.

Stress

Katrina and Rita evacuees whose homes were declared unlivable after inspection were entitled to either rental assistance or rent-free trailers from FEMA. By law, benefits were supposed to continue for up to eighteen months after the individual moved in.

When we left Memphis for Baton Rouge I think it was probably the early part of September, and we applied for the FEMA disaster relief help, we requested a trailer. I let them know that I had land available for selected trailers to be put on. From September to the fourteenth of February, nothing about the trailer has happened. In late November, days before Thanksgiving, the people came and did what they call a site inspection and put down a little pink flag. That particular person said that within ninety-six hours they had to deliver a trailer. They promise you to get this or that, but you know it just didn't materialize. The people FEMA had to hire had little or no training evidently.

According to Sheila Crowley, hurricane survivors occupied 87,824 trailers as of March 2006. However, Louisiana had requested 40,000 more trailers than were supplied.[6] Nearly every day for months, the Porters would call FEMA or drive out to their property in Sunshine to see if their trailer had been delivered.

Since Katrina prices skyrocketed. We're paying one thousand dollars a month for this house in Baton Rouge, and the most FEMA gave us for housing assistance was a total of twenty-three hundred dollars a month for three months. At one thousand dollars a month, you understand, the arithmetic won't permit that to add up.

FEMA's an organization that's just undermanaged, over-taxed, and overwhelmed with a thing they never expected to handle. I sympathize with them because the task that they're faced with is very enormous. They don't have a precedent to work by, because these other hurricanes just didn't match up to Katrina. The flooding made the big difference.

A Small Business Association (SBA) loan would be helpful to a point but there's a danger in it too because it's a thirty-year loan. The property you've had for fifty years or one hundred years, you wouldn't like for it to get lost in the next whatever number of years you've got left. The idea of losing the property that has been in the family forever, the product of so much hard work, scrimping and faith pains me.

Irvin was worried about how he could make the monthly payments of seven hundred dollars for the next thirty years on a fixed income. Defaulting on the loan, however, would have jeopardized the land he offered as collateral.

New Orleans was a pain and a puzzle. The mortgage company early on had threatened to foreclose on the mortgage. There were so many different proposals going around in the fall of 2005, until I really didn't know would they let you rebuild or would they bulldoze the place down. I decided I would pay off my mortgage with the proceeds from the flood insurance. So that's in the works and they should get it done in the next couple of days. To this day I don't know if it was a bad decision to use the insurance money to pay off the mortgage, but if the bank went into foreclosure, there would be legal fees and whatever I had to pay.

This week's newspaper had an article about was the "toxic soup" a myth or a reality. Different reports are conflicting and which one you have more faith in might help decide what you're going to do.[7] Alice's already bought into this health hazard, because they suggested that people in our age bracket refrain from going to the area. With the levee rebuilding situation as it is, whether it will be redone and whether it will be redone right, or whether it will be designed again to relieve one part of the city at the expense of another—all those things are still up in the air.

Reclaiming Existence

Our son Alvin Porter works for the Sewage and Water Board in New Orleans. My son's job was relocated here to Baton Rouge from New Orleans. He was on a committee to find housing for those workers and by doing that type of work, he was able to come across this place for us. Baton Rouge is like home. The way Baton Rouge has treated the people to me seems very good.[8] The people I know who are living up here, they like it, and some have even bought homes.

Baton Rouge doesn't want to become a New Orleans. I find that admirable. They welcome the people, but some of the things that went on in New Orleans, and the way some New Orleanians live and want to live, they don't want it. An example is selling liquor at the games. Baton Rouge said, "That's fine in New Orleans, but you're not welcome to change us into a second little New Orleans. You can't overrun us and impose on us a different style of life."

For the time being, I'll just return to the Sunshine area. If something else develops, I'll monitor the situation in New Orleans from there. After the storm, I decided to fix up my daddy's house on land that has been in our family since 1903. The country too is changing, but it's a more free feeling. Sunshine served the area of what we call Cancer Alley, so-called for its high levels of toxic emissions from the chemical plants and the electrical plants that made the population grow.[9] It provided jobs for those local people, and instead of people moving away like I had done, second and third and fourth generation people are coming back and populating the place. So twelve or thirteen years ago, there was enough people to call it a town.

Conclusion

Katrina provided opportunities for good and for evil to come out. In my mind, it allowed things to happen that people had been wanting to happen: people wanted to get out of town or people wanted to change the makeup of certain areas.

We're supposed to be the land of the free and opportunities and have rights that are not available to people in other countries. The American dream is to have your own place and be able to go into business and rise as high as your abilities will allow.

I guess faith's the thing that keeps you going. Knowing that you might not perceive what's going on and why it went on, but knowing that it's going to turn out all right in the end.

FIVE

Leonard Smith

   Born in Clarendon, Arkansas, on November 17, 1938, Leonard Smith was the oldest
of seven children reared by sharecropping parents. He never had enough money growing up,
and he found the omnipresent racism intolerable. "This ain't the place for you," Leonard
recalls his grandmother advising him in his youth. He moved to Pine Bluff in the middle of
his junior year and lived with one of two uncles. Four days after he graduated from Merrill
High School in Pine Bluff, Arkansas, he joined the U.S. Navy and traveled all around
the world. During his fourteen years of service, Leonard saw active duty as a Navy medic
in Korea, the Vietnam War, the Formosa Straits, and during the Cuban Missile Crisis.
Leonard fought and was wounded in Vietnam while he served with the Fleet Marine Force.
In San Diego, Leonard worked for decades as a carpenter and cabinet maker. During his
second retirement, he began his love affair with New Orleans, where he traveled once a year
for Mardi Gras until he relocated there permanently in the late 1990s. One of his first acts
was to join the Zulu Social Aid and Pleasure Club, a multimillion dollar organization especially well known for its annual Zulu Ball and its community outreach efforts.[1] In 2004,
he was elected "Mr. Big Stuff," the Zulu who carries the crown for the king during Mardi
Gras. Before Katrina made landfall, Leonard made sure his wife, Dorothy, was evacuated
from New Orleans, but stayed himself to protect his property and to safeguard, he thought,
his fragile health. He was the survivor of many life-threatening situations, including cancer
most recently, and was not intimidated by a hurricane.
   This interview took place in Waterproof, Louisiana, on November 8, 2005, in the
living room of Leonard's in-laws' home, where he was staying with his wife.[2] A tall, lanky
man, Leonard was wearing black dress slacks and a matching button-down shirt with a
gold scorpion medallion dangling from a gold chain around his neck. He immediately got
down to the business at hand and told his story with very little prompting from D'Ann.
   Leonard tells his story of being precariously positioned on his rooftop for three days
while he waited to be rescued. He evaluates the rescue effort with a more demanding stan-
dard than civilians might use because of his wartime experiences. He was eventually airlifted along with his neighbors to a staging area on Causeway Boulevard in Metairie and
then driven with a police escort to a shelter in Houma, Louisiana, where he received critical
medical care. He dedicates this chapter to the people of Houma, Louisiana.

I was born on November 17, 1938. I did one tour in Vietnam with the Marine Forces. I was doing logistics for six months, and then I was with the Fifth and Seventh Communication Battalion combined. I was a senior petty officer, so they took me and gave me six more corps men, and I had to set up over there.

I came up to retirement in the military, but I came out with a disability. I settled in San Diego, California. A few years later, I got it upgraded to a one hundred. After that, I was a carpenter. I retired twice, first from the military and then from my other job.

I started coming down to New Orleans for Mardi Gras. I stayed with Mack Slan (see chapter 15), a friend from San Diego days, for two or three weeks in the early '90s. In 1996, I said, "I'm going to make a change." I was still single then.

I married Dorothy when I came to New Orleans. My house is on a one-block street right off of Morrison Road. It's in eastern New Orleans. I live in a development called Kenilworth: beautiful neighborhood, beautiful street. Very quiet, very clean. It's a one-story house with three bedrooms. There were eleven houses on my block, and I think five of them had two stories. What I like about it is at night there's no cars on our street. Everybody's car is in their driveway.

The first week I got here, I joined the Zulus, because Mack's cousin was a Zulu member, so he sponsored me. I know a lot of people in New Orleans, because not only are there the five hundred members, but then we got our associate members over six hundred.

What I like about being a Zulu is that you have some of everybody in there. In fact, the whole elected officials of New Orleans were Zulu: the chief of police, the fire department chief, city councilmen, and the district attorney. It's an influential group. What happens when you walk through the doors as a Zulu, everybody's the same. You're not referred to as doctor, sheriff, lawyer, or whatever. You're just Jimmy, Richard, Johnnie, Leonard.

Katrina: "I Was There in My House When It Went Down"

I heard a week or so in advance that there was a hurricane building. It didn't bother me. I've been through a lot of stuff. That's the reason why I didn't even come out. I sent Dorothy out in the truck. I decided to stay because I believed we could have survived the storm, which we did. My thing was I really believed we was going to get some water, but I really didn't believe we was going to get flooded out.

The storm was a lot of wind, a lot of rain, and across the street from me, one tree came down. In my back yard, I have two trees. One of them went down, the small tree. A couple of limbs came off the big tree. That was the first day. People had been calling me and asking me how was I and everything.

I was in the house. I was just doing some things. The telephone went out. Then I started looking out the window. My street drains real good, and that's when I noticed the water. The street had water. Then I started noticing water coming up on the front lawn. Now the street is probably a foot or so below that lawn. Then I figured something was wrong, but I didn't know what. This is in the daytime. After I saw the water coming on the lawn, I said, "I'm going to have a little water in the house," so I got all of our clothes and everything from the closets and put everything on the three beds.

Meantime while I was doing that I guess the water was coming faster, which I hadn't anticipated. By the time I got through doing all this, and I looked out my back, where I have a patio enclosed with plexiglass, the water was five feet up over the plexiglass, and I knew it was very dangerous.

I had already checked the attic and pulled the ladder down. I had some things I was going to take up in the attic, but I didn't have time to get them. I didn't have my keys and my glasses or nothing. By the time I got to the ladder to get to the attic, the water just exploded. It sounded like a bomb. And everything just went straight up in there: the couch, the TVs, the refrigerator, and the freezer. The freezer just exploded. I had less than ten seconds between when I realized how dangerous it was and when I got up in the attic.

Now I'm in the attic and the water's rising so fast. I hadn't figured out how to get out of the attic yet. I had three bottles of water. No lights. I found a tool box in the attic, but I couldn't knock out the roof. Then I found the vent. It's got a vent which is about 18 × 24. So I used a 2 × 4 and knocked that out. The water kept rising, rising, but now I know I can get out through the vent. I'm looking at the water. The water stops about one foot from the attic. So now I'm a little bit more relaxed.

For the next three days, I was in and out of the attic on the roof. You had to go on the roof when the helicopters were coming, so you can see if they can see you. There were two ladies, school teachers, diagonal to me. They were on their roof. My neighbor on the corner was on his roof, a guy behind them, and a guy behind him. I called out to them. One of the ladies on the roof, her name was Katrina. All day long we would communicate with each other. At night the helicopters would stop, so we would go back to the attic.

In the meantime I started figuring out things because my neighbor's house next door is two stories. Everything was just floating up to the attic. The TV come up to the attic. I took this cord from the TV, because I wanted a lifeline. I was going to swim over to his house, and go up higher, if the water continued coming.

About the next day the water receded some. I went down in the house. The water was up over my chest, but I could walk around through the house. All the food was ruined because the water had came up over the cabinets.

It was a feeling of desperation. See I had an operation in August 2005. I started getting dehydrated while I was on the roof. Those three bottles of water I had? Two of them rolled off the roof. I was hungry. I got infected. But I got through it.

I think they actually had about eight helicopters for that whole area, but most of them wasn't rescue helicopters. That's all I saw in the eastern area. There was one helicopter that was always in the same place, but they wasn't picking up nobody. They only had one Navy helicopter out there them three days I was out there. We would talk to each other, and I'd say "Look, that helicopter is doing its job three times faster than the rest of them." I didn't realize until long afterward that they had a lot of people on the roof from the high-rise apartment complex down from me on Morrison Road.

The thing about it is no boats came. It just amazed me. After being in the military, I know there is nowhere the military can't go: water, land, sea, or whatever. They can do it. It was quite different from some of the places I'd been. I was in Vietnam with the Marines. I've been overseas. I rode ships. It was real different. When you stand there, and you see no reinforcements with the helicopters, and you look out there at all this water! You're saying to yourself, "There is no safety net for me. Where am I going to go? Unless a helicopter comes down and gets me, there is no other way I can go."

I was very fortunate our house didn't have any big trees. The people in the helicopters saw me, because they came back and got me at night, and when they came back, they came over my house, and shined their light down. My neighbors and I had communicated with each other: whoever house they come to, make sure they got all of us. Someone in a helicopter came and got me first because my house was a little bit more open. They seen me better.

I was impressed with the people who rescued me. Those guys came down on the roof and got across that roof just like spiders. We would have called them spider man. They was doing their job. They just didn't have enough help.

We went to a staging area off of Causeway Boulevard in Metairie underneath the overpass. At the staging area, they had some oranges and water. There were a lot of military people and policemen at the staging area. They tried to be as nice as they could. They probably didn't have no water and stuff. Whatever they had, they tried to share it with us. I was treated with respect. Ain't nobody going to disrespect me in a situation like that. You won't like me after I get through talking to you, but you're not going to disrespect me.

And they had three buses there, and they were waiting for the last bus to fill. I guess we probably stayed there two or three hours. You just get on the bus and go. They took us to Houma, Louisiana, where one of our Zulu member lives.

Houma: "They Was Overbearing Nice"

The shelter was real good. It was a hall. It was the second part of the convention center. They have two convention centers there in Houma, and this was the smallest one. About three hundred and fifty people were there. They had a wall and two sections. At the shelter, they had TVs but there were so many kids in the shelter, so they wouldn't show us what you were seeing.

We tried to keep the morale up. We'd joke and talk about different things. When we got to the shelter, we all put our shelter cots together. With me being so tall, the ladies found me some things to wear, and they found me a couple pair of tennis shoes. The first day and night, we had to wash up. But then someone made an outdoor shower. We rotated around. The womens in the morning, the mens in the afternoon. And we stuck together. We became like a family.

The first night, all the pets and things, was in there with us. But the very next day they took the pets somewhere else. Then they started fixing it up better and better. They start making a dining room section—it's still in the sleeping area, but they tried to take the senior peoples and the sick people and make sure that they served them. So that shelter was excellent. They kept something twenty-four hours for you. They brought us home cooked meals.

They had a medical department and that was good. We had some really bad off people, and they made beds for them. The cots weren't sufficient for their condition, so they went and got mattresses and beds and blankets. They took care of the people real good.

I was fortunate enough, because they had to send me to the hospital, so I was out in Houma a little bit more than the other ones. We had a lady in the shelter on dialysis. Everyday me and her would go to the hospital.

A lot of people started coming around us, because they could see how we were. And this one young man that had been on this high rise near Morrison, they took his wife and kids, but they didn't take him in the helicopter. I guess they had boats over there too. But they wasn't taking the most physically able people, so he had gotten separated from his family. When he found his wife and kids, they were at Southern University. Those blacks who lived in Houma on their lunch break would come over to the shelter and talk to us. And this one guy, he say, "I found my wife and children, but I can't get no way to get out of Houma." The other guy say, "As soon as I get off work, I'm going to bring some more of my buddies over here, and we're going to take you." So I went to use my ATM card, and got some money out, and I gave the guy twenty dollars. I say, "You don't know whether you're going to find your wife right away or not. When you get there, you're going to need some water and stuff." I gave Katrina and the other lady some money too.

The staff really bent their backs for us. I met this police man. I sent a thank you card to him, because he helped get me in the computer for my family to find me. I didn't remember the phone numbers or nothing. And I'd never been to Waterproof, Louisiana, where my wife's people are, before.

Waterproof: "Fine, Just Boring"

My family found me. One of my wife's brothers lives in Houston. They called him, and he said, "I'll go get him." He drove from Houston and came and picked me up.

Things in Waterproof, Louisiana are fine, just boring for me. I think the population is under two thousand. If you drive around the lake, they've got mansions on the lake. Thirty or forty years ago this all used to be white. So the kids moved away. Then the people die off, and the kids don't come back. They just don't care no more. I understand you can get property around here pretty cheap. Just go down there to City Hall, and pull the tax records. My income is stable, my retirement benefits and all. I make fifty thousand dollars a year now for just waking up in the morning.

It's just that I'm so far away from the hospital. I have prostate cancer, and there's a follow-up procedure for prostate cancer. When I went to the hospital in Jackson, Mississippi, I didn't have an appointment with Urology. They did make a space for me. In fact they tried to do as much as they possibly could. I had lost my glasses. So they did my glasses, and they did give me a urology exam. To redo my teeth it took six trips up there. I got my shots and everything. But I got my medication straightened out, and I'm going to be alright. I tell them, "If you don't do it, just give me a statement why you're not going to do it, and I'll send it to the Disabled Veterans Association."

I've been back. The house still has some stuff in there, and I'm sure it's really in bad, bad shape. My headboard went and broke up just like a cracker. You have to remember that water stayed in my house for over two-and-a-half weeks. I lost everything. Everything! All the papers: the marriage license, the deeds, the insurance papers, everything! They tell me my house's a complete disaster. There will be no inspection by an insurance adjuster.

I'm going back to New Orleans definitely. I'm not going to rebuild. I'm just going to rent. I told my wife, "Next time we hear a storm is coming up, we can just put everything in the car and go."

Conclusion

FEMA participated in the Hurricane Pam exercise at LSU in 2004. Hurricane Pam simulated what would happen if a Category 4 or 5 storm hit New Orleans. Included in the simulation was the fact that approximately 110,000 people in New Orleans did not have access to private transportation. The simulation predicted that if a major hurricane hit and broke the levees, thousands might either drown or be stranded for days. The results of the exercise, published in the Times-Picayune, *were widely known by officials and the educated public.*[3]

I've always said that we knew this storm was coming. I was really disappointed because I thought that the government had a lot of things in place in the city of New Orleans. After Wednesday, August 24, we still could have did a lot of things. We could have closed the schools. Made sure all the school buses were filled with gas. Either move them out of the city or move them to a higher place. A lot more things could have been in place, especially equipment, food, and water.

The response should have been different. I think about this a lot. Don't you know I know how fast the military can move? I was called on Friday to go to Camp Pendleton, Monday to go to FMSS (Field Medical Service School) to go to the Marines. Thirty days later I was in Vietnam. It happened, ok? The military can take you anywhere in a matter of minutes. If you can't walk it, the U.S. Air Force can drop you. I guess I'm an optimist, especially having been in the military.

Pete Stevenson

Pete Stevenson was raised in the Upper Ninth Ward. He dropped out of school after the eighth grade to help support his single mother, Rosemary, to whom he dedicates this chapter. He worked as an auto mechanic for most of his life, with a ten-year detour in prison. On the eve of Katrina, home for Pete was an eight-block area that encompassed the Lafitte Housing Development, and extended to North Claiborne, a compact world easily navigated on bicycle.[1] In the projects, he had many friends and lovers.

The setting for this interview was Pete's efficiency apartment in Cullman, Alabama, known nationwide for the trial of Tommy Lee Hines, a mentally challenged black man convicted of allegedly raping a white woman in 1978.[2] At 8:00 a.m. on December 30, 2005, he opened the door to a stranger, the interviewer.[3] He was both ill at ease and eager to share his troubled thoughts with someone. The interview was conducted in the combination kitchen and living room furnished by the church people of Cullman. A table for two resting on eighteen inches of linoleum was all that separated the kitchen from the tiny living room, with its faded brown carpet, worn couch, and old TV.

He is one of the poorly educated individuals who left New Orleans before the storm only because of an enormous extended family network. Above all, his example shows how much courage it took to leave everything familiar at a moment's notice with little cash, or in his more colorful expression, "on a wing and a prayer." Pete also graphically illustrates the extent of the secondary trauma to the people of New Orleans caused by the mishandling of the Katrina rescue. In exile, the shortcomings of his earlier education complicated his life. He remembered neither the last names nor the addresses of most of his lovers and friends, none of whom had cell phones. Pete's experiences demonstrate three competing tensions in the post-Katrina debate: First, he missed his community and his friends in Lafitte, who gave meaning to his life. Second, although he didn't feel in any danger, he had grown weary of the omnipresent gunfire. And lastly, without his community in Lafitte, his disability check alone was not sufficient for him to find housing in New Orleans.

I was born in Decatur, Alabama in 1945. I had five sisters and three brothers. I'm next to the oldest. My daddy, Thomas Jefferson Stevenson, was dead. My mama says I look just like him. I was raised by my uncle. My uncle used to work at a paper mill here in Alabama. Then my mom came and got me when I was seven years old. I went to New

Orleans. I was raised in the Desire Project. It was good. I never got in no trouble with nobody, or anything like that. I was raised the way my mama taught me: Baptist.

The original Desire Project opened in 1949 as a response to the postwar influx of individuals born and raised in the rural South. It was home to over ten thousand residents. By the mid-1960s, it was the most densely populated area of New Orleans with an average of six residents to a unit. The schools in the Desire area were overcrowded, requiring students to attend in two shifts. The dropout rate was one out of two.[4]

School was nice. Math was my favorite subject. I got to the eighth grade and then I dropped out of school to make some money working fulltime. I been doing mechanic work since I was 16 years old. I learned the hard and easy way, I picked it up and I learned it. I never went to school for it. I was doing hotel work on the side when I was younger. When my mama had her twin daughters, I paid for all her hospital expenses. I started working as a hopper when I was eighteen. We dumped garbage out of cans into the garbage trucks. The pay was about $2.50 an hour. After about three years, I got hurt on the job and had to have an operation on my back.

I got married when I was twenty-five and my old lady, Yvonne Stevenson, was twenty-three. We rented a one-bedroom apartment on Abundance Street. The day that we got married was my birthday, May 14. One month later, on her birthday, she died. I haven't remarried since then. I like the single life. I stopped drinking in '75.

After my wife died, I moved back in with my mama. She moved all over New Orleans, and I moved with her while I was still getting on my feet financially. I liked being in Uptown best.

Then I got in a little trouble. I shot my baby mama's brother. I was working at this filling station doing mechanic work, and he told me he was going to come in and rob me. I shot him in the side. He's still living. He asked me when I got out of jail, "Why did you shoot me?" I said, "You said you were coming to rob me, so I shot you. I'm going to shoot you before you shoot me!"

I did ten years. It was rough. I did it at Angola, Louisiana's big state penitentiary that sits on the Mississippi River.[5] I didn't want to stay in New Orleans at the time I was getting out on parole, so they paroled me to Pasadena, California, where my brother was. I was fixing cars and worked in filling stations. It was the best place I ever went to. You could do what you want to do out there, if you set your mind to it. Have fun, live happy.

I came back and visited New Orleans from time to time. When my mama told me she needed me to be back here, I came back to stay in 1989. I also came back because of my son, Anthony Brown. I just did the right things from that point on. She was living in the Iberville Project by then. I had enough money to do what I wanted to do when I worked on cars independently. I was able to keep change and stuff in my pocket. I was living near my mama.

Before the storm, I was staying in the Lafitte Project. Now I'm totally disabled. I suffer with seizures. I takes my medicine every day. Medicare pays for it. I have to put a few dollars with it. I'm on disability. It's never enough, but you have to make do with what you get. After the bills get paid, believe me, you don't have nothing left.

Everything was good for me in New Orleans before the storm: the people that I knew, this friend that I had, the things that I used to do. Everything was just smooth. I had a lot of women friends in the project; we did a lot of things together. People I met while I was in the neighborhood. What I liked best about living in New Orleans was being able to go places, do things. I got around by bus. If I didn't catch the bus, I'd take my bicycle and ride around.

I didn't mess with too many dudes. I just mostly stayed by myself. I didn't go to bars because it's not safe. You go to a barroom, next thing you know you might hear gunfire going off. Matter of fact before we came here, that Thursday before the storm, it was a dude followed another dude and shot him seventeen times in front of his door. You always heard gunfire. Nothing but gunfire. Shooting, killing. All useless stuff. I got tired of it. And I didn't want to be the cause of somebody else being sent to prison. I didn't mess with nobody, so nobody messed with me. I felt safe for myself.

Cullman: "I Came Here on a Wing and a Prayer"

I heard that Katrina was headed toward New Orleans about a week before it happened. The weatherman got on the news and said, "Everybody take precaution. It's best to evacuate." I decided to leave New Orleans on Sunday, August 28, with my whole family. We all thought we were going back the next day. I just packed a shirt and a pair of pants. I had my wallet, but most of my papers was left at home. Everything else was left on the first floor in the house in New Orleans.

According to Pete's oldest sister's granddaughter's mother-in-law, Velbert Stampley, better known by her nickname "Sugar," the drive on the highway took hours. Her oldest son led the caravan of fifty people in ten cars. None of the fifty knew anyone in Alabama.[6]

The cars made it. The little money that we had, we ate off it. We survived, we got here. I had about three or four dollars in my pocket when I first got to Cullman, Alabama. Sugar's son got everybody into the Hampton Inn in four or five rooms. When we ran out of money, we went to a shelter at the McGukin Civic Center.

Shortly after Pete and his extended family were displaced from their shelter by a gun show, a local woman assisted them. On September 2, 2005, Alabama Governor Bob Riley announced a state initiative called "Operation Golden Rule." The initiative was designed to assist an estimated 10,000 displaced persons to find long-term shelter in the state.[7]

Everybody treated me fine. There were some rough spots for my family.[8] I ain't had no run-ins with nobody. So far I haven't. When they came and gave to us freely, they was helping us out. They gave us clothes, they saw to it that we was being fed.

I got this apartment by the help of the lady that owned it. She called my sisters and they came. They've been very nice. They did us good justice. I can't talk bad about them because they did what they can for us. They helped with the furniture too. I bought the little record box right there.

I don't have a bicycle here; I have a car. The people helped me to get one. The church is helping me. It was so hard for me to get a driver's license when I was in New Orleans. The written test, I know it's going to be harder out here. I'm not very good on reading too much. I wish I could get back into school.

New Orleans: "They Meant to Kill Us"

I can't go back to the city. I don't want to have nothing else to do with New Orleans. Anybody with sense should be fed up. The blowing of the levees was meant to kill the blacks and the poor whites.[9] The two people that helped blow up the levees killed themselves.[10]

A friend in the city text messaged Sugar that the levees had broken shortly after it had happened. The friend described the loud boom and the gossip on the street. Sugar relayed the news to Pete and his family watching TV at the Hampton Inn.[11]

I been knowing people in New Orleans didn't like blacks. I first became aware of racism many years back. Like when they call you "boy" and all of that. I don't too much worry about it no more. I done growed up, and I done learned to let things go by that are not supposed to be. I'm nonviolent until somebody threatens me. Then I have to do what I have to do.

The people ain't back there. They got Lafitte fenced all up, even though it only took on twelve inches of water.[12] The doors are all boarded up.[13] They don't want nobody in there no more.

Pete's worries were confirmed by comments such as those of Rep. Richard Baker of Baton Rouge, who was overheard by a Wall Street Journal reporter saying to lobbyists, "We finally cleaned up public housing in New Orleans. We couldn't do it, but God did it."[14]

"This Is a Lonely Trip"

I got a family. They all good people. My mama's by my little niece's house in Cullman. She's eighty-two years old. My sister and them, we split up now. Everybody's spread all around the world. We talks to them over the phone. They all came for Christmas.

What makes me feel the worst is the things that I done seen. Dead bodies floating up and down the streets haunt me. I didn't see it personally, but seeing it on TV was bad. Yes indeed! Seeing people floating down the streets of New Orleans and not knowing who they were. I still can't find several friends that I used to be with. They were just friends and they'd come by and visit. Still I'm worried about all the friends I had, that they died or got washed in the river. Thirty feet of water, you know some of them got lost. I had one that was a friend of mine, her name was Sandra Williams, and I ain't heard from her yet. Another of my friends, her first name is Beverly but don't know her last name. I think she stayed about four doors from where I was staying. That's about all I know about her. Couple of dudes that I knew was good friends of mine. I ain't heard from none of them. I had a friend, Carl. He stayed on Claiborne around me.

I done been to the Red Cross with all of them. I put their names in the system. Somebody's got to be missing, because the Red Cross ain't never told me they found none of them yet. I misses them. My mind would be at ease knowing that they are alive.

Best Western behind I-157, they got a clinic in the back. I guess I'm telling you I need to talk to a psychiatrist, because as the days go by after the storm, and things that happened to me since the hurricane and everything, it starts coming back down on me. I feels it. I think about it. It hurts. I'd like to go to that mental health clinic and see if I'm stable enough to hold myself together. The burden that's coming down on me, it's kind of getting to me a little bit.

Conclusion: "I'd Like My Life Back"

Family can't do what I need. I just don't have nobody that I can say is mine. My whole life is interrupted now. I ain't got nothing going for me, I ain't got no one to be with.

I sleeps alone. I don't have no one to talk to. I don't have no talking friends to talk to here. I've met some people who's been very nice to us since we've been here. But they don't have the same history.

What I need most is a friend, I'll put it like that. And something just to keep me going, keep my life together, and help me the best that I can be helped. But I also need financial support. I guess that's the most important thing.

I'm not going through other storms. Next year they have another storm, might be the same thing. There ain't no sense in going back.

I likes it here better than a bigger city. It's peaceful, it's quiet. You don't see policemens like you would in New Orleans. You don't hear gunshots. I can't get around because I don't know nothing about Alabama like I should. This is my first time coming back to Alabama in about fifty-three years. I'm learning the best that I can.

The good Lord, the people in the community, and I guess my health, life, and strength has gotten me through to this point. I would tell the younger generation to be cool, be righteous, and stay together.

SEVEN

Parnell Herbert

Raised in the Lafitte Project, fourteen-year-old Parnell Herbert started washing dishes in the French Quarter in 1962 to help supplement his single mother's wages. He dropped out of school, which he found boring, and took to the streets. He joined the Navy during the Vietnam War, and while on leave from the Navy in San Francisco he was introduced to the Black Panthers, a party founded in 1966 in Oakland, California by Huey P. Newton and Bobby Seale. Parnell never joined the organization, but he sympathized with its values from that time on. On his discharge from the Navy, he got married, had children, landed a technician's job with BellSouth, and bought a house in Gentilly. Parnell's work with BellSouth eventually took him to Pensacola, Florida, where he realized how much he loved New Orleans. Known as the "Poetic Panther," in retirement he began to publish poetry and give workshops on racism around the country. He drove from the city by car to Virginia, where he watched the disaster unfold on cable news shows.

This interview took place on January 11, 2006, in an Urban League office in Houston. Parnell is a tall, muscular, and youthful-looking political activist.[1] His heritage was prominently displayed in his two silver African bracelets. The simplest of questions were answered with detailed, eloquent, and impassioned discourses: a cross between poetry, storytelling, and theater.

A common thread of Parnell's life is his attempt to understand the genesis of New Orleans's street culture and the impact of its drugs, alcohol, and violence that affect so many members of the community. He places his childhood, adolescence, and coming of age as a black man into a political context. The youngest member of the retirees, he has enjoyed a comparatively easy financial transition to post-Katrina life because of his BellSouth pension and his computer-savvy daughter, who guided him through the FEMA maze. Although his primary residence is now in Houston, Parnell is actively involved in efforts to rebuild New Orleans along racially just lines. The Poetic Panther's story traces one strand of black activism from the 1960s to the present day. His story is the journey of an African-centered artist into a leadership role that continues in the Katrina diaspora.

My father disappeared on me when I was two years old. He popped up again, right after I'd done my first cruise to Vietnam. My sister threw a surprise party for me. We went to this house, and this man walked up and hugged me, and said, "Look at my baby."

I knew that was him. And I just froze while he hugged me. In a few minutes, he was gone. He's a stranger to me.

Someone described me as a self-made man. I developed myself by emulating different characteristics of men that I saw because my father wasn't there and my two brothers were gone. I took what I saw from different men that I liked, and I applied it to myself. I liked his walk, I liked the way he dressed, I liked the way he talked, I liked the way he combed his hair. It's like putting together this puzzle.

My mother was living in the Lafitte Project when I was born in 1948. Two brothers, two sisters, my mother, my aunt, my grandmother, my uncle, and me in a one-bedroom apartment. We had segregation right there in your face. They had jobs, and we didn't. They lived in a project also, because you had the Iberville Project behind Canal Street built in 1941, but it was better maintained.

Mom was like thirty-five when I was born. She didn't know much about what was going on in the streets, but she was always well respected in the community. I used to see her as naïve but I found out that mom used to help feed Black Panthers while I was away in the Navy. She was the very first elected president of the Tenant Council of the Lafitte Project. She was very special to me, and to our community in general. All my friends' mamas referred to her as Miss Marie.

Unless she had an extension cord in her hand, she was just a quiet old lady. Mama was good with the extension cord (her disciplinary weapon of choice). She was a little bitty lady and I was the baby in the family. I was spoiled and pampered (by all the women in the house), but I caught my share of whippings too. From me up to my oldest brother, she'd do all five of us. She'd take us into the bathroom, turn the water on so my grandmother couldn't hear, and she'd fire us up. I think I was fifteen years old when I got my last whipping from her. I remember I brought my hand up to catch the extension cord just to keep it from coming down on me and she thought that I was getting ready to fight her. She said, "Oh, you want to fight?" She threw the extension cord on the floor and squared off at me. After I convinced her I was not fighting back, she picked up her extension cord, hit me a couple more times, then sent me to my room.

I remember the last thing that she normally would have whipped me for. She sat me down on the sofa, talked to me and cried. She said, "I try to do the best I can and you're too big for me to whip. I just don't know." The extension cord had nowhere near the kind of power those tears running down her face did. I was like, "Ma, take the extension cord. Whip me. It'll make you feel better." Somewhere along the line she got smart on me, so all she had to do was look like she was going to cry, and I'd melt.

My mother used to make five dollars a day cleaning white folks' houses. My mother also ironed white folks' clothes that they would bring over a couple of times a week. These little white kids called my mother Marie. I was like eight or nine years old when I started noticing this. And I said, "Man, this is not cool at all." That's when I really started developing a dislike for whites. As well respected as my mother was in the projects, these little kids would call her by her first name.

I was aware of slavery. Just a couple years after slavery, my grandmother was born, so her parents were slaves. She would tell stories that her parents had told her. I never considered myself patriotic to this country. I always had a little bit of anger in me.

Street Life: "You Feel Kind of Invincible"

I was a pretty busy kid. I was active out in the streets, hanging out with the fellows. I used to get high: I smoked weed, popped barbiturates, drank alcohol, smoked cigarettes. All this before I was fourteen years old. I never did heroin or any other hard drugs. I never injected myself with any drugs.

The first three months of tenth grade I was suspended three times. After the third time, I said, "Why go back? Every time I go, all they do is put me out again." School was really boring to me. I dropped out and went to wash dishes at Brennan's French Restaurant in the Quarter. I'm making like eight dollars a day washing dishes. I gave most of it to my mom. I would keep maybe two dollars as cigarette and beer money for me.

When I was coming up, everyone had a job. If you didn't have a job, you went to jail. It was called vagrancy: no visible means of support. You could be the biggest drug dealer in town, have money and clothes, but if you didn't have a check stub to show the police, you went to jail for vagrancy.

I only washed dishes for maybe six, eight months, and then I decided, "I'm going to go back to school, because I don't want to be a dishwasher for the rest of my life." I was at Joseph S. Clark High School, and they wouldn't take me back. So I had to go to Booker T. Washington. By then the streets kind of had me, and I dropped out again.

It was exciting. You get to exercise a little freedom. The police in New Orleans today have got the elders scared to go out. But they don't really scare the young people, because the young people feel like, "I can either outsmart them or I can outrun them." We used to stand out on the streets in the project and when the police passed by, we'd throw rocks at them because we knew we could outrun them. They couldn't catch us in the projects. That was one way of beating the system.

Street life was all about hanging out, getting high, and being cool. It is an addiction to the lifestyle, to the allure. You were defiant and girls always liked the gangster. (Someday I'm going to write a book about ladies and thugs.)

The streets offer night life. A whole different world being able to prove that you're tough. These kids out here today are killing each other to establish that reputation. Once I tried to stab a complete stranger. I was so young and so dumb, the knife was the wrong way, and it closed and cut my hand. This guy had on a coat that was two inches thick. And he was running away. I bled all over his coat, but I had the rep because everybody thought I stabbed this guy. "Don't mess with Herb, he'll stab you." Once you had that rep, you were alright and your family was safe. You feel kind of invincible. It was a lot more challenging and interesting than those books they tried to make you read in school. In hindsight as I look at the young brothers today, and I look at myself back then, I feel like it was really fear because we were more afraid to try to succeed in this world with the deck stacked against you. Whereas we could stay in our own little community and be a big fish in a little pond.

Our people in New Orleans have been failed by the system. We teach our children that this system is created and designed to protect them. Then someone kills your brother and reports it, and the police don't do anything about it. Now you got to go back to survival instincts. You know that you're going to see the person who killed your brother every day. You and your whole family now are set up as chumps. They may kill another of your family members, so you have to do something to recommand the respect that you or your family once had. You've got to have it known that there's a price to pay.

"I Was Safer in Vietnam than in the Streets"

My best friend Charles Harris, wherever he was, I was. Charles told me that they were going to do this robbery, and this guy had a gun. We never had guns in our group, because one thing back then the establishment did what they could to keep guns out of our hands because they knew that we would use guns on them. They had just opened this theater on Tulane and Carrollton, so they were going to rob this theater. I said, "I can't go man, I asked mama to warm the beans. If I don't go back to eat, she's going to be all over me." They went, and I went inside and ate my beans and a couple of hour later Charles knocked on the door and told me that they did the robbery, and some white boys chased them. Charles stayed with the guy with the gun, and he just fired back into the crowd, and didn't hit anyone, but Glen went in a different direction and they caught him. Later that same night the police picked Charles and the guy with the gun up.

I looked at how closely I had come to being a part of that. At the time I was sixteen years old. I had two brothers in the Navy and they both had been trying to get me to join. So I felt like the timing was right. I had to wait until I made seventeen. My mother was against it.

I did three cruises to Vietnam. I worked the flight deck on an aircraft carrier, fueling planes, loading bombs, and mixing napalm. I wasn't political at all then because I had no idea what Vietnam was. While I was in the Navy, I did get involved in a robbery in 1967, just one day after my second cruise. I had spent nine months overseas, one day back in the states, and I was in jail, the Santa Rita County Jail in Oakland. We jumped on this guy just because he was white, just for the hell of it. The guy had about ten dollars. During the time I was in jail, that's when I learned about the Black Panthers. Huey P. Newton (originally from Monroe, Louisiana) was in that same jail. None of this was out when I went overseas nine months earlier, but there it was all over the place. I was ready to join but they wouldn't take someone who was in the military. Meeting Huey Newton gave me a purpose. Before I just didn't like white people; after Newton, it was white people are my oppressors. After forty-seven days they dropped the charges, and let us both go back to our ships.

We did some violent things in the streets but I've never killed anyone. If I had done something like that back then, it wouldn't have bothered me. But after changing, going through what I've been through, and becoming who I am, I know that would eat me alive. In Vietnam, I was an accessory: I loaded the bombs, fueled the planes, and mixed the napalm.

From the Streets to BellSouth

When I got out of the Navy, I got married right away and I got back into the streets. My wife at the time supported us. My wife went to work and brought the babies to the nursery. I stayed in bed, woke up, and got high. She'd come back and pick the babies up, and I'd hit the streets. We'd moved out of the projects into a house in Carrollton/ Hollygrove in Uptown.

Wake-up calls hit me periodically at different stages of my life. I decided I needed to do better. I went to Delgado Community College. It was like a trade school then. I went there for electrical construction. I failed all my math courses but I aced the drafting course.

After a couple of years at Delgado, this company came back there looking for young draftsmen and I was the one, naturally, because I was married, had a child, was responsible, and needed a job. I quit Delgado and became a topographical draftsman for Urban Transportation and Planning Associates. I went from a draftsman to chief draftsman to traffic engineering inspector to chief traffic engineering inspector in five years. I didn't have the time to hang because I was also going to school at night. I had a wife and baby at home.

In 1973 we bought our home in Gentilly, a lakefront area drained and developed in the 1920s and '30s. We had to get a court order to remove a clause from the deed covenant that would not allow the seller to sell to "Coloreds."[2]

I hired on with BellSouth. BellSouth transferred me to Florida. My plan was to stay there long enough to get ten years so I could get that vested pension and then I would come back to New Orleans. I got out there and started liking the job. In New Orleans I had been a service tech going into the houses. In Florida I was a lineman. I enjoyed the freedom up there in that bucket.

It was, however, like culture shock because Pensacola was 14 percent black. New Orleans in the '90s was 60 percent black. I was the only black in a thirteen-man crew. I used to feel so out of pocket being around all these white people. I felt like I represented the black race to that group. I would iron my uniforms before I'd go to work. I always wore a white shirt. I always made sure I did a good job. I would carry an extra clean shirt, that way if I got dirty during the day, I would put that clean one on before going back in. I would shine my work boots. They could say, "The black guy does his work, and he stays sharp when he wants to." I represented my people well, and I made whites feel uncomfortable.

I would always show up thirty seconds early for work because I didn't want to sit around and socialize with them. When the O.J. Simpson trial came along, not once did any of them ask me what I thought. I knew they were curious as hell, but I had them kind of intimidated. Being from New Orleans has its rewards. Word is, people from New Orleans shoot you.

Four years ago I transferred back to New Orleans, and I was back almost a year and I injured my back. That pain was real. After that second surgery, they put me inside on light duty. When I made fifty-five, I officially retired, which gives me the independence to do the things that are important to me.

A Passion for New Orleans

I developed a passion for New Orleans when I went away from it. I came home three weekends a month. I'd drive four hundred miles in one weekend. I missed seeing black people. I felt great just about the time I would hit Slidell, when I'd start seeing black faces in cars. I still like going to clubs and seeing my sisters dressed up and all decked out. I started getting involved in more and more of our cultural activities.

I became Afrocentric when I was in Pensacola. I had this room called the African room. It was all decked out in African masks and carvings. Then it went into the next room, and pretty soon it took over my whole three-bedroom house. I've even gone so far as to take the DNA test. I know I'm descended from the nation of Cameroon, the Bamileke/Fulbe people.

I enjoy being in what we say is the most Afrocentric city in America. I feel like it could be a much better city, if the white establishment weren't pitted against us. It's like when the Haitians had the nerve to declare themselves free from slavery after the Haitian Revolution. Not only did other slave-trading nations place them under an embargo but the recently emancipated slaves were forced to pay reparations to their former masters, the French, whom they had thoroughly defeated.[3] Similarly, back when Dutch Morial was elected first black mayor in 1978, the city started changing. White flight occurred in the '70s. It was like they said, "OK New Orleans, you want your black mayor, black chief of police, black fire chief, black city council, and black school board? We're going to take our money and go." They did everything they could to starve us out, to marginalize our leaders, to minimize our wealth, and to break us down. They watched our levee system and our school system go down.

"The Last Militant"

At work I was a union rep. The little struggle I had in there as a union rep was nothing compared to what I saw out there on the streets of New Orleans. I'm making thirty dollars an hour, and complaining and fighting. I get out here and look at the people who were really struggling. I used to go to these people's houses every day, but I never really considered what they were going through, because I was so burned out from working so many hours.

Once I left BellSouth I looked back at my life, and there were so many blessings bestowed upon me. These blessings weren't for me to lay on the beach in the Bahamas. I said, "I need to give something back." That's when I started getting involved with the undoing racism workshops with the People's Institute for Survival and Beyond organized by Ron Chisholm. We go to schools and talk with kids. To these young brothers now I say "You are displaced warriors. You don't know who your enemy is. Your job is not to terrorize your own community, but to protect it." When you look at the way someone in the 'hood walks, he's got a certain lean to his walk, a certain strut. That's how a warrior used to walk around a village. Fear is nothing compared to respect. I try to present myself in a way that those black kids are going to pick something up from me and take that and apply it to themselves.

In June 2005, we got started a local organizing committee, the African American Leadership Project (AALP). The AALP is a network of about fifty African American community, business, and religious leaders and representatives from New Orleans. We focus on policy analysis, strategic dialogue, and consensus building to defend the interests of our people.

During our June Siege against Racism, all these grassroots organizations came together with the New Orleans Mardi Gras Indian Cultural Coalition. We demanded that the mayor, chief of police, city council, and seven police captains all be there. They showed up. Big Chief "Tootie" Montana had a heart attack and died.[4] He said, "I want this stuff to stop," and he keeled over and died. We had them at that point. Then we asked the people from the Essence Fest not to go into the French Quarter for one day, during the July 4th weekend. We had a march on Bourbon Street, where we had pickets in front of Razzoo's Club and Patio, which caters to whites. We were protesting the killing of a young black male tourist who had been killed by a white bouncer on December 31, 2004.[5] We had people who hadn't spoken to each other for years starting to work together.

Katrina

I never evacuated before. I was at my house and my daughter who evacuated to Natchez, Mississippi called and said, "Dad, this thing is a Category Five. It's coming down the mouth of the river, and it's going to bring this wall of water. You've got to get out of there." My sister and I made this pact, if I ever left for a storm, then I would get her too. I'm still invincible, so I can survive anything. But now I've got a sixty-two-year-old sister over there depending on me. That changed my flow of thought. We drove out to Virginia. My son is stationed out there.

On the highway we'd heard it came through as a Category Three. My sister and I were ribbing each other: "You're running from a Category Three hurricane." When we got to Virginia, we watched CNN. That was a nightmare: water was flowing out of my eyes, the way it was flowing through that levee. I knew my house was under from the TV.

Something that really surprised me was the number of African American people in New Orleans who had large American flags in their homes. Were they flags that once draped a loved one's casket? Why did the survivors begging for help from the U.S. military forces feel it necessary to wave the American flag, as opposed to a white sheet or blood stained towels?

One friend of mine, her mother was in a hospital bed, and they were rolling her to the Superdome before the water got too high. The helicopter came, and they lowered the harness and pulled her up. The daughter thought they were going to take her up too, but they said, "No, we don't have room." They flew off and she's never seen her mother since then.

Losses and Concerns

I lost my car and my dog; that was my biggest regret that I left my dog. He's a one-hundred-and-fifteen-pound, twelve-year-old rottweiler like me. He's going to take care of himself too. I owned my house in Gentilly for thirty-one years. It took so much out of me. You accumulate a lot of stuff in thirty years with your kids and photos. All my clothes are gone.

I try to leave FEMA alone as much as possible. My daughter's all over that. "Take care of my daddy." My daughter called me with instructions on how to get stuff started. She was on line. I got into the system with them. Because I have a source of income, I didn't have to worry.

I didn't save anything but this one African mask I took with me. That was my favorite piece. I've been lugging it around from New Orleans to Virginia to San Diego to Houston. That's going to be the first piece I hang in my new apartment.

Our elders are dying. You hear everyday an elder died, but it's happening at a rapid pace right now. I believe that the stress of being displaced like that has gotten to them. We've got our kids here in Houston fighting the Houston kids. Houston has gangs. The gangs have certain turfs and areas where people move off the sidewalk when they walk through. New Orleans kids are not getting off the sidewalks for anyone. So that's a fight.

Ideally, I'd like to see this government create programs to train these young people to rebuild their own city. Take this young brother off the streets and put him behind the wheel of a bucket truck, a dump truck, or a backhoe. Teach him construction. Teach some to draw up the plans. Train some carpenters, electricians, and plumbers. Teach them to do the work to rebuild the city, work they can do anywhere in the world. These old houses in the Ninth Ward that's down, I'd love to see them go in there and tear the rest of them down, but rebuild them with the same people who were in them before.

I'm concerned that some of my people will use temporary accommodations provided by FEMA as a vacation instead of focusing on getting themselves together, and being prepared when this time is up. I'm really worried about our people being caught off guard at the end of this period, and we're going to have a bunch of homeless people out there.

Politically, we at AALP had things going really strong, and I think that's why Katrina took so much out of me, because we had accomplished so much, and we were on the verge of accomplishing so much more, and then they poured that water on us. All fifty of us just scattered all over the place. And it was like, man, I don't have the time or the strength to do this all over again. While we were gone, in New Orleans they terminated the teachers, changed public schools to charter schools, and did a lot of things that we fought against.[6]

Vision

I love New Orleans, I love the people of New Orleans, I love the culture of New Orleans, hell I love those raggedy buildings. The French Quarter, I've got no love for that. We can take our culture anywhere. As long as I know that my people are alright, I can be anywhere.

Our ancestors were forced over here. Their bodies are buried here, but their spirits were never enslaved. Most hurricanes are gentle breezes off of African shores here to the land of descendants of European thieves and horrors. So they're coming from Africa, building up this rage across the absorbing throes and this watery path where so many black lives were sacrificed at sea. It's tearing this land apart, a vengeance of Mother Nature and the Mother Land combined.

I look at my ancestors, how they were before slavery, how they must have fought trying to escape slavery, and how broken and humiliated they must have been during the period of slavery. I try to carry myself the way I think a strong warrior would have done in a village.

I've never been to Africa. That's on my front burner right now. All I'm doing is packing, getting ready to go. I feel it's going to be a very spiritual thing for me, to set foot on African soil, knowing what people I'm descended from. I want to go through the Door of No Return from the opposite direction.

At first I said, "I'm not going back to New Orleans for them to allow something like this to happen." I believe that there was a deliberateness to allow tactical portions of the levees to deteriorate so that they would be the weak points. They couldn't beat us otherwise. So that's how they drove us out. I told Strong Buffalo, "When Americans came to your land, they tried to enslave your people, and they couldn't do it, because you could walk away and you know the terrain. So they tried to annihilate your people and they went to Africa and they stole my people. Now we're not working out, so they need to get rid of us, and now they're going to South America and get another people."

But then I felt, "They want us out? We coming right back! We're going to show them. They're not just going to take our city! This is the land of our ancestors. We built this." After that, however, I had a lot of time to think, sitting out there for a month in San Diego watching palm trees and humming birds. I said, "What was my original intent in doing this work? Was it to make sure white folk don't take the city? No. My intent was to make life better for the black folks of New Orleans." And there are many black folks of New Orleans whose lives are better because they've gone on to better places, and they've gotten better jobs and better housing.

Right now I feel like in 2006, I'm going to be here in Houston because there's nowhere to live in New Orleans for me. Maybe 2007 that entire year I want to be in the New Orleans on the ground fighting. In 2008, I want to go back and watch palm trees and hummingbirds.

AT THE HEIGHT OF THEIR CAREERS

EIGHT

Harold Toussaint

Born in 1950 in the uptown community of Pigeontown, Harold Toussaint chose his family name in adulthood to disassociate himself from a reminder of slavery. Although he feels most at home in France, he is quick to say that he has never forgotten his time in the Upper Ninth Ward, where his family moved when he was six. A self-taught, free-spirited sommelier who has won contests in Paris, San Francisco, and Boston, Harold took a sabbatical from his craft when he returned to New Orleans in 2000 and became involved in his church. Deacon Harold, as he is known, takes pride in his pre-Katrina role as a caregiver to the oldest generation and a mentor to the young people of his church and community. He witnessed the first five days of Katrina and its aftermath from Bayou St. John near City Park. Friday and Saturday he spent at the Louis Armstrong International Airport for twenty hours, and until February 2006, he stayed at a Howard Johnson hotel in Decatur, a suburb of Atlanta, where at least eighty-seven other Katrina evacuees were housed.

A short, slender, and impeccably groomed man, Harold was wearing perfectly creased dress pants and a short, beige, leather jacket with a hand-knit, striped, wool scarf stylishly draped around his neck on the day of his first interview in the lobby of the Howard Johnson on January 6, 2006.[1] He carried himself with aristocratic grace.

Harold is representative of the New Orleans' "gumbo": the unique mixture of racial and ethnic groups, languages, and history. Harold's narrative sheds light on the complicated reasons intelligent people of all classes decided to weather Katrina in the city and the complexities of race relations during and after the storm. Evidence of the spiritual wounds inflicted by a military presence more preoccupied with crowd control, it seemed from Harold's vantage point, than rescue, haunt this otherwise upbeat account. Acts of humanitarianism, generosity, compassion, and resourcefulness, however, ground this testimony from start to finish.

I traced my roots back to Senegal, Martinique and Haiti. The whites in the family I traced to France. My father's grandmother's people intermarried with Houma Indians, but came from the Black Forest area: France, Germany, and Switzerland. We all see someone else as a victim of slavery, because I see myself as a free thinker. I'm actually a descendant of Shuloushouma (Joseph Abbe Couteau "Tacalabe" Houma), who was a chief of the Houma. Many of the people were at Biloxi when Bienville arrived in 1699.[2]

I was born in 1950 in the Carrollton (Pensiontown) section. It's near the river bend of the Mississippi river, not too far from where Mahalia Jackson was born. They tell me she was born in a house that was built on the river on stilts. We moved downtown after about age six, but the childhood up there was almost like being in the country. I do remember my mother and the people bathing us in those large tin tubs outdoors.

My father worked in a warehouse with agricultural products. When there was a surplus, my dad would bring home bananas, or whatever fruits and vegetables. He tried to volunteer for the military, but they wouldn't let him because he had asthma.

My mother was a maid; she was what they called a *first-class maid*. She worked in the kitchen of whites, some poor in the Florida Projects, but she also worked for some very wealthy people. She cooked foods like Oysters Rockefeller, and she would bring us some home. My mother knows all the old recipes and has old techniques passed down from her mother. Some of them were classical French items, and some were very much African and Caribbean in influence.

My education is piecemeal. I dropped out of high school to help my mother. My parents had separated and I decided that I would be the breadwinner. When I got on my feet financially, I gave her some money to open a beauty shop. When I dropped out of Carver High School, I was reading things like Marcus Aurelius's works and meditations, Greek literature, and the classics.

From '69 to '78, I lived in Boston. I got married there, so I stayed in Boston. When I went up to Massachusetts, in 1970, I took the GED as a walk-on, I didn't study, and I passed it. And then I took some French extension courses at Harvard University in the evening—I wanted to be fluent in modern day classics. I had a B average as a parent.

In '78 I went to San Francisco, because at that time that was the gastronomic capitol of the states and I went out there not to be left out. When I went to San Francisco in the late '70s, I took some hotel restaurant courses. In the '80s I went back to Boston to get my daughter in Brookline High School. That prepared her to go to Barnard College.

I'm self taught, but I feel so gifted of God, because learning comes easy for me. With no formal education back in 1980, I took a sommelier exam and scored the highest in the country. Because I've worked in fancy French restaurants, I was able to excel. And I pursued it with a passion. Knowledge is something people can't take away from me.

New Orleans

New Orleans is one place where people will feed you. When I was a small child of maybe four or five years old, I remember I didn't enjoy going to my mother's mother's place for one reason. I mean I liked them, but when it was dinner time, they fed you so much. We had to overeat. The table was filled with cornbread, yellow grits baked in the cast iron skillet, mustard greens, black-eyed peas, and gumbo. We never had a hungry day, and we didn't know we were poor because everybody shared. Our economy was not trickle down but circulate around.

When I was away from New Orleans, I thought of that song by Louis Armstrong so many times—"Do you know what it means to miss New Orleans?" I missed the smell of fresh roasted coffee beans in the air and the warm greeting people have.

People like my family are the reason people come to Louisiana. We have a joy for life, but we also know when to be serious. We celebrate life. We were taught as children in the churches to cry when a child is born, because they're going into trials and tribulations, and to rejoice when they pass away because they've gone on to better things. We know this is not our home. This is a marketplace. And that's why we don't fight for the soil, but for the soul.

I came back to New Orleans in the year 2001 because I needed community. I stayed there because I've been involved with a Holiness church. Through praise and worship, an atmosphere's created that drives out negativity and puts us into a dimension of tremendous peace. I know that the new leadership will come from the church, grounded and rooted in true love. African American leadership has always been from the church. Even in our harshest times we were able to have happiness, because under the church we had a spiritual covering.

The people in my church look at me as in a position of leadership. The men try to mentor the boys, and the women try to mentor the girls. They all call me uncle. They have much respect for me, and I can't disappoint the children. Especially as a Black male, I enjoy helping young Black children. "Yes, you're of value! Yes, people love you! I'm doing it. You can do it too."

Katrina

I made a vow that I was going to stay during the storm to help the senior citizens, mothers, and children who would be left behind. As a deacon in the church, I know it's my responsibility to help the people who cannot help themselves. If I died, let me die serving the Lord, and serving Him is serving His people.

A lot of people look at the people left behind as if they were stupid, but many of them were elderly and didn't have the resources to leave every so many weekends, as my sister was doing. There are storm warnings every few weeks in the summer. The wages were very low there, but the cost of living is very low. Others stayed out of a basic heart commitment. Some people voluntarily stayed behind to look after extended family. My landlady is an eighty-year-old woman; she's my second mother. I didn't leave my side of the double shotgun we shared in Bywater, until she had left with her grandsons at 6:00 p.m. on Sunday at curfew time.

"The Ritz-Carleton of the Storm Situation"

When Katrina came, I was at my sister Carolyn's apartment on the fifth floor of a seven-story, semiretirement building called Park Esplanade over by City Park right on Bayou St. John. It's on the site of a first Indian settlement right where my ancestors had settled in New Orleans in 1708. It's a fortress with over three hundred luxury apartments. I was put there by God to help a lot of senior citizens, the majority of whom were white. I do love all people, especially senior citizens. The people in the building became very fondly and respectful of one another. Once the hurricane struck, all the class barriers broke down, as far as the people relating to one another. New Orleanians have a sense of unity in crisis; the racial barriers break down. We were too busy in survival mode to hate. We didn't think about how they would have treated us otherwise.

Morale was very strong. We all acknowledged God. The whole building prayed one prayer to the most High God, the God of Abraham, Isaac, Jacob, and David. We called on the God that recognizes all of us. As soon as we finished praying, a woman's son she hadn't seen in years showed up. He had some political connections to the National Guard.

I went to the manager of the building and said, "I have a background in hotel and restaurant administration/operation," and I offered my services as a coordinator to inventory the people in the apartment complex, and to organize some young men to be of service to the people. The general manager of the building herself said, "We're not going to turn anybody away. And we're not going to kick anybody out." So there were black people and people in general that came in the building from outside, and we gave them an empty suite overlooking the park.

I assigned a floor captain to each floor to inventory people and to find out what the special needs were of anybody, like if at evacuation time they needed to be carried out. We found out who were the ministers, doctors, and nurses in the building. My younger captains were a motley crew. When I first inventoried, I had at least forty able-bodied men in the building, not counting senior citizens. When helicopters came, guess who fled first? The able-bodied men. That left about four or five of us including myself; two of them were white. We bonded very close. One person had ADD (Attention Deficit Disorder), another person had a drinking problem. Everyone was a little "throwed off" in one way or another. I looked around and I said, "I feel like Joshua and Gideon in the Bible. We don't need a lot of men. We just need the vigilant men, the ones that forgot about themselves." They were the ones I could count on in a crisis.

There was a restaurant and a grocery store in the building, so the restaurant owner was kind enough to empty his inventory and cook with a generator to make sure that everyone ate. There were people who needed someone to come and check on them, and to bring them food. So we made sure that everybody had something to eat and drink. We did this from Monday morning when the storm hit to Friday afternoon, when we evacuated by the Chinook helicopter.

"Get Back! Get Back! Get Back!"

We were looking for the National Guard. I was in Hurricane Betsy in 1965 in the Ninth Ward, and we were flooded, but the National Guard came within a day or two. In 2005 we first saw rescue people on Thursday. We were fortunate because there were a lot of whites in the building, 85 percent at least. We knew that they had friends and family in positions of influence. So we knew that they would get help reasonably early. We had supplies through Thursday.

One afternoon I was on the bridge in front of the building by City Park on Bayou St. John and Esplanade, and we saw the federal troops pass by in some kind of black uniforms. (It said, "Federal Police" on the back.) There were three or four of them in a bulldozer. A lone National Guardsman from the building who was playing "soldier" said to me, "Harold, go and tell the Federal Police that they need to come over with a bulldozer, because we might have to knock down some posts or something to make room for a helicopter landing." I had on a short-sleeved shirt and my pants were wet. I had no place to hide a weapon. As I approached them, I just waved my hand out and said "The guardsman over there; he needs to talk to you." The Federal Police pointed their M-16s at me and said, "Get back! Get back! Get back!" I raised both of my hands up and said "No, I'm trying to tell you that the Guardsman says he needs your help." Again, "Get back! Get back! Get back!" I felt they were ready to shoot me. They wouldn't let me come within ten yards of them. What if somebody was dying and needed their help? I felt that right there explained what was happening elsewhere in the city. These people see us as enemy combatants. They're not here for us. All these federal police saw was that I was black, and blacks are criminals. That's what I got from them when we needed them most. It was very discouraging to be treated as an enemy combatant rather than someone who needed to be rescued.

Evacuation

The people who rescued us came Thursday, by helicopter, and landed in City Park. We didn't pull our people out of the building until the helicopters came. We got the poorest health people out first, and the senior citizens. We had different crews, like an upstairs crew carrying people down five, six, and seven flights of stairs in their wheelchairs. One of the guys in the building commandeered a boat that somebody had left behind. After we carried them down the stairs, we got them across the bayou in the boat and up to the bridge at Esplanade and Bayou St. John. We were able to evacuate people into the boat because there was still four feet of water to go through to get to the bridge, where there was a little van driving them over to the helicopter.

There was a man whose heart just gave out on him as soon as he got across Bayou St. John. We were dancing and singing in the rain, when he passed. It was fortunate because he was a loner, and he didn't die facing four walls. We felt a sense of peace that this person at least passed away in the sunshine.

I started to stay behind, but then my spirit said, "Harold, if you go back, they're going to mistreat you. You'll be just another nigger." I knew when the whites left, I was in danger.

I arrived at the Armstrong Airport sometime late Friday afternoon. Some of the senior citizens had been evacuated Thursday afternoon. I found out my people were still there and that the young able-bodied people were being moved out upstairs. So many senior citizens in wheelchairs were sitting around.

It was like a war zone. The whole place was densely crowded. People were just moaning, and they were on military cots. When I first got there, there was a handful of triage people that did put a sticker on the people with a health need. If it was really urgent, they were sending them upstairs at the airport to "the half-morgue." There people were dying and some were dead already. Probably there were some paramedics. On Saturday afternoon, I did find one doctor.

The military people that came from outside treated us like we were diseased or contaminated. They came from various states. They didn't want to communicate with us. I felt like, these people think they are in Haiti? They're so stand-offish. There was a total disconnect. They were there for a semblance of order, but you didn't feel a patriotic or missionary impulse.

When I arrived there, the first thing I tried to find out: who was in charge, because I have some administrative background, so I need to go to the top. It took me a long time to find out who was in charge. When I finally got to the man in charge of the operation, I had to play a race card. (He was white.) I pointed out the senior citizens I was responsible for. The majority of the senior citizens I was responsible for was white. I was outside at the airport talking with a nun about God. I'm Protestant. We were talking about how every time we talk about God, something miraculous happens. Soon as we came back in the airport, one of the persons came up and said, "Deacon, Deacon, come here. We got a plane! We got a plane!" The guy that I gave the notice to saw all those good people and he gave us a heads up.

And so we rounded up about forty senior citizens in my group. Some of them went upstairs and went to try on their own, but I said that we should all stay together. I felt bad. I had to be mean to those folks. But there was a window of opportunity, because the line was thinning out and I didn't want there to be a gap and they closed the door. I ran and rallied all the senior citizens together. I'm ordering them like they're young people. "Get up! Move! Move! Move!" I made sure that the people had their medication more than anything. We finally made it to the plane just in time. We got out on Saturday because Al Gore had commandeered an American Airlines airplane he paid for with his own funds.

Former Vice President Al Gore had offered to pay for two flights, but the founder of California Pizza Kitchens stepped in and paid for the second flight. The original intent of the rescue missions was to airlift patients from the city's hospitals, especially Charity. Gore had received a call from a physician at Charity who had saved his son's life asking for his help. Greg Simon, president of FasterCures, who helped to coordinate the logistics of the rescue, wrote that forty non-hospital patients were mistakenly put on the plane, although given their age, he guessed, need for medical care could not be ruled out.[3]

I was talking to him just like I'm talking to you. He showed a big heart. He was a concerned citizen, saw a need, had the resources, and jumped in the fray. Even the people around said, he doesn't want any credit out of this. Otherwise we would have had to wait another day. We were only there about maybe twenty hours.

I had to go because I was responsible for the people from Park Esplanade, and this one lady was ninety-something-years old and in very frail health. You know it's like a parent, you stay with the weakest child. My people said, "You have to come with us."

We were flown to Tennessee near Knoxville. On the plane, they assessed the people's needs and sent them to five or six different hospitals. Gore's people already had the papers to go to designated places, so that was very organized once we got there. From the airport, we ended up at Blount Memorial Hospital, where they checked us out and gave everyone treatment. I ended up with my adopted mother, the oldest lady. When we got to Maryville, I found out they sent her to a hospital by herself, so I volunteered to go along with her. She told the doctors I was her son. (She's a white lady.)

I tried to contact my sister Carolyn. She took the second to last helicopter from her building. I found out that she was at some air force base outside of San Antonio, Texas.

I was only in Maryville one day, because I have a sister, Vanera, who lives in Powder Springs, a suburb of Atlanta. The lady that I was responsible for had went back to the hospital, because she had an arthritic condition. Her niece and her niece's husband were arriving at the same time my sister arrived to pick me up. The lady was close to tears when she saw who was looking after her great aunt. I felt comfortable that she was in the hands of her family. In my heart I wonder where they are, and sometimes I wonder how their spirits are.

"Transitioning One Another through the Transition"

Two days later Carolyn did arrive. Everybody's focus then was getting the rest of the family together. Making sure we knew where everyone was, everyone was safe, and nobody was out of the loop. At my sister Vanera's house, we had a big reunion: my mother, my father, my two older sisters, and my youngest brother. Other family members were at my brother's house; he lives in the same town a little further over. So there was eight or nine adults, plus the daughter of my sister. It's important to see their faces, to know they're all right, to be able to see and touch. Even if we fuss and cuss at one another in between, that's fine as long as they're healthy. I was at my sister's about a week-and-a-half.

I heard about Red Cross and FEMA through the grapevine, public service announcements on television and in the newspaper. On September fourteenth or fifteenth, we went to the Red Cross Center in Latonia, which is at least forty miles from Powder Springs. It was a giant warehouse that was packed, but we were grateful for whatever we got. It was as organized as it could be. We had to go back and forth. They gave us a ticket, come back in a few more days, come back in a few more days. None of that mattered because we had a place to stay.

The Red Cross processing center in Latonia assessed everybody's needs, and I told them I needed a hotel. A friend of ours that we know in the hotel had about twenty or twenty-four of them all in one house. Tremendous stress and they've just come out of the storm, so they really need to break down in their own space.

Under FEMA regulations, displaced people were entitled to stay at hotels and motels as they looked for temporary housing or prepared to return home. The American Red Cross initially orchestrated the placements, which were paid for by FEMA. FEMA covered 85,000 rooms a night at the peak of the crisis.[4]

The Howard Johnson Hotel in Decatur seemed to be the only one left. As soon I got here, I saw the need. I realized that the church is where the people are. It was God that touched my heart to be here, because in the building there were about eighty-five rooms of evacuees, and there was no coordinator, no one to glue things together, and no counselor. I had to step into that role.

I linked up with the church that meets here in the building: Love and Grace Full Gospel pastored by Michael Haynes. They were very receptive. They were my first impression of Georgia. They saw me as a true Christian. They said, "Harold, we want to use you as a liaison to help your people." They had a supper to draw the people together, and a clothing give-away the first time. They didn't give any scripture or doctrine. They just met the basic need. And then the second or third time, we started nurturing to the emotional and spiritual part of the people. The church helped to get people together at the meetings, and then some people told their stories and that was therapeutic, and we discussed our needs and our situations. The church here gets nothing from the government. They've been very helpful to me and to my community. Sure we'd like to be funded, but when the funding isn't there, we're going to do it anyway.

Looking back, I categorized three waves of people that exited the hotel. Wave A were the people who already had position, education, and resources. Within two months, they were able to get themselves up on their feet. The one fellow was a contractor with his own business. Some of them were school teachers. They were the people who knocked down doors back home. The first group had cars. Wave B were people who didn't have as many resources, but who had wherewithal. Some of them had cars, but I bet it's only 20–30 percent. They were able to get up and go in months three and four. Now from month five going into month six, Wave C, the people that remain were the people that were left behind anyway in New Orleans. Those were the people that really need social workers to help walk them through the process. Of the people left behind, one fellow is illiterate and people take advantage of him. They're human beings who still have need of shelter.

Atlanta is spread out in comparison with New Orleans. I went to Lawrenceville from Decatur by public transportation to try to scout out a job prospect there. The distance is 27.5 miles. I just rode the bus to look in the area and didn't get off. It took almost seven hours just to get out there on public transportation.

You know some people are very much in the shock of being displaced, just like every exile and refugee goes through. I've been in France, Germany, Switzerland, and Belgium. I've been going there since the early '70s, at least several times a decade. I speak the language, and I get along with them. It takes me six months just to get over the culture shock of a new environment. What about the people who have never left the community? In New Orleans, everybody lived around the corner from one another. There are people that never left the zip code. They're in culture shock.

They're definitely traumatized as well. I was talking to a little boy here. We were playing football. It casually came up that he was on the rooftop for two or three days. There's one young lady in the church who was on the overpass in New Orleans for eight days in the sweltering heat with barely anything to eat the first couple of days. Sleeping out in the elements!

With the first installment of money I got from FEMA, I bought a Toshiba laptop for eight hundred dollars and a portable printer/copier. (My computers all died in the flood.) Since that was the only funds available to me at the time, I used it to start the "Louisiana Community Newsletter." So I was able to access all the FEMA information as far as helpful resources, download all of that, and put it in the newsletter and communicate it in the building. I go to bed three, four, five in the morning because I'm online doing FEMA research.

The hotel people were connecting me to any group from outside that wanted to help the hotel. They connected to me, and I'd connect them to the people. We loaned each other money. I gave some of it away to people who didn't get their FEMA help yet, because it was very slow for a lot of people. Some people didn't have food. Others didn't have clothing. We had to purchase these things for one another with that first batch of funds that we had.

I set up an office right here in my room and I had people come in. My job is to take care of my people and not to get into somebody else's business. If I see something's not right, I step away real quick. I worked with the Lutheran Services so people could try to find jobs, apartments, and childcare while they're working; tried to find people counseling and therapy; and helped people go online and check the status of their FEMA applications. I didn't initiate any applications for them. I tried to speak in the administrative language of FEMA for the people who didn't know the process or didn't feel that they knew the vernacular that FEMA uses.

It took FEMA months to get an outreach branch. When the outreach people are here, they show up like a thief in the night. They call the night before and they say they're showing up tomorrow. And then I have to hurry and do the legwork and post signs everywhere and say, "FEMA is coming in the morning." There were three of them yesterday, and they were behind a *locked* door with no signage! I asked one FEMA lady, "Why do you have the door locked?" She said, "It's kind of rough around here." I said, "I know the people in here. They're not going to bother you." She wanted to help, she was kind enough, but she couldn't get over her fear, her disconnect with black people. She was a white Californian. She asked me to walk her to her car to get the catalogue. She said, "I wanted you to walk me to the parking lot because I was afraid." It was the middle of the day. High noon!

D'Ann, one of this book's authors and a German American, spent two nights at the hotel in late January 2006. At no time of the day or night did she feel in danger in the hotel or the surrounding environs.

I've experienced racist attitudes by some FEMA workers. You don't have to be called a nigger to be treated subhuman. I've been made to feel as if it were my fault that the hurricane happened. They can tell you have a black voice. It's masqueraded now, but you can feel it on the emotional level from the tone and the attitude.

If I were running FEMA, I'd make sure the people were thoroughly trained. I'd set up an academy with manuals to train them. I definitely would include cultural awareness training. I'd make sure that they learned the nondiscrimination policy.

New Orleans, October 2005

When I first got back to New Orleans in October with my sister and my nephew, we felt estranged. We felt like strangers in the place where we had deep roots. I have traced my lineage back to no less than ten generations. The blood of my ancestors is embedded in the rock of this soil. The soldiers were an occupational force.

The National Guard remained in the city until the late fall. They were assigned to help with the clean up and to augment the New Orleans Police Department, which suffered attrition after the storm.

These people claim to liberate us, but they've taken over. The soldiers claim to help us rebuild, but nobody lifted a finger. They were standing around with their guns.

They look at you like, "You're still here? What are you up to?" We felt criminalized. I said to myself, "You bastards, do you know how long my lineage is? How much blood I have in this soil?"

They took the middle part of the French Quarter by Jax's Brewery right outside Jackson Square and the Mississippi River. They set it up like a club house with bright lights, steak, potatoes, and such. I tried to walk along the river as I often did to smell the fresh roasted coffee beans, the night blooming jasmine, and to feel a true sense of place, the real freedom that New Orleans has, because New Orleans has unconditional growth on the soul level. We couldn't feel that any more. I had to sneak in there to walk along the river. It felt dead, soulless, and neglected, like nobody cares.

You can feel it in the air that they haven't buried the dead. One hundred days later, they found people in a house in Gentilly. They didn't care about retrieving the black bodies. They're still finding them. Had it been white bodies, you know the premium they'd place on recovering them? When I came back to look at my place in the Ninth Ward, I saw on it: "9/16 (was entered), one dog dead." I buried my dog in the back yard.

Residents were discouraged from reentering the city during the draining of the flood-waters and the reconnaissance of damage. Military personnel entered each house, checked for dead bodies, and left painted marks on the exterior of the house indicating the number of bodies discovered. Some of the dead were found by family members immediately inside the front doors, leading skeptics to wonder how thorough the searches had been.

I don't think there will ever be a New Orleans like there was before, and I love Louisiana. I call it my country. I don't think it can be put back together the same as it was. I think it can be made into a Disneyland. I think they can sterilize it by keeping out us Blacks.

I've accepted that I've been put into a new land to develop a new paradigm and a new way of thinking. It was the hand of God that moved us out of that situation and exposed it to the world. I accept that there's a new way. There were too many people in hard-luck circumstances that need to be exposed to a new way of seeing, a new priority, and a new song.

Conclusion

The military didn't bring down life vests, but they brought M-16s and they pointed them at me. That tells me they didn't even intend to rescue. They thought it was an insurrection because of a few guys running around with TVs and tennis shoes. It felt like they said, "That's horrible, let's go put a stop to that." Rather than put a stop to the waters that were going over our heads. Yes, it's emotional. We felt laid low. We felt like we had no value, and the land was no longer ours.

I'm relatively at peace, because I got the lesson from the storm. I see Katrina as a catalyst to encourage dialogue and to give a 360-degree perspective about the plight of the poor and the disenfranchised. As Richard Pryor said, "If you talk to a person long enough, you fall in love with them." I'm at peace because I'm willing to change.

I would like my daughter to learn that not only are there physical Katrinas but there are personal Katrinas. And we all get hit by storms in life, and those storms come to wake us up, to get us out of situations that we don't have sense enough to get ourselves out of.

I had a dream when I was sleeping. I saw the letters: "post-Katrina." Why am I thinking post-Katrina? The work doesn't stop with simply the rescue part. We can't say, "Now that we've gotten the people out of the water, now that we've gotten them into apartments, our work is finished." Now we need to fix the bad conditions that Katrina only brought to the surface. Because if we don't fix that, there'll be political Katrinas rather than weather and wind. Why not fix the lives that have been bombed over here emotionally, socially, and economically?

Cynthia Delores Banks*

Cynthia Banks, the second oldest of eleven children, was raised in the Desire Project in the 1950s and early '60s. She attended college in Long Island, New York on a Martin Luther King Jr. Scholarship, became an administrator at Hofstra University, married, and started a family of four. At the urging of her parents, she returned to New Orleans after the death of her husband. Working two fulltime jobs, Cynthia purchased a home on Lake Carmel in New Orleans East before the storm. Her eldest son, Jermol Stinson (see chapter nineteen), took a bullet to the neck at the age of twenty-two. Extrafamilial networks of caring and creativity made it possible for a widowed mother of four to keep Jermol out of a nursing home in New Orleans by allowing her to orchestrate around-the-clock volunteer homecare. Four years before Katrina, Cynthia started her own daycare service for children of poor working or incarcerated parents, "Free to be Kids."

On a wintry Sunday afternoon in early January 2006, Cynthia spoke openly about the impact of Katrina on her family.[1] The setting was a sparsely populated waiting room in Kindred Hospital in Dallas, where Jermol was recovering from an operation necessitated by his stay in a nursing home in Texas. Having come straight from church, Cynthia was still wearing a neatly pressed pantsuit with stylishly pointed leather mules. She wore her coal black hair in a three-quarters-inch-thick Afro.

The sensitive observations and retelling of the traumas and tribulations of others are particularly valuable in this account of a woman positioned in an uptown hospital during and after Katrina. Many hospitalized and disabled residents had no means of obeying the mandatory evacuation order called the Sunday before the storm's landfall. Their survival, with the help of families, volunteers, and some officials, is a story of persistence and triumph over difficult circumstances. Having lost both her home and her business in the flood, Cynthia talks of starting over in Lancaster, Texas.

I was born in '45 in the Sixth Ward. I grew up in the Ninth Ward, the Desire Project. A lot of really resourceful people came out of Desire. We opened Desire up after we lost our house to fire because of faulty wires. The Red Cross moved us. It was supposed to be temporary. My family left around the time I graduated from high school in '63.

In our building there were four families with eleven, ten, eight, and six children respectively. The families were very close. We shared food; if I didn't like our supper,

I would try and trade with another family. We had sock hops, flower gardens, and live bands.

My dad, Richard Gullette Sr., was a chef/cook. He cooked morning, day, and night. At first he worked Ochsner Foundation's Brent House for the doctors. For thirty-five years he cooked for them. He cooked at one of those French Quarter restaurants on the weekends, and in between he cut all the neighborhood kids' hair to make extra money to raise eleven of us.

My mom, Delores, must have loved my dad to have all those babies. I remember an image of my mother at 4:00 a.m. on her hands and knees stripping hardwood floors, varnishing and polishing them. My mom was a very strict disciplinarian. She needed to see us in her line of view as she was cooking supper at 5:30. We had to be ready for our baths at 8:30 and in bed by 9:00. She had all eleven of us marching to her tune. As she grew older, my mom became the most gentle woman. She spoke softly. We were not being beaten with a belt because we wouldn't do what we were supposed to do. We were being talked to.

I was second to the oldest. I did not have a lot of responsibilities. My grandfather on my mom's side was one of those ministers or preachers that had a heart of gold in the French Quarter on Burgundy Street. His congregation was a lot of the French Quarter people, the jazz players and all of that, and he spoiled me rotten. Whenever my mom said I had to watch the kids or I had to do something, I would always find a way to ease and call him, and he would always say, "Let Dirty Face come by me." He practically raised me. I spent most of my time with my grandparents. On Sunday, Wednesday, and Friday, I would be in the French Quarter at church. My granddad lived in the Third Ward on Lopez Street, right around Parish Prison.

The culture made New Orleans special to me: the second lines, the music, the Africanity. The districts offered something different, yet they came together to be able to create a sense of everyday life joy.

I graduated from Carver Senior High School in 1963. There were five of us from around the United States who held these Martin Luther King Jr. scholarships and went to New York. I went to school in Long Island. That's where I met my husband, Edgar. He was a classical singer. That's where I had my kids. I got my LPN license in 1969 at New York State University, and got a job working nights. I graduated from the New York Institute of Technology in '74 with a B.S. degree in Criminal Justice. Upon graduation, I was urged by my counselor, who was a Caucasian, good looking guy who influenced me to apply at Hofstra University for an administrative position. I got it. It was a day of real celebration, because one of the things Hofstra was up against was the fact that they had to hire someone of color, because they were totally white and male. It was a manager over this Public Safety Department, which gave you the position as liaison between the president and the university. I loved that job. I completed a master's degree in Counseling and Education in the early '80s. The week of graduation, my husband died during open heart surgery.

So I had the opportunity to live in New York twenty-three years. Then I moved back to New Orleans in 1985 at the insistence of my parents. I worked two fulltime jobs that allowed me to make a monthly mortgage payment on a house in a subdivision in New Orleans East.

"Ma!"

My oldest son, Jermol, is thirty-six. He was a longshoreman, and he loved it. I always wanted him to go to college. When he graduated from high school, he said to me, "Mom, I just graduated, and I know you want me to go to college, but you didn't know this. I have a girl friend at Dillard University and she is pregnant. Ma, all I know is that I need to go to work." He met this older guy. The guy must have become very attached to him, because the guy said to him, "I'm going to take you down to the union, and we're going to get you on a ship." I didn't know until after the fact. He said, "Ma! I met this guy, and he took me down, and he helped me register. I need you to do one big favor for me. I need you to buy my gear, because I'm leaving out of here tonight at 12:00." And the next thing I know I was buying gear, and he was gone.

His son, Jermol Stinson Jr., was born with severe brain damage. I kept the baby when Jermol was at sea. Whenever he came home for three or four months, it was just him and the baby. He took the baby to Children's Hospital, and he said, "I will not leave here until you teach me how to care for my baby." He'd put the baby in a little sack in the front of him, and he'd be gone. Such a beautiful relationship he had with this child that did nothing but sleep.

Two years later, Jermol came home. And he said to me, "Mom! I love the water." He said, "I got to get you to Africa. Mama, it is so beautiful." He said, "The one thing I've learned in the two years that I've been on the ship is that I've got to go to school. I'm taking a leave." The year he was shot, he had just come home.

On July 4, 1992, Jermol was coming out of a K & B drugstore in Gentilly, right up around Dillard University. My nephew had just come into town to do some recruitment for the Marines, and they were together. My nephew got in the car and he saw this guy walk up to Jermol but he thought the guy knew him. And the guy asked him, "What's up?" Jermol said, "What you thinking, doc?" My nephew said that the guy pulled a gun and shot him right in the neck.

Jermol was in surgical intensive care for seventy-eight days. They couldn't get him off the respirator. The doctor said to me, "If we don't wean him off that respirator, he's going to have severe brain damage." We prayed and we believed God. The nurse called me at work one day, and she said, "Ms. Banks, you got to get here. We weaned Jermol off the respirator for thirty minutes and he didn't have any problems."

From that point, he went to rehab for a week. At rehab I was so angry. It took the rehab physician, medical doctor, surgical doctor, rehab director, and physical therapy nurse to say to me: "There are no resources out there for him. We will have to put him in a nursing home because he is total care." I said, "I'm looking at all of this knowledge and money sitting around this table, and you tell me my son's got to go to a nursing home. I don't think so. His emotions are worth the world to me right now, and you will not destroy him by putting him in a nursing home." They said, "We can't let you take him home because you won't be equipped to take care of him." I said, "It is my decision where he goes."

I brought him home, and between myself and his brothers and his sister, we worked around the clock with him. I came to the conclusion, "When we go down, we all go down together." We finally got the tubes out of his stomach, the trach out of his throat, and all of the other stuff that he had. We are best of friends, partners at war together against this cruel system.

When I was at work I would always wonder, "God, there's no more money. I don't know what to do." There was a two year determination period before he could be eligible for Medicare. So when he first got out of the hospital, I couldn't work for about four and one-half months.

But it got to the point where you say, "This thing is bigger than me. I just don't know what to do." The nurses from Charity Hospital would visit us. "Ms. Banks, we could come over here two-to-three hours for you. You don't have to pay me."

In the process of that he made so many friends—nurses and nurse's aides—and they would all come over. All of his high school friends would come and take him out for lunch. They learned how to operate his van. He's got a power wheelchair that he operates with his head.

I'd do things like buy all kinds of goodies to keep in the house and the refrigerator. His friends would come over, "Do you want me to go in the kitchen and fix you something to eat?" My food bill was enormous, but the medical bill would have been ten times worse.

"Free To Be Kids"

I founded an organization called "Free to Be Kids." I created an early childhood program that worked with kids from six weeks to seventeen years old. Under that umbrella, we created a program called "Children of Promise, Children of Incarcerated Parents." It was an after school program. We were open from 6:00 in the morning until 12:00 midnight. We were in the process of expanding to twenty-four-hour service when this whole thing hit. It was my life.

Before we started Free to Be Kids, I literally pulled together ten people out of the communities that had worked in the school system and the Department of Social Services that had knowledge and resources. We worked for about a year, every month meeting together, hammering things out at a table in a raggedy building, trying to figure out how we were going to make this thing work. The biggest part was training those individuals how to have a change in their mindset as it related to engaging with people. We tried to penetrate the bitterness and anger that exists in both parents and child.

My church, Christian Faith Ministries, had a building and they leased it to me. It was a huge, beautiful, brick building in New Orleans East along the I-10 service road between four apartment complexes lined up. We got all the service from all of those complexes.

I loved being able to fight the system for the kids and their parents who have no idea how to navigate through a system that is so cruel, and being able to say, "They did what? They turned you down? Let me get on the case. I will do everything in my power." Parents were denied assistance far too often. Getting beyond the biases of workers was difficult. Individuals were judged by their appearance and communication skills.

Then I'd go to the agency and I'd say, "I have a lady and this is her story." They would say, "There's not nothing we can do for her." I would win them over by being nice in spite of their being insulting and nasty. I would say, "I know you're so overworked. There's just not enough help for the needs that's out there, but for the needs that we can meet, maybe I can help you do whatever it is you need to do in order to make this happen." They'd stand there and look at me; then they'd go out of the office, and they'd leave me sitting there for a half an hour. They'd come back, and I still would be sitting there. And I'd say, "God, if you guys just had more workers, you'd probably be able to get more done." Then they'd say, "Ms. Banks, I got something for you." That was just so fulfilling for me, because you were able to touch not only the parent, but the workers and the kids for a big piece of pie, like a Christmas toy for a kid.

The impact on the kids was just phenomenal. My godson's wife's got this little boy, Kaylor, she cared for because the mom was incarcerated. And he's got these play men. "Hold my man, hold my man. I'm going to kill your man!" And I say, "Oops, you can't kill my man. I don't want to die. You got to make peace with my man, you can't make war." I said, "Say 'I love you' to my man." He said, "I love you." I said, "Make them hug." He made them hug. I said, "Say, 'you got to build community.'" He's on the floor playing with the little men, and he's going, "Build community!"

We were in our fourth year. We were blossoming with kids and were looking at new funding. We were creating a huge mentoring program to address the needs of these kids of incarcerated parents. I was in the process of creating a lot of our literature.

The Aftermath of Katrina in Kindred Hospital

Before the hurricane hits, my godson Ricky calls me from Dallas and he says, "Just keep straight on I-10 and you'll find yourself here. Whatever I need to do to make it happen, I'm doing it. You need to know that you're not without a place to go." I said, "I wish I could."

I called my sister right before I left that Sunday evening to go to the hospital with my son. I said to her, "Gwen, Jermol will not feel like he was deserted by the entire family on tomorrow. So I'm going up to the hospital with him."

Jermol had gone to Methodist Hospital before the storm for some evaluations. He was moved to Kindred Hospital, a private hospital, after the tests because he developed a fever. In the process of that the hurricane hit. The week in the hospital was one of desperation. Nobody knew what to expect. Kindred was located in the back of Touro Hospital.

I think the greatest concern was the one that nobody really had before the hurricane, the flood. We knew that there was a hurricane coming and we needed to prepare to have food and water. We knew that we needed to be able to make sure that the places and the areas that people were in the hospitals were safe, that there wouldn't be flying glass, but *nobody* dreamed that in the week to come the situation would become worse. So we all prepared for two or three days.

The staff showed so much compassion for the patients in the process of trying to prepare them for the move from New Orleans and at that point they didn't know where they were going to be evacuated to. The hospital had already negotiated for the patients to be airlifted out and taken to another hospital that Wednesday.

I think the hardest part of that process for me was when the National Guard came in. I think that's when even more confusion ensued. They arrived that Wednesday, and said, "Nobody will be airlifted. You have a generator that's giving you a little light. We have places that have no generators at all." We understood that. We had no idea that it was going to take from Wednesday till Friday. The busses that they had negotiated to come in were on hold. The federal marshals wouldn't let them through.

We scraped and really scrapped to save water for those that really needed the water. There were patients that were in ICU there that had no water, no food. Those individuals of course came first. You had so many people there: workers and the workers' families, other patients and those patients' family members. Everybody wants something to eat.

Everybody is in fear because when night comes the hospital is dark. Everybody is asking the question when. People are becoming anxious and impatient, and they just want out. They no longer want to hear that we've got a plan, and we're implementing that plan. So we had people lying on the floor from Wednesday till Friday not knowing how we were going to get out of there. At this point my major concern was, "Lord, don't let me have to put my son downstairs on that floor in that doorway of the first floor, where people are constantly in and out. His intravenous is not going to work there. It's too unsanitary." They wanted to move him down, but I just refused. They were also very empathetic with me. I kept him up there in that room until the emergency units were not only there but they said they were ready to go.

Interstate 10

One of the hardest things was literally watching people on I-10 that Thursday evening. My second son, Dwyan, he'd come back to find me and his brother to see if he couldn't get us out of there. I said to him, "We're going to be fine, and I think two things need to happen. You need to stay with us because I understand that National Guards out there are pretty bad, and the city out there is pretty bad." So I said, "Let's try to drive through and see if we can't get to New Orleans East." I wanted to see if there was anything at all we could salvage. I was in a car with Dwyan because my car was still way up in the parking garage. I don't know how long it took us just to get to the interstate through all of the uprooted trees and all of that debris.

When we got to the interstate and I saw all these people—that was heartbreaking. People were begging and crying to be picked up. At that point, I didn't even have a bottle of water to give to anybody. And I said to my son, "Let's turn back. This can't be real. Where are all these people coming from?" Then I learned that there were people who had stayed behind to wait out the hurricane and had somehow gotten up on the interstate to get out of the water. It was devastating to think that there was no such thing as privacy. Personal needs have to be met.

Evacuation

At that point the hospital didn't have any food left. There was a little water that they kept for the people that was dehydrated, on medications, and things of that nature that had to have water. I'm going to be very honest with you. There's a Wal-Mart that's not far from the hospital. Some of the guys had been there and said, "There's a Wal-Mart that's wide open, and people are in there taking whatever they want." So I went in there and I got as much milk as I could carry out of there. So I gave milk out to everybody that wanted milk. There was no food.

I let them take Jermol down, and when they took him down, they took him right into the emergency unit, which was a little more sophisticated than an ambulance, because it had enough equipment in it to sustain hours of travel by land. When they began to move those emergency units out, all they could do was put two patients in there at a time.

But the thing that touched my heart so was one of the National Guard people said, "Let's check a nearby hospital to make sure that nobody was left there because all of the people at the hospital had just left, and they found twelve people in the corridor on the floor that were left behind to die there. One of the men was barely breathing. The emergency unit guys pulled him out and they did CPR (cardiopulmonary resuscitation) on him, put him in one of the units, and took him with us. The rest of those people died on that floor. It was like, "God, this is a battlefield. How do you look at people just die right before your eyes, and there's nothing that you can do?" They were older. They were mixed: black and white, male and female. They were alone.

The National Guard that was at the hospital, I must say, was very, very human. They were firm, but they were very human because they took the time to explain and talk about what happened. They explained to us that when they left, there were stories of all kinds of looting and stuff that was going on in the city. They explained that they would find routes that would avoid having to go through the water. We took River Road.

River Road is a pre-World War II highway out of New Orleans that passes through Baton Rouge. It runs along a natural levee that withstood the flooding.[2]

It took us hours to get to Texas because of the route they had to take to avoid water, a lot of trees and electrical lines that were down in the street, and things of that nature. It was getting dark. But once we got into Pasadena, Texas, those people at that Kindred had food ready, drinking water, and bath water to bathe the patients. They had a welcome party waiting for those people. The young lady that came into my son's room earlier tonight, she was one of the people in Pasadena. When my son and other guys came in that were completely handicapped, even though she knew it was time for her to get off from work, she said, "We need to bathe these guys up and get them comfortable so they can have a good night's sleep." So they brought comfort into a situation where people were just physically, mentally, and emotionally exhausted.

Trauma and Losses

We find that people that have been through any kind of disaster end up being retraumatized continuously by an insensitive system. I did some volunteer work at the Dallas Convention Center. I assisted with feeding the people in the shelters as soon as I arrived. We were feeding them simple stuff like hot dogs, coleslaw, and chili. Kids were saying, "I don't want that coleslaw on my plate." The lady in charge of that part would say, "This is not the Sheraton!" (They were sleeping on the floor in the shelter at the time.) "If you don't want it, just throw it away." At first I didn't say anything, and I just listened to the people's orders, and if a child said, "I don't want coleslaw," I didn't put it on the plate. If he said he wanted two hotdogs, I gave him two hotdogs. She became very upset with me. "You cannot customize the requests of everybody. There are too many people." I said to her, "I'm not customizing. I'm simply making the plates as they come up. I just eliminate whatever it is that they don't want."

My sister made the decision to stay in New Orleans in her apartment, even though she had a car to leave. She's an RN. Two elderly women were left in the apartment complex and had no way of getting out of there. She said she couldn't leave the two old ladies, so she stayed there to take care of them. She took them to her apartment upstairs. She got about two inches of water in her apartment, and she lives facing Lake Pontchartrain.

New Orleans is shaped like a saucer. Large masses of older earthen levees surround Lake Pontchartrain at the edges of the city. This land closest to the lake is higher than the land in the middle of the saucer.

We couldn't find my sister for almost two weeks after the storm. When we finally heard from her, we learned that when she was outside on the street just looking to catch somebody's attention to get the two elderly ladies out, the National Guard had come and insisted that she leave with them, because she attempted to try to contact somebody to pick up the two old ladies. They insisted that she leave by helicopter too. The children of the two women met them at the airport somehow. My sister was at the airport almost twenty-four hours. Then they made her get on a plane and go to Cape Cod. They all had weapons. And she said she felt like a prisoner. She had no say about her destiny at that point. At Cape Cod, she ended up on a naval base for weeks. We didn't know what had happened to her, where she was. She had no pocketbook, no ID, no nothing. Finally she was able to contact her sons, and they got her from Cape Cod.

There are some members of my family that I don't know where they are. The rest are all scattered: Arkansas, Georgia, and some parts of Louisiana. My family members are devastated, and don't know what to do. They call, "Cynthia, do you think we're going to be able to go back home?" And I just don't have answers for them. The young people look to us to help them find some kind of closure to this thing. My niece Diamond said "Auntie Cynthia, I am so lonely for my cousins and for my family. I feel so bad living way out here in the country."

My ten-year-old grandbaby, Jwyan, a published poet, said to me, "Grandma, I was feeling really bad about living in Baton Rouge, but I can't live in St. Bernard Parish now." She's been back on two to three occasions, and she's had the opportunity to at least begin to bring some closure to a lot of her feelings. That's what it's going to take to overcome some of the shock she's had about living in what she calls "the ghetto."

I think about how my children are all separated now to the point where all of their conversation is about the pain that we've gone through and what they've lost, instead of being able to move forward with the plans that they had. My second son, who is so devastated by all of this stuff, wandered the streets for weeks trying to figure out what move he's going to make next and how could he help the homeless on the street. My daughter, Cheryl, had to say to him, "You're homeless. Did it sink in yet?" They were all moving forward before the storm.

We've had experiences with people saying, "We have given you." I say, "Wait, wait, wait! Let's start this conversation over because it's not coming out right. I'm a hardworking citizen to the tune of sixteen-to-eighteen hours a day. I worked every day and took care of a family. I'm a homeowner. You're not giving me anything. You've taken away everything that I've worked for and that my parents worked for not because of Katrina but because of negligence and corruption. Now we can start this conversation over and I can say to you what my needs are right now. And let's see if you can assist me in getting them met."

I had pictures of my mom who is dead now that I'll never be able to get back. I had a Bible that I've had fifty years with all my sisters and brothers in it, my parents in it, my grandparents in it, my great grandparents. It's gone. I'll never ever be able to get back all my children's hard work, their awards, everything from school, their degrees, their pictures of their childhood, or love letters from my dead husband.

I can only look at these things as an experience that tomorrow will serve to be rewarding as I begin to rethink my values and look at what's most important. I promise you that it is because of my handicapped son that I need a home. My home is in my heart and my children. Now that my home is gone, I would just want to travel and work with the people that have been so devastated. I would say to them, "Something good has definitely got to come out of all this."

Disaster and Disability

I was told that the fastest way to get services for your son is to put him in the nursing home for forty-five days, because it will take you forty-five days to go through the paper trail. And if he's discharged from the nursing home to home, he will have completed the required paperwork, determined whether or not he's really eligible for state benefits, and he'll be able to get services right away. Otherwise, he's going to have to go on a two-year waiting list.

It has been almost ninety days now with my child in a nursing home. It was hard on him psychologically to be in the nursing home. He said, "Mama, there's five different levels of urine in here you smell from the time you hit the door till the time you get to my room. I don't want my friends to come in here." He was ashamed. It is psychologically, emotionally, and socially challenging. He had to have an operation because he developed a bed sore on his hip that got infected. He wasn't being turned. He couldn't eat the food because there was no taste to it.

Jermol's home care in New Orleans worked with him building relationships and people pitching in to help. That has placed a lot of uncertainty in our now. All of that has been disrupted, and he keeps saying, "Ma, what are we going to do?" I think he's got an inner fear that presents itself in our conversations sometimes of going back into the nursing home from this hospital. He keeps saying, "Mom, you worked so hard for twelve years to keep me out of a place where I've now ended up in." I keep saying to him, "Jermol, God was ever present in the beginning, He's present now in the middle." We just keep moving forward.

Conclusion

Many of the Free to Be Kids parents have called me. One parent called me from Mobile, Alabama, and she said, "Ms. Banks, I don't know what to do. My son won't eat." He was in the hospital a lot, so whenever he would come out of the hospital, we'd always make this great big deal at school. We gave her his teacher's number. His teacher told him that as much as she wanted to come and see him, she couldn't because of the hurricane. But that everybody was safe, nobody was lost, and she loved him a lot. She wanted him to do her a big favor: show his mom how he can feed himself. His mother called me back. "He's eating!"

I would be one of the ones to say, "Lord, I'm really tired." I've got to the age where I should be retiring now. I'll be sixty-two next year. I should be tending my garden, and doing the things I enjoy doing. I love reading, driving around looking at the different neighborhoods, the communities, meeting people, and going out to eat.

When I look at Katrina and my son, part of what you do is you put your emotions on hold. You're in survival mode. You look at everybody around you that tends to be falling apart, and you make yourself very busy to work with them through their issues. At some point, I had the time to work on the anger I was feeling about what happened, how it happened, being left in that city until Friday after the hurricane, and watching the devastation of what was occurring and looking at the lives of people just being totally destroyed right in front of your face. I mean watching twelve people left in the hole dying. How do you begin to internalize that stuff?

There has been a great awakening for those who didn't know that these kind of inhumane mindsets exist. The way people were hovered over with guns like they were criminals, rather than victims. Something good has got to come out of this because it has been exposed to the nation. So they're poor. They don't need to be treated with dignity? I could not be rich, if it had not been for the poor. I had the opportunity to watch people hide food in America: during the storm, after the storm, right now. They don't know if they're going to have it tomorrow.

My daughter, Cheryl, said to me: "I keep saying to my brother Kyle that the soul of music is in black folks because they've been through a lot. That was the thing that gave them the endurance that they had to get through much of what they've gone through." She's twenty-six. I asked her, "How do you experience something that you only had the opportunity to read about?" She said, "listening to you sing." During my times of greatest concern of what tomorrow's going to bring, I always hum old spirituals and I find so much relief. It's like new energy.

It's the same kind of built-in resilience from experiences that will bring the citizens back to New Orleans to begin to rebuild and to work through the issues that they're facing right now with the system and its own agenda. We're all seeking for the same identical purpose: to be heard, to be treated fairly and with justice. I look at tomorrow as being very promising, because it leads to learning new strategies of being able to overcome.

TEN

Denise Roubion-Johnson*

In 1972, Denise Roubion-Johnson was a debutante and the homecoming queen of Joseph S. Clark High School. She rose from meager economic circumstances to direct a cancer-screening program in the outpatient clinic of the Medical Center of Louisiana at New Orleans (Charity Hospital). A wife, mother of four, and a grandmother, Denise led a busy life of family, work, and education before the storm. Denise, her husband, Ronald Johnson Sr., who suffers from sickle-cell anemia, and her sixteen-year-old son, Ronald Johnson Jr., rode out the storm at University Hospital, where she spent the remainder of the week giving emergency care to newborn infants, mothers, and ICU patients.

On a hot, muggy Saturday in late August 2007,[1] the fifty-three-year-old health care provider was wearing a vertically striped, button-down cotton shirt with white pants. Even on her day off, Denise fielded phone calls from colleagues and patients. The interview took place at her custom-built Eastover home, a gated community of executive homes located at the far edge of New Orleans East. One entire wall of windows looks out over the gated community's pond and golf course. Downstairs her console is adorned with pictures of her with Bill Clinton, Jesse Jackson, Ray Nagin, and other politicians. Upstairs, prints by Alfred Gockel line the walls.

Denise's relationship to her family, patients, and community afforded her a unique perspective on the acts of race-based discrimination she witnessed. By October 2005, she had resettled her family in Houston, reconnected with some of her missing cancer patients, and returned to New Orleans, where she worked as a nurse practitioner to help the large number of women who lost their health insurance after Katrina. Her narrative describes the situation of women survivors who are unemployed, uninsured, and diagnosed with cancer. The insistence of these women to return to New Orleans despite the lack of adequate medical care is an example of the draw of the city.

I was born in 1954 and raised in three different housing projects (before they became dangerous): Lafitte, St. Bernard, and Desire. I was here for Betsy in the St. Bernard Projects. My mother's relatives were from the Lower Nine. New Orleans always felt like home. Growing up, you'd see everybody sitting on the porch, and you'd be waving at somebody, "Hey now, how's your mama and them?" You can give the greeting and feel so good about it.

My kids have had advantages in life that I never had coming up. We didn't even go to the beach because my mother, Marlene Roubion, worked constantly. She worked so many fifteen-hour shifts because she was a single mother raising five kids with the help of her parents.

My mom was the most excellent nurse. She was an LPN, and we were going to school together to be RNs. She always used to say, "Niecey, treat every patient as if that was your family member regardless of their race, their religion, and their ability to pay. You treat everybody as if that was mama, Papa, or Grandpa Roubion in that bed." That's the kind of nurse my mom was, and I promised that I would be the same kind of nurse. She died in 1984 at the age of forty-seven at Charity Hospital. She had a one-day surgery, and a resident in anesthesia accidentally punctured her heart, so she died. That was a funeral like none other. She had so many doctors and nurses there. She touched a lot of people.

When my oldest daughter, Rhonda, graduated from nursing school prior to Katrina, they asked me to be the guest speaker at her nursing program. I said, "I want every one of you to go out here and refer to your patients as Mr. or Mrs. So-and-So, Mrs. James. Never refer to your patients as the lady in bed ten, or the man in bed number nine because that could easily be your mother or your father. And when you're alone and working the night shift keep in mind that you have a guardian angel looking over you, and her name is Marlene Roubion."

I started as an RN, moved up to be an advanced practice nurse with an M.S. in nursing, became an adult nurse practitioner, and finally worked on a clinical doctorate. In 2001, I became affiliated with the Louisiana State University School of Public Health's Cancer Program. We had a cancer center for uninsured patients. I screened women with cervical cancer. We had a clinical nurse specialist who took care of women with breast cancer. I would send them to her, and she would do the workup, because the mammogram machine and everything was at the Charity campus.

My family lived in New Orleans East in the Kingswood subdivision. I built that house and stayed in it for seventeen years. We were such a close family. We would do stuff every weekend. We'd either go to the movies or go to dinner. We did so much together as a family. My husband, a retired bus driver, had his first stroke in 2001. He's almost completely blind. He has sickle-cell anemia. He's far exceeded the statistics. Males with sickle cell usually don't live much over forty-five. His sister died at twenty-six with sickle cell. I built the house in Eastover, because I needed a handicapped-friendly house for my husband. We were only in that house for nine months when Katrina hit. I had just finished getting my home perfectly decorated.

Landfall

I'd never left for a storm before. I wasn't obligated to stay. I decided to stay because it was hard to get on the road with my husband because he was sick, and he was having some mild pain symptoms at the time. I was afraid to get on the road, because I thought he would go into a full-blown pain crisis. The doctors whom I ended up living with once I moved back to New Orleans after the storm, Mary Abell and Don Erwin, were in Vermont. They were like, "You got to get out of there D. We're looking at the weather channel. You're going to get hit hard in that area where you're at." I said, "I don't have anywhere to go."

My daughter Renee, her husband and three young children evacuated to Florida the Saturday before the storm and were faced with severe price gouging along the way.

I left Eastover but decided not to leave New Orleans. I called the hospital administration at University Hospital and said, "I'll come and volunteer my services, if you all can provide some place for me to put my husband and my sixteen-year-old son." They knew he was sick, so they gave me some place for him and I went to work, not knowing I would end up working the way I would be working. I only brought two pairs of underwear and scrubs. That Monday we were like, "Oh good, the storm has past. We're all going home."

Aftermath: "Then It Began"

Later on that day we saw water coming up from nowhere. The Central Supply people had a portable TV with them. I knew Eastover was flooded. Everybody was like, "Oh God, if Eastover with all those million dollar homes is flooded, can you imagine."

We were on the second floor and up. There were a few hundred patients, because we had the whole newborn nursery there. A lot of employees had their immediate families with them because the essential employees that needed to be there had their families with them.

At the very beginning, we tried to coax people in because you could see people physically drowning in the water because they couldn't swim. The water was over their heads. A lot of the homeless people from the church around the corner (that's where they usually are) were let in. After the water stopped coming up, you got to see the bodies floating.

I've been in the medical profession for over twenty something years, and I've never experienced anything like that in my life. We lost power on Tuesday. The generators were in the basement—the water was up to the first floor. Things you take for granted like lights, air conditioning, and ventilators for the Intensive Care Unit: There was none of that. God, it was stifling hot. We were going to the bathroom in plastic bags. Everybody: patients, nurses, doctors. The smell of the feces and urine was unbelievable. It got worse and worse. There was no running water after a few days, so there was no bathing. You'd just wipe off with normal saline.

We never ran out of food; we just didn't have hot food. We had canned peaches and cold ravioli. We never ran out of water. We just had hot bottled water to drink. About the third or fourth day, the Coast Guard or somebody did a drop to us of water on the roof.

Some guys rode up to the hospital in the hot tub they had stolen like a motor boat. They wanted to come in but we didn't let them in. They thought we had drugs. They started shooting at the hospital. I only slept a little, because we were so afraid. People were breaking in at night. They did steal one nurse's purse. We had to barricade ourselves in at night to keep the homeless people from coming in and stealing from us or possibly hurting one of us.

I spent most of my time throughout the whole hospital. I mopped floors because there was water coming up. I worked in dietary. I worked in administration for a while. I helped out the nurses that were in the ICU. You just went wherever you were needed.

We all bound together. Colleen Lemoine, a nurse specialist, is the nicest person in the world. She goes to her locker, and she says, "Come Denise, come with me." Who goes in their locker and pulls out brand new pairs of drawers still in the pack? And she goes, "Here's a couple of pair of drawers for you." And she had all these little soaps and toiletries from when she goes to conventions and conferences and stuff, and she's divvying it out to our friends. Trust me. Six days in the same underwear? It was a very big deal.

At University, we controlled our own space. The administration was pretty good. Dr. Dwayne Thomas and Dr. Cathi Fontenot were pretty good about saying, "We're working on it." Blacks and whites worked together. The whole hospital was banding together trying to keep everybody safe and get everybody out. At University, even the prisoners got good treatment. Because University is a teaching hospital, so they had prisoners in there.

We found a cell phone that worked and I called my oldest brother, Eric, who was in Alabama, on his cell phone. He said, "Where are you? CNN is reporting that everybody's been rescued from all of the hospitals." Nobody was gone out of any of the hospitals. The VA (Veterans Administration) was still there. Touro, Tulane, Methodist. All the hospitals' staff and patients were still there.

About the fourth day, people started losing hope because it was like a feeling of helplessness. There were days when we just thought we weren't going to get out of there. You would just think to yourself, "This is the United States. Why isn't anybody coming to get us?" They kept telling us, "They're coming tomorrow." Tomorrow comes and night falls again. We heard helicopters. We saw helicopters. That's what made it so bad. They were just passing us up.

On Friday some local people came by in small boats. We sent mamas and babies to Tulane Medical Center, a private hospital, by boat. They had the Medivac, a helicopter with supplies and equipment on board, already picking people up. Most of the mothers were black. In the nursery, I might have seen two or three white babies, if that many. So the babies that were really, really sick and in ICU, and a couple of really sick people from the Intensive Care Units, we sent over to Tulane. Tulane sent the mamas and the babies back. They said they didn't have any room for them, so they sent them back. The mamas, the babies, they were all crying. We were all crying. Why would you turn these women away? Instead of letting the people get on the helicopters, the staff was getting on the helicopters and leaving. They wouldn't take the Charity patients with them. That was our little glimmer of hope and it was gone. It was devastating on the mothers because they thought they were going to be saved with their babies. It was hard on everybody.

We didn't get out until Saturday morning. I'm standing on the walkway between the hospital and the administrative office, and we could see the convoy of boats, hundreds and hundreds of little boats, the ones with the big fans on the back of them, coming in droves. It was like somebody was leading them all in. We also saw a helicopter, and I said, "Hey y'all. There's a helicopter, and this time it's for us." There were men in army clothes or Coast Guard clothes. So we started putting the mamas and the babies on the helicopters.

While I was up on top helping to evacuate the mamas and the babies, they were taking our families out down below on these big boats that came in. That's how I lost contact with my husband and my son for days. My son got a crash course in getting used to taking care of his father. They said, "Oh, don't worry about it. You'll all end up at the same place." I wanted to make sure all the patients were gone. I got out on a later boat with the second-to-the-last group to leave. As we're leaving, the bodies of the homeless people from across the street near the church were bumping against the boat. They didn't make it up in time, I guess, and they drowned.

Myself and about ten-to-fifteen other hospital personnel went from the boat to the Hyatt. The majority of them were white. By us being one of the last groups to leave, we ended up with the people from the housing projects from all over the city. They were coming from the convention center and the Superdome because the Hyatt was right next to them. The men with the guns were routing them to the bus. But when the bus came, eventually, they said, "You're all going to get on this bus right here." We weren't told where we were going. We were told to shut up and don't ask any questions. So there were ten or fifteen of us in the front of the bus, and the military placed people from the different housing projects on the publicly provided bus after us. They had not been able to afford to leave the city on their own before the storm. That's a different class of people. I'm not knocking people from the housing projects because I grew up in a housing project, but some of them were hostile and cursing, because no one was being told where we were going. One of the white nurses kind of joked with me and said, "We're going to die!" But because I was black, I could understand their frustrations, even though I did not agree with some of their actions.

Texas: "We Were Herded and Treated Like Cattle"

I witnessed discrimination by the military. Once we got to Texas, it was a noticeable big deal. The mere fact that we were pushed around: "Don't ask any questions." I feel like if it were a group of white people, we wouldn't have been treated so poorly. But we were labeled as "poor black people," and we looked the part. We were smelly, so it didn't matter that I live in, this Eastover house that I built, and I drive a Lexus RX 350, and I make a substantial income. None of that was relevant. Katrina was the great equalizer. I got frisked just like the woman next to me from the housing project. I had to stand in line in the hot sun, just like everybody else. I had to go to the bathroom, just like everybody else. I got yelled at to shut up, just like everybody else.

I ended up in Tyler, Texas. Getting there in itself was the most embarrassing ordeal you could ever go through. Texas had makeshift processing centers where you had to stop to use a bathroom. It was like you took that vacant lot across the street and you set up a tent and a port-a-potty. At the first one, there were church folks with a sign, "Welcome Katrina refugees." They gave us water, Gatorade, and sandwiches that they had made, because we had been on the bus for a long time. At that checkpoint everything was all lovey-dovey. Dr. Fontenot and I were together. She's a white woman who is a very kind and down-to-earth person. We went to the potty together. I stood outside the door as a guard for her, she stood outside the door for me. We were fine. We got our Gatorade, got our snack, and got back on the bus. But at some point she had to leave because her father was dying, so her husband met us half way between Texas and Shreveport. They let her off the bus.

Once we left the first stop, it was like going to a concentration camp. If you had any weapons, you had to declare your weapons before you got to the front, or else you were going to be arrested by the Texas police. They only took ten people in at a time to be frisked. At these processing centers is where you gave the Red Cross your name and social security number. After we were frisked and declared we didn't have a weapon, we got back on the bus.

I don't remember seeing the white women from University Hospital who boarded the bus with me get that kind of treatment. They were kind of like pushed up to the front of the line once we got to Texas. We were driven to the Houston Astrodome at 6:00 at night. Then we were told that there were too many people there, so they turned the bus around. I wanted to get off the bus in Houston, because the plan had been that Dr. Fontenot's cousin was going to pick me up from the Astrodome and take me to his home. I was not allowed to get off. No one was.

We had state troopers that were following us and in front of us. Every now and then, the troopers would stop and would get on the bus. You got all these irate black people on the bus, so they would stop every so often. The bus driver would be like, "These people are asking questions. They want to know where they're going." And they were like, "You all just shut up. You'll get where you're going when you get there. Don't ask any questions." That went on for pretty much twenty-four hours for us. We didn't get where we were going until Sunday morning.

Reclaiming Existence: "Kindness of Strangers"

After the Astrodome, we were back on the road another six hours. When we got to Tyler, we were brought to some church. We were still smelly. There was no shower. The church let us use their cell phones, but the people we were trying to call were God knows where, because they had been transported out of the city as well. Husbands came and got their white wives in Tyler. None of the black hospital employees had the means to have anybody come get them.

I was so distraught because I couldn't find my family, so a hot meal meant nothing to me at that time. I told my story to some of the women there. I said, "I'm a nurse practitioner and I was one of the last few people to get evacuated from the hospital. I need to find my very ill husband and my sixteen-year-old son. Dr. Fontenot gave me eighty dollars. I don't have any ID, and I don't even know where they are." (My son had my purse because I didn't need it on the floor at the hospital.) So they did some typing into a computer, and I gave them all my family's names, social security numbers, and they found them in San Antonio, Texas.

The church people called to see how much would be a one-way plane ticket to San Antonio, Texas. And it must have been like 150 dollars. So the people in the church and some of the people from the hospital who had money on them gave me money, which, when added to my eighty dollars came out to be enough to get the plane ticket. The police officer from Tyler took me to the airport and explained to them that I didn't have any ID because I was a "Katrina refugee." Those words weren't as hurtful at that time. I was dealing with too much. So he brought me to the airport, and he waited for me until they put me on a small commuter plane that took me from there to San Antonio. The army had somebody pick me up and then brought me to the Naval Base or whatever it was that my family was at.

I walked in there, and wasn't there five minutes, before I said, "Oh, no. We cannot stay here." I'm not a snob, but there was no way in the world that I was going to have my kids with these people with their pants sagging, babies walking around with poo-poo, pee-pee diapers on. There was a place to change their diapers now. There were shower facilities in this place, because it was like a military base. And there were hundreds of people in one big room with cots. The police and the guards were not friendly. Here you were with "mostly poor black people" from New Orleans. We didn't get any special treatment. I got on the phone immediately and I called Dr. Abell. She called this preacher and his wife Mary Ellen. They're Episcopal. Once again, white folks are rescuing me. She said, "Give her a call." So I called her, "Mary Ellen, I can't stay here. It's just too overwhelming for me." Six days of being closed in, of worrying about if you're going to be robbed or shot, of seeing bodies floating in the water. I always take pride in taking care of people. And now I needed other people to take care of me.

Mary Ellen and two taxi cabs carried me and my family to a hotel. All of us were in one hotel room, because the hotels were booked up anyway. This was more like a motel. It was so terrible that when I pulled the cover back to lay my head down, there was a cockroach dead on the pillow. I hadn't slept in days. I just brushed him off, laid down, and went to sleep.

Next morning Mary Ellen sent this lady named Peggy McGaughy, and I will never forget that lady as long as I live. She was a very gentle, white woman. She and her husband Kent had a ranch in San Antonio, and she was there for the weekend. These were wealthy people. They all went to the same church. She rented us a car to get us to Houston, where I had a brother.

So my daughter, Rhonda, and her four kids, my son, my husband and myself, we all went to this house in Cypress, Texas. Sally Hardin at the church did drives for us to get food and blankets and stuff you need to start a household. We were not welcomed by our neighbors with open arms. Their attitude was more like "How dare these poor black people move into our affluent community." I stayed with them almost three weeks getting my family settled.

"Taking Care of These Patients Is My Life"

My boss, Donna Williams, heads the CDC (Center for Disease Control) funded cancer program for Louisiana and she found out where I was. She started e-mailing me. She set me up with gas cards to help patients get to and from their appointments. I was finding our patients in all the cancer hospitals. We had a list from the Cancer Registry of colon, rectal, any kind of cancer. The patients were so relieved when I actually found them. This let them know that we didn't forget about them. We got what records we could, so that we could match patients up with their doctors in New Orleans. Most of the doctors had left, but we could find them through e-mail. Even though I had my own problems to worry about (I lost my house, my husband's ill and in the hospital, I can't find my other daughter Rochelle, and still had not heard from my daughter Renee and her family), we still did what we could for our patients.

I left my family and my seventeen-year-old son in Houston to come to New Orleans and take care of these women because I knew that they needed me. My kids said, "Mom, you're going to leave us here." But I had to come back here to take care of my patients. I told them, "You have me. If I'm not there with you in person, I'm there with you in spirit. Some of these women don't have anybody else. They lost family, they lost it all."

Dr. Abell immediately said, "You can come and live with Don and me." She was a resident in medical school when my mom was a nurse, and my mom took care of her. So Dr. Abell returned the favor by taking care of me. I lived with her and her husband for ten months.

The CDC said, "If you can get your staff back, we'll continue your program." Donna Williams found everybody and pulled it back together. Then we started seeing patients in November 2005. That was continuing the breast and cervical cancer screening program, only now I was running the whole program by myself. I was the only clinical person. The St. Thomas Community Health Center in Uptown gave us a medical home to take care of the women.

My work is needed more now because so many women don't have insurance, and the stress of what's going on is making women sicker. All the patients were without insurance. That number increased dramatically because we had a new population of uninsured.

In the fall of 2005, the Louisiana legislature took over the Orleans Parish school system. Seven thousand teachers received letters ordering them to resign or retire. Many lost their insurance benefits as well. Professors and adjuncts were laid off or permanently downsized at most of the local colleges.

People who once had jobs were coming in with large breast masses. These women were coming in with III and IV grade breast cancers. They'd go to a private doctor, sit there and wait their turn, and then they'd say they were there because they had large breast masses. When they told the admitting nurses they didn't have insurance, they were turned away because that wasn't considered an emergency. We diagnosed five and six women with cancer a week. Some women are so savvy. They go on the computer, and they find us. I've had a woman who was a college professor who came in with a large breast mass. Her daughter found us on the computer. She said, "You know my mom has this mass, and I researched it. Would you take somebody without insurance?" I said, "We take you regardless of your ability to pay."

Surprisingly there were a lot of people left in the city after Katrina. Others were still trickling back in, because they just didn't want to leave. You were seeing more and more people coming back, and more and more people showing up sick. We would say, "If you're sick, don't come home." That's why my husband and my son ended up staying in Houston a year. There's no health care in New Orleans. But they wanted to come back home.

Both University and Charity Hospitals were located in the center of New Orleans and suffered major flood damage. By November 2005, of the hospitals within New Orleans, only Touro had reopened its doors for emergencies. The only three fully functioning hospitals in the greater metropolitan area were outside of the city limits.

We had to figure out what to do with these people. I oversee the clinical aspects, get the workups done, and do the biopsies on some women. I worked until 10:00 or 11:00 at night trying to return the calls of people getting back and wanting to come in and see me.

I had different organizations like the Key to Life Foundation that were helping me pay for mammograms, where other hospitals were turning our patients away because they didn't have insurance. But the American Breast Cancer Foundation, the Susan

G. Komen Foundation—all of these foundations gave money to the program to help keep us running.

Over the past two years, we have managed to have a full breast clinic up and running again. We have diagnosed over sixty something women with breast cancer. All of those women are getting treatment. We get Medicaid approval within one or two days for these women.

What I love most about it is hearing a woman come back and saying, "Thank you for saving my life." Last year I got the American Breast Cancer Spirit Award. It means a lot to know that you can help somebody who, had I or the program not been there, would have had nothing. They would be dead. We lost a lot of people, but we saved a lot as well.

Conclusion

Katrina wiped it all out. All that new furniture I just bought? It was gone. So I had to start all over again. This house has a good view for my husband who's home all day long. That's why we got the big window so he can sit there and watch them play golf. It's got something to make him happy, although nothing makes him happy without his grandchildren here.

We had a FEMA trailer on our property in Eastover, even though it was not hooked up. The bylaws didn't approve it, but we did it anyway. About forty percent of the residents who came back in 2005 and 2006 had trailers. We felt like if we're paying all this money to live back here, we've still got to pay our Association dues, then we should be allowed to put a trailer here while we rebuild our homes.

My contractor, Alvin Masters, lived right next door to me, so he worked on my house and his house at the same time, but it took me ten months to make my home livable again. My insurance wasn't very good. We didn't have enough to pay for the actual repairs out of what they gave us. But my contractor really worked with me to keep me within a budget that I could afford. Dr. Erwin helped me write a letter to my insurance company, because at first they didn't give me the full 250,000 dollars that flood insurance caps at, and I had more than 250,000 dollars worth of damage. I didn't get anything from the Road Home, so for me it was a Road to Nowhere.[2]

I feel like we've been robbed of so much of our life. I lost so many friends and family members who died or moved on to other states and are just not coming back. We didn't ask for this to happen to us. Nobody asked to lose all of their pictures: pictures of my mother, pictures of my children as babies, pictures of my grandparents. We can't get any of that back, because everybody else lost theirs too. We didn't have any out-of-town relatives. We do now.

My son graduated from high school in Houston. I flew in for my son's prom and graduation. The biggest loss is not having my children here, but I talk to them on the phone as much as I can. My daughter's kids: I miss them. They're going to be in Houston for a while. My daughter worries about her dad dying and her not being here to be with him. There's not a lot you can do for a sickle cell. I just miss the closeness of family. We don't have that anymore because we're living too far apart. It's not like they can come over every Sunday and just hang out.

One day I smelled a smell that reminded me of the week in the hospital, and I just walked out and cried. I was like, "God, what is wrong with me?" I still have good and bad days. If it rains really hard, I get really nervous. I was in the St. Thomas Clinic about a month ago and it rained real hard and the power went out. I can't be in the dark when it rains. I waded through that water, went across the street, got in my car, and made it out of there. If I hear a certain kind of knocking, it reminds me of the bodies knocking on the boat. My daughter in Texas doesn't want to come back here because when it rains hard, her kids freak out. The six-year-old when it rains asks, "Are we going to die? Are we going to see more people floating in the water?"

For a whole year, I found myself just crying anytime I heard anything about Katrina. We were treated so poorly. Going through those processing centers! My son said, "My daddy had to stand in line in the heat for hours in pain to get frisked." To be such a rich nation and to be called refugees. Regardless of our socioeconomical status, we were Americans! The word "refugee" took on a whole new meaning almost as hurtful as being called "nigger." We were productive, working, tax-paying Americans being referred to as refugees.

I tell my kids to be humble. I was herded like cattle like everybody else regardless of my ability to pay for services. Nobody cared. I was just as low on that totem pole as that lady who was on welfare with five kids. If they take anything from this, be thankful. I lost all of my material possessions, but I got my family back! My dream for the future is that my kids can all come back and live in New Orleans while my husband is still alive to enjoy the grandbabies.

The neighborhoods are not any better off than we were two years ago. The American people have pretty much forgotten about us because they're not here seeing all these empty houses. If you go down certain streets, there are so many blighted houses. You go in the Lower Ninth Ward, and you see concrete slabs. I never go and look out the window on the other side of my house, because who would think in this million-dollar community that there would be a slab next to a house like this. That slab next door to my house has been there since Katrina's waters knocked it down and the owner has never returned.

There are people still struggling for some place to live, sleeping in cars or sleeping in those contaminated FEMA trailers, or in gutted-out houses.[3] I saw an article in the newspaper one Sunday about this eighty-something-year-old lady who was living on St. Roch Street in this gutted out old house with no heat, no walls, no electricity. I took my daughter and I said, "Rochelle, let's go look for this old lady." We went driving down St. Roch Street, because they didn't give you an address but they showed you the whole block. I was going to put that old lady in a hotel for that cold snap of a week. Or she could have come and stayed with me and my daughter in the trailer. But we couldn't find her. I told my daughter, "This is so hurtful that I can't help." This is the one that's going to worry me. I wonder how that old lady's doing.

Kalamu ya Salaam

Kalamu ya Salaam (Pen of Peace) was born Vallery Ferdinand, III, in 1947 and raised with his two brothers in a modest home built by his father on St. Maurice Avenue near Law Street. He is recognized as one of the Lower Ninth Ward's most talented artists, as well as a revered community activist and a feared political critic. A leader during the civil rights movement in New Orleans, Kalamu attended Carleton College in Minnesota for two quarters on a partial scholarship, before becoming overwhelmingly homesick for the black culture of New Orleans. Kalamu was a leader of the Black Arts South Movement and a participant in Free Southern Theater. From 1983 to '87, he was the executive director of the New Orleans Jazz and Heritage Foundation. Before Katrina, he lived in Algiers with his wife, Nia, ran the cultural listserv e-drum, and codirected Students at the Center, a writing program for high school students. Like his grandfather Noah Copelin, he has influenced generations of New Orleanians.

On July 10, 2006, the bearded activist, writer, and educator opened the red front door of his home for an interview.[1] He was wearing his customary attire: black jeans, black t-shirt, and black shoes. Only his red-rimmed glasses challenged the ensemble. The phone rang continually for the entire four-hour session, but remained unanswered. The interview was held in a small room painted deep red and lined with framed African art, at a dining room table decorated with a bold kente cloth runner.

Kalamu is an example of the dynamic, creative, hardworking, and stubborn talent that once was common in his neighborhood. In the spirit of the Lower Ninth Ward, he was a bookworm who even read a book on how to give "licks" in order to defeat a bully and earned a reputation for toughness that gave him license for his artistic and intellectual expression. He observed Katrina from Houston, where he drove before the storm. During the fall of 2005 while on a cross-country tour, Kalamu videotaped seventy-five hours of frank, emotional interviews by displaced survivors for a project called "Listen to the People" in an attempt to memorialize in words, sounds, and images the devastated people and places he loved, and to stir a nation's conscience as a secular prophet. The spirit of the Lower Ninth Ward lives on in Kalamu's relentless will: not only to survive, struggle, and win but above all to create.

as i mature
i wonder
how we made it

we being
the men in my family
the various
african-american males
who colorfully crossed
past bold confusions
intentionally engendered
by the infamous hidden alabaster hands
of america's human marketplaces

...

if only i can embody that black eloquent
strolling through the spaces I move
returning home at dusk
from the workplaces/the social
slaughterhouses with nary a drop of blood
messing up my mean cleanness, no malice
on my mind, and just a grinning wide
with some kind of alligator tossed
casually cross my shoulder

I was born in '47. My father, Vallery Ferdinand Jr., was in the Korean War. While he was away, we were living with my grandparents on Lizardi Street. Uncle Dewey lived next door, and he came home one day with an alligator across his shoulders. It was an alligator tail actually, now that I think about it. They cooked the alligator that day, so I had alligator.

My grandfather Noah Copelin, a minister, always encouraged young people in the church. If they played an instrument, he'd have them come up and play as part of the service: drums, trumpet, whatever. And different of us, and as I think back on it, it was mostly the males at this point, but he'd have the announcement of the order of service, and the reading of the scripture. He'd have a young person do it—get up behind the pulpit. Get experience. So I had a lot of that kind of experience. I think there was a general assumption that I was going to follow in my grandfather's footsteps. And it just wasn't to be. I left the church at fifteen.

I have four distinct memories of Theresa Copelin. First, when she administered a whipping, you had to go get the switch. Second, when I was very, very young, like four or five, she had chickens in the back yard. And I remember her wringing the chicken's neck. The third big memory I have is when my oldest child, Asante, was born, we lived in the house on Lizardi Street with my grandmother for about a year. There's a picture somewhere of Theresa holding Asante, her great grandchild. The last memory is not really of Theresa, it's just the situation in '85. We went to Nicaragua. I was there when the Witness for Peace people were kidnapped by the Contras. I was supposed to come back after a week, but I wanted to get to the East Coast of Nicaragua, Blue Fields, and we kept after them. So we went part of the way by cab, part of the way by bus, part of the way by ferry. Nobody knew where—I didn't know where I was. And there was just no phone service. My grandmother died at that time. I didn't find out till after I got back.

I think our temperament clearly comes from the family as a whole, and our parents in particular. My father is very, very, very encouraging in all kinds of ways. *Our Women Keep Our Skies from Falling* is dedicated to him.[2] There's a picture on the back, he and I, and he was so proud of that book. You would have thought he wrote it.

I remember my father read *Reader's Digest*. And I read it for a while, because he read it. We lived in the Lower Ninth Ward, which was a long bus ride—five miles— and there weren't no express buses. Every two blocks it's stopping. By then there was the Galvez bus across the Canal. I remember he would read the *Reader's Digest* back and forth to work on the bus. Most of it was politically narrow, but it still covered a lot of different genres and interest areas, so I think I maintained that kind of curiosity.

I think we were all reared to follow our own mind. As parents, they stayed out of our way. They would be very stern with us at different stages about things of principle, but as far as what we wanted to do, you do what you want to do. It's worked for us. It requires a certain element of toughness that a lot of people don't have, because they didn't grow up that way.

Across the Industrial Canal

The appeal of the Ninth Ward was that it was relatively inexpensive land, and you could build any kind of way you wanted to build. I can remember helping neighbors to build houses on a weekend. When we moved on that block, there were the Masons and the house on the corner. Across the street was the Townsends, and there was a white family that lived diagonally across from us where the Donates lived after a few years. The Donates—that's the house I remember helping to build—what he did was as his family grew, they would add on. I think they ended up with nine children. We would go over there and help. We'd build a house. We thought we were having fun chopping down trees, burying trash, and all that stuff. So there were four homes on the block, and then our house made the fifth.

It was a really rural area. I don't remember any other part of the city being like the Lower Nine. When we moved to the Lower Nine, the next block was a farm. It wasn't just a house with some land that was large. I mean people had horses, pigs, and chickens. Because we were separated from the rest of the city, and because it was "so country," the Lower Ninth Ward has in many people's eyes a stigma, but that encourages people to just really be tough.

We fought with oyster shells on the streets. But we didn't consider that fighting. Two fights I remember. I got in a fight with the neighborhood bully, and he told me not to throw a punch, and I threw it. He blocked it and knocked me down. I never forgot that one. From that I learned how to fight because people coming from the Lower Nine don't have the ability to back down in you. Another fight I had at Rivers Frederick. I didn't want to fight this person, but he just kept picking on me. So I went and bought a book on judo—read the book, studied the book. And there were a couple of techniques. The fight lasted for about four or five minutes. We circled around each other. He tried to throw a couple of jabs and so forth, and I surprised him with a lick and it was over. After that there wasn't a need to engage in fights.

I got run out of a couple of parties in the Seventh Ward because I was from the Ninth Ward. Literally we were running down the street, and people were throwing bricks at us just because we were from the Lower Ninth Ward.

Betsy happened to the Lower Ninth Ward, which was then a completely isolated part of the city. But everybody in the Lower Ninth Ward had somebody somewhere else in the city, so that was a way to make it through that, which is very different from the way things are happening now. And also the Federal Government came in with the SBA loans. And they were 1 percent, 2 percent, something like that. I think it was a traumatic event, I think it was a watershed, but I don't think it made that big a difference to the city as a whole. The Lower Ninth Ward population wise was just a small percentage of the city.

Whether the levees were actually blown in 1965 or not, I don't know. I know historically that the power structure has the will and has demonstrated the behavior to do it. I'm not a paid up member of the Negro Conspiracy League. I don't think we need conspiracies to explain just brutal class warfare. Ultimately the people that run things are not the diehard race-first people. They haven't been for a long time. I'm not confused about that at all. I'm also not confused about the fact that there is a major race issue, but often the people who are pushing the race issue the hardest, both black and white, are not necessarily the people who are in charge of anything. The reason I caution against buying into the conspiracy theory with Betsy is because even if the levees were blown, it was the power structure that did it. It wasn't the Klan.

The Ninth Ward is part of who I am, but it's not all that I am. It's a root orientation and not a future direction. So I don't have to live in the Lower Ninth Ward. But I carry part of it with me always.

"I Support Revolution"

My junior and senior years in high school, I was very, very active in the civil rights movement. I was extremely blessed to experience the civil rights movement because this helped me to learn to love Black folk. When I was sixteen, I spent many Saturdays going door to door doing voter registration and voter education work. My job was both to convince people to register and to teach them how to correctly fill out the application. While we often worked in the poorest neighborhoods of the city, I was culturally enriched because all of the houses I entered viscerally taught me aspects about my people and myself that I had not previously known, particularly the blues. While I taught mathematics, grammar and spelling, they taught me a music which literally wailed its defiance of status quo propriety. There is nothing as defiant as the blues ten a.m. on Saturday morning, cranked up loud and reverberated by the wood of those row on row of sparely painted, if painted at all, shotgun houses. Blues as tough as that woman who answered her door in bra and brown skirt, cigarette at an angle in her mouth, an angle which complemented the comb in her hair. She continued combing her hair as I tried to convince her to register to vote, neither smiling nor scowling at my naïve attempts to bring what I thought would be an improvement in her life.

In retrospect, I understand that my parents actually encouraged us. They never made any speeches. They never said, "You have to do this." But if we decided we wanted to do something, they would encourage us by not making it a problem to do it. And with my father, it went a step further in some senses, because he would supply transportation sometimes for demonstrations and those kind of things. My father kept a shotgun.

I think as far as I'm concerned, consciously, I got my activism from my father. When he came back from Korea, he applied for a job at the Veterans' Hospital. And they wouldn't give it to him. He fought it. He went all the way to Washington, DC and came back. So he was the first Black person who was nonjanitorial to work at the Veterans' Hospital. He trained a lot of people. In fact the two years I was director of the Lower Ninth Ward Neighborhood Health Center—that would have been '70 to '71—some of the people that worked and came to work there knew my father from the Veterans' Hospital.

By the time I got to St. Aug, I was so active in the civil rights movement. The civil rights movement was nonviolent, but it was also confrontational. It took a lot of courage to do some of the things we did. But it wasn't like you were throwing licks. We didn't carry weapons.

I also found out later that my mother was very, very, very active in the Teachers' Union. The one thing I do know is that there was a major teachers' strike in '68 or '69 when she walked out alone. Her sister Neomi Foy was teaching at that school and didn't walk the picket line. My mother was initially the only teacher from Hardin Elementary School who actually walked the picket line. One of the times I was arrested, she said, "Whatever you decide to do, think about it. If that's what you really want to do, be prepared to do it alone because you cannot count on the other people." There are things you know in the abstract, but when you get a direct statement like that from someone you respect so deeply, you think about it in a different way.

I'd known Dutch Morial from NAACP days. We were demonstrating on Canal Street: picketing, organizing a boycott of merchants for jobs and accommodations in terms of restaurants and what have you.[3] That boycott lasted for over a year. I mean we were demonstrating every day, picketing in the rain and what have you. We'd go there every day of the week except Sunday. At a certain point, the merchants said they were willing to negotiate, but we'd have to stop picketing. The Youth Council of the NAACP who had taken the leadership on this said that we would stop picketing once some negotiations were done. They contacted the adult chapter and Dutch Morial was head of the Adult Chapter of the New Orleans NAACP, who in turn contacted the National Office. The National Office sent one of the field secretaries out, and they read us the riot act, which basically was, "If you don't stop demonstrating, we're going to put you out of the NAACP because these merchants are ready to negotiate." This was in March of '64. We used to meet at Mt. Zion church, and I was sitting there before meeting on the steps. I quit the civil rights movement right there. I felt betrayed by the adults because except for Mr. (Llewellyn) Soniat, adults had not been active on the picket line like we youth were. And they were going to tell us how to do it. I knew that Dutch was genuinely interested in advancing the cause of our people, but he was also an accommodationist.

I was in Korea during the Vietnam era. After I came out of the army, I'd been trained with weapons. I've already gone through the civil rights movement, so I'm not afraid. I had a carbine in the closet. Not my mother nor my father ever said, "You all shouldn't have any guns." I would assume that some of that had to come from their parents. I know that my father's father got run out of Napoleonville. My grandfather Ferdinand had gotten into an argument with a White man and was going to shoot him with a shotgun. To me, this was a definite plus.

I was one of the leaders of the takeover of Southern University in New Orleans. We just literally took over the campus. When we got to Dean (Emmett) Bashful, he said he wasn't going to move. And he sat in his big chair. I told them: "Wheel him out in the chair," and they actually—I mean we had big cats—wheeled him out in the hall way. We called Baton Rouge and told them we had taken over the school. Whenever they were ready to negotiate, we'd be willing to negotiate. We ran the school for a couple of months. We kept a couple of the professors.

Early on, the administration tried a tactic of calling a meeting with the parents.[4] And they were going to try to get the parents to put pressure on all of us. My grandfather had a stroke when he was speaking. I don't remember what he said because I was in the back. As far as we were concerned, it was a charade.

Noah Copelin was in the midst of an emotional appeal to the parents not to blame the kids but rather themselves for being so submissive to "the man" for too many years. Their passivity, he argued, necessitated more confrontational actions by the students.

When we kidnapped the governor, I think that's when they got scared. Governor (John) McKeithen came to New Orleans to speak at a church on St. Charles right off Canal Street.[5] That church only had the front door and the side door out into an alley. We were wonderfully organized and were able to mobilize in the matter of an hour four hundred students. We surrounded the church. He could not get out. He had said he wasn't coming to the school. "You ain't going to the school? You ain't going nowhere." And I remember when he finally decided to come up. He had the state trooper, security, plain-clothed. And they said, "The governor says he's going to go to the school." We said, "no! We're going together." So he went to dash to try to get in his limousine, and one of the guards tried to stop me. He pushed me, and I pushed him back. And he looked up, and they recognized that they couldn't move, because the students had the car surrounded. So finally two of us got in the car with the governor, and we went to SUNO. On one side of the stage, an American flag, on the other, the flag of black liberation. He said, "What's that?" I said, "Sir, that's the flag." That's what had been the big issue, because we took down the American flag on campus and put the liberation flag up—we always flew it wherever we went.

The red, green, and black liberation flag was a symbol of black self-determination; it also represented contemporary African decolonization efforts. At SUNO, students felt the campus was a remnant leftover from Jim Crow times, and they refused to accept an education they considered outdated and unequal.

He said he wasn't going to speak on the stage with that flag. I said, "Take all the flags off the stage." So we had no flags on the stage, and he spoke and he said that something needed to be done to defuse the situation, and he was going to send some people down to negotiate with us. And that's when the negotiations started. What they did was not allow about twenty-four hundred students to come back in September. That was the only way they were able to break us.

We had a shortwave radio, and we were listening to the police, and they were trying to catch one of us by ourselves, so they would kill us. They would talk about it on the radio. But we always traveled at least with one person, and as we moved around, we let people know where we were going. My life has been threatened numerous times. I always took it seriously, but I never let it stop us from doing what we had to do. They started following me after the flag. It was on until we took over City Hall. It was constant virtually from '69 to when Dutch was elected.

The monument to white supremacy, a twenty-foot obelisk recorded in the historical register as the Liberty Place Monument, was erected at the foot of Canal Street near the French Quarter in 1891, one year after the Louisiana Redeemers had rewritten Louisiana's constitution to codify white privilege.

The standoff with the international conclave of the Ku Klux Klan was at the white supremacy statue downtown at the foot of Canal Street. Our slogan was, "They're not going to be the only ones shooting, and we're not going to be the only ones dying." And the police knew it. The FBI knew it. The night before, the Klan had a rally in Algiers, and they shot at the white police. Somebody called the police and the police tried to break up their rally. It was the largest demonstration of armed Black folk I have ever attended. We sent out the call and even the street people responded. Nobody messed with Black people that day. The Klan folk changed the time of their demonstration. It was the most beautiful demonstration I'd ever been a part of.

When we took over City Hall, Dutch decided not to confront us. We had a team of people whose only job was to call city hall numbers and tie up the phones. When we went in, it happened so quickly, they had no time to react. And our tactic was to make sure that they didn't just ignore us—we had whistles. "Blow the whistle on Dutch." And you could not work in that building because we were in there blowing our whistles. We were fighting police brutality.[6] At that point, it seems negligible now, but back then there were thirteen people killed in a twelve-month period by the police under all kinds of circumstances, including the killing in old Algiers, and we said that was just too much. The takeover of City Hall, which lasted three days, was the culmination of a lot of those struggles. I drew some serious conclusions about the future of organizing after we stormed City Hall, because there was nowhere else to go. I could see that people were divided. This was the first Black mayor. They wanted to give him a chance.

I considered myself a supporter of revolution because I haven't made any revolutions. I was in the People's Republic of China in '77. I went to Nicaragua. I went to Cuba twice. I went to Tanzania. Surinam, the coup attempt, I was down there. You can't make a revolution without realism. I'm not a romantic at all. I don't define myself as revolutionary.

Sophisticated people don't slaughter their own food. They prefer to buy it and prepare it. My grandmother taught me that. I saw it. Never liked it, but I understood. You've got people that have no stomach for revolution. I never will forget Julius Nyerere, the first president of Tanzania. In '74, I interviewed him. He said, *"All* governments are conservative."

Coming out of that, I made a decision to start a writing workshop, and so essentially what I decided to do was to work with younger people. I recognize now that all my adult life except a brief period, basically the '80s, I have worked with young people. When I came out of the army, the first thing I did was hook up with Morris Jones at the Lower Ninth Ward Community Center. I was teaching photography, and then Mrs. (Olga) Jackson started what was called a Boy's Club, an after-school program for young kids, in her house. And I started working with that, then Ahidiana. (*Ahidiana was a Pan-African nationalist political formation which had operated a school since its inception in May 1973.*)

I got involved in '96 with Students at the Center. So I continue to work with high school students. Some have bought into the American dream; some are looking for other things. My task is not to tell them where to go, but to make them strong enough to make that journey. That's what my folk did for me. I just try to make them strong enough to understand what it takes to do something. And be an example, day in, day out, day in, day out, day in, day out, day in, day out.

I've been doing e-drum since August of '98 every day. It doesn't cost anything to produce, but it takes a lot of time and effort. I made a conscious effort to always include a lot of information that is not "directly Black," because I think we need to be involved in the world. I know that e-drum is influencing people and has impact all over the world. You just have to persevere. It's time to model consistency of work and getting people thinking about the issues.

Katrina: "We Intend to Overcome All Obstacles, and Continue"

There are two ways to deal with a hurricane: either hunker down until it blows over or run and get out the way. I've learned to pick my fights and from what I saw coming, I wasn't up for tangling with Katrina.

Kalamu was part of the evacuation to Houston on Sunday morning. His home in Algiers was only lightly damaged; but he had lost many of his possessions kept in a storage unit in New Orleans East: over five thousand CDs, over three thousand books, historic papers, and equipment. Eventually he moved with his wife to Nashville, Tennessee, where she was transferred temporarily for her job. He spent most of the fall on the road. Although he had only ten dollars in his pocket, he immediately began an ambitious plan to document the tragedy in the words of the men and women most affected by it.

I felt I had to do something. In a nutshell my goal is to travel around from shelter to shelter, community to community and work with people so that they can document their own lives and share their views with the world via the Internet. In the end, I conducted over thirty video interviews, over seventy-five hours. That was what was in my power to do. I didn't have to wait for money. We got one grant to Students at the Center, and we earmarked ten thousand dollars of that for "Listen to the People." Everything else was paid for with money I earned from speaking engagements and the like as I traveled. It took a lot for us to do what we did. But that's the way change is made. If you're looking for a funder for the revolution, you can forget it.

We missed e-drum just one day after Katrina. I do that 365 days a year. Connectivity was the issue. At one point in Nashville I had to pay four dollars to sit in Barnes and Noble for two hours. We didn't miss a single issue of "Breath of Life," a conversation about black music.

A guy interviewed me today, and he asked me, "When you got back home, what did you think?" I told him, "I never got back home. I'm here, but this is not home." The Lower Ninth Ward from Claiborne to Florida Walk, it's so obvious that you've got to tear all that down. The East is irreparable. I don't even think they're going to put up new condos. The infrastructure was already stretched. The city was crumbling as it was. There's not the political will to rebuild. They cannot rebuild New Orleans and the Gulf Coast and maintain Iraq and Afghanistan. I think the plan is basically death by attrition.

Somebody asked me, "What do you want to see in the future of New Orleans?" I said, "What do you mean? What is the future of New Orleans? Who knows what's going to happen?" The reporter followed up: "What would you like to see?" So I said, "Bush's head on a stake in Congo Square and Nagin's on a stake right next to it." (Congo Square is the place where my ancestors, even as slaves, crafted and created a space where they could be free on Sundays. They kept African music and dance alive.) And he looked at me. He didn't understand the historical reference: Congo Square is where they used to do executions of African Americans. And it was the site where they executed the leaders of the largest slave revolt in American history with the 1811 revolt.[7] The rebellion involved approximately five hundred slaves. When the soldiers put the revolt down, they cut the heads of the leaders off and put them on stakes. Then I told him, "I'm not that interested in Bush's head per se, but I know in order for that to happen, there would have to be a revolution, and that's what I'm really interested in." The '80s have convinced me that reform will not work in America.

Conclusion

People don't really want to know what's going on in New Orleans. It's like when you ask somebody, "How you doing?" You're not really asking them. We as Americans are junkies. We're addicted to materialism and consumption, and anything that threatens that, we don't want to hear about. You cannot organize junkies as long as they have a supply of dope. It is going to have to be a disruption of the supply of dope. I firmly believe that the environment is going to directly disrupt the American economy to a significant extent. And nothing is worse than a junkie down on their luck.

Since the 1930s, Louisiana has lost 1,500 square miles of coastal wetlands. Wetlands have the capacity to diminish the power of the storms and their surges.

In Louisiana, the problem is erosion of the coastal marshlands. In fact a significant portion of Plaquemines Parish is no longer dry land, so there's nothing slowing the hurricane. The environment is the key issue here. It's going to take a minute for people to connect all the dots, but these are historical developments. And after a while, if enough things that never happened before keep recurring, people are going to start asking questions. When they ask, "What can we do about New Orleans?" I say, "Elect an environmentally conscious leader, whether the dog catcher, mayor, head of the Rotary Club, whatever."

Keith C. Ferdinand

Born in 1950, Keith Copelin Ferdinand is the son of a third-grade teacher and a laboratory technician. In 1965, Hurricane Betsy severely damaged his family's home and took his grandfather's life. After two days on the roof of their home and one night on the roof of Hardin Elementary School, the Ferdinand family temporarily moved in with relatives on Louisiana Avenue. Keith went on to excel at St. Augustine High School, and in 1968, he entered Cornell University on a full academic scholarship, financed by L.L. Nunn, who in 1910 founded the Telluride House where Keith lived while in Ithaca, New York. In front of Cornell's Willard Straight Hall, he decided to devote his life to providing medical care for the community that reared him. Sixteen years later, he and his wife, Daphne Pajeaud Ferdinand, then a registered nurse, opened Heartbeats Life Center, a cardiology clinic, on Poland Avenue in the Upper Ninth Ward. During its twenty-one-year history, more than ten thousand patients were treated at Heartbeats, with its board-certified cardiologists and full-service work-ups.

The first interview of Keith took place on January 3, 2006, in his crowded ABC office.[1] He was suffering from depression from the loss of Heartbeats and his lakefront home from the floodwaters, but he became animated whenever he talked about the Lower Ninth Ward of his youth. His descriptions of the people and the neighborhood that, in his heart, he never left served as a way to honor his community after Katrina made serving its medical needs impossible.

Keith's recollections of his parents and the community of his youth remind us of the second generation of black pioneers who homesteaded the Lower Ninth Ward when it was a black homeowner's paradise. The detailed descriptions of Betsy offer us insights into the pioneers as they responded to what they believed was mainly a man-made disaster. In his narrative, Keith explains why such a dedicated, experienced professional is reluctant to return to practice medicine in the city he is proud to be from.

I was born in New Orleans, Louisiana in 1950. My father was Vallery Ferdinand Jr. Vallery Ferdinand Sr. was a self-taught preacher, a longshoreman, and an occasional plumber. He was married to Julia Bennett Ferdinand, who died young. My oldest brother is Kalamu ya Salaam. For the longest time, I was known as Kalamu's younger brother. My other brother's Kenneth. I'm the baby. Kenneth is a gentle guy, strong and tough like my daddy, but really kind. Kalamu was ostensibly tougher than we were because his personality was gruffer. I just can't perceive that somebody could roll up on me. Unless you're ready to take on those other big guys, why would you mess with Keith?

Our children have a lot better life than the three of us Ferdinand boys had, but we were raised to be part of the warrior class. We were raised to challenge the world and not doubt ourselves. Control the space you occupy. Don't be a sucker.

Ferdinands are doers. You didn't get gold stars on your forehead for doing what you're supposed to do. I mean it wasn't a warm, touchy-feely, "You're doing real good baby. You're out there struggling." If you came back and complained somebody called you a name or said you were dark, my parents would probably look at you like you're crazy. "Why would you even listen to anybody who would say something like that?"

There wasn't any whining in my house. My father ran it like a sergeant. My mother was a sergeant of the Ninth Ward teachers. And we never were told not to do this, don't go there, don't hang out. We could do whatever we wanted, just so long as we came home safe. My closest relationships were with my parents.

A healer and worker, my father was born December 12, 1919. His story, and I believe it, is that he walked from Napoleonville to New Orleans at the age of nine. A self-described "poor boy," he ate stale French bread for five cents and drank milk to survive. Ninety miles south of New Orleans, Napoleonville is basically an indentation off the bayou into the sugarcane field. It's obvious that the purpose of the town is to serve the cane field. The white folks said, "Build your home here. There is where you all are going to live. Go right out there and work, work, work." I can see how my daddy would not want to stay.

My father did see combat in the Pacific Theater during World War II and the Korean conflict. He rose to become a sergeant. He only talked a few times about having to stick bayonets in people. I remember a uniform with medals on it.

He said, "You've got to respect a man's job. Treat everybody right and the hell with them," meaning that you don't bite anybody, but you don't make large efforts to please people. You do what's right, and then, if they don't like it, then to hell with them. You move on.

There was no public transportation to the back of the Lower Ninth Ward. The closest bus connecting us to the rest of New Orleans went down St. Claude Avenue, which is twelve blocks away from Law Street where we lived. Every morning my father would get up and walk to St. Claude, and he would catch the bus to go downtown to the VA Hospital. So in order to get to the VA for 7:30, he'd get up and leave at 6:00.

My uncle, Sherman Copelin Sr., had Louisiana Mortuary, where my dad assisted with autopsies. If he wasn't going to his mortician job, he would come home at 6:00 p.m. Weekends, that's when he did his second job most of the time and when he did most of the grass cutting and brush clearing trying to get the neighborhood together. I've seen my daddy work ten, twelve, fifteen hours outside in the summer sun. As a kid, he chopped sugar cane, so it was no big deal for him to go outside in ninety-five degree weather and work. My father would say, "I need three volunteers: you, you, and you. Let's go!" He was premodern fatherhood, where you would bargain with your children. He would say, "Get up! Let's go cut the grass."

I remember our daddy going to Sears and Roebuck with a broken washing machine, a lawnmower, a blender, or some appliance that he purchased. When you go to these places, especially as a black man in the '60s, the powers that be reasoned you should be happy that they allow you in the store to buy something. He went back to the guy and he said, "I bought this thing, and it's broke. I want another one." The guy would say, "Fill out this form." And he said, "Nah, I want another one." The guy would say something else. "You can eat it, because I'm not taking it back." My father got his new machine. I don't remember my parents with a situation where they backed down.

I remember my daddy specifically saying, "Negroes don't own a tooth-pick factory," meaning that we didn't occupy any of the means of production in this society. There was a black-owned pharmacy on Galvez and Caffin and a little movie theater that eventually closed after integration. There were a few corner stores. To go shopping you had to go out to the front to St. Claude. There was an A & P grocery store there. When we were kids, these older guys and my daddy would meet in the front room and talk about opening their own businesses.

My mother was born March 12, 1918. She went to Dillard University, became a New Orleans public school teacher, and taught third grade. She was only about five foot four, 110 pounds, but she was really, really tough like Joe Pesci. My mother chose the toughest kids. If a kid was overgrown and a juvenile delinquent, she'd say, "Give him to Inola."

When the school bell rang at 3:00, my mother wouldn't move. She lived in the neighborhood, so she could walk home. She would keep after school the ones that today would probably be placed in juvenile delinquent homes, and make them do things like decorate the board by hand or help her prepare the lessons. I later became a physician in the Ninth Ward, and I've had several patients who would come and say that my mother taught them how to read.

My mother went to Nicholls State University in Thibodaux, where she got a Master's in Education. She drove sixty-two miles by herself and took her classes. She helped to form the teacher's union for which she received a lot of flack. Her tires were slit one time because of it. At that time the black teachers weren't unionized.

My mother died in 1975 of Hodgkin's disease, before I graduated from med school. My father was like Clark Kent. Then he died from an infection from a rare strain of pneumonia in 1987.

Lower Ninth Ward: 1950–65

My predominant loyalty was the Ninth Ward. The young men who are from the Lower Ninth Ward feel good about it because it kind of meant you were edgy, dangerous, different. We used to say, People "from the Nine don't mind dying." We were proud of it. I think it was a combination of factors, because you were from a particular area, but also because we were pioneering type people, not people who were in long-standing historical neighborhoods, which were comfortably defined as New Orleans. Some of our areas were swamp lands or the rough edges of the city, where you had tall grasses, marsh, trees, bushes, and shrubs, so they had to be claimed as a place to make a home. So the neighborhood was built by people who were fairly ambitious in terms of trying to make a homestead for themselves.

My first recollection of a substantial childhood was in the Ninth Ward. After my father returned from Korea, he built a house at 2517 St. Maurice Avenue. It was one of the last streets in the city of New Orleans, one block from the corner of St. Maurice and Law. It's the corner of the city. After that you're in the swamp or you're into St. Bernard Parish.

Most of the people back there from the early beginnings were homeowners. The land back there must have been pretty cheap, because it was all filled in swamp. My daddy had a contractor. These are usually black men who may have a license and some basic skills, and they will fix a porch, brick your house, put down a floor, and put up walls. My father used to expend huge amounts of energy cutting and trimming grass, cleaning out lots, and cutting down weeds. He did it for everybody, and he wasn't paid at all. If he was cutting grass, he would keep cutting until there was no more uncut grass in his line of vision. My daddy thought that our block should be the best block in the world.

We had a living room, dining room, kitchen, and two bedrooms. It was a box, but it was our box. It was brick veneer with a garage. There were a lot of reasons to go to the Ferdinand house. People came for the good food, for the stable environment. We might have had one of the few televisions that worked. There wasn't any hollering and screaming in my house. My mama had strawberries she planted in the front. In back, we had Japanese plums, tangerines, pomegranates, pecans, and figs.

Our back yard was the swamp. We used to do war games with sugarcane reeds and bottle tops. You take the old snap-off bottle top and hammer it into an arrowhead. When we were young, kids from Desire and the kids from the Lower Ninth Ward would war with little shells and weeds that grew almost like bamboo. So we had little "gangs," and there would be little territorial fights. We were nine, ten, eleven, twelve, when we had these fights. Nobody got hurt, and we never did anything really terrible.

The community took care of itself. There wasn't a lot of random crime; old people could walk home. You could leave your door open with just the screen door locked so a dog wouldn't come in. Guys would fight each other if they were drinking, or if you went to a certain barroom or pool hall, you might get in a fight, but the random, crazy stuff we didn't do.

The rest of the people, if they weren't a teacher or weren't in the ministry or a longshoreman, became postal workers, a very good job, or mechanics. Some people picked up trash, did odd jobs. It was all about work. There wasn't a lot of lying around doing nothing. I mean there were always some fellows around the corner with brown bags, drinking, hanging out. But that wasn't unique to the Ninth Ward.

September 19, 1965: Hurricane Betsy

I was fourteen years old. I had one of these boom boxes, but I was kind of a strange kid. I actually would listen to the shortwave and the weather radar. On the shortwave radio, you could hear some of the fishermen and people in outlying areas chatting, and they were saying the storm's coming. I was up when the water first started coming. It came very, very rapidly. I jumped into the water, and went from house to house, knocking on doors, telling people, "It's flooding! It's flooding!" So people started getting up and they would scramble out of their houses. A lot of people came over to our house, because it was higher than some of the others.

During the life-changing trauma of Betsy, Kenneth and I became permanently linked. Ken and I together darted into the perilous hurricane-force winds with torrential rain, flying debris, and sparking electrical lines, only to turn around, because we could not reach grandpa. For many, many years I blamed myself for my grandfather's death.

The water was up to eighteen to twenty feet. I remember my daddy took a 2 × 4 plank and kind of busted the ceiling. We scrambled onto the roof of our house. The next day a lot of people started coming, private people that had boats to go shrimping, crawfishing, and fishing. Some helicopters came and some of the people were fighting to get on helicopters. But my mother said, "Just wait." When we were sitting on the roof for two days, there was no moaning and groaning. We waited until an amphibious boat came from the military and took us to the Claiborne Avenue Bridge. We walked across the bridge, and then we got on buses and then they brought us to City Hall. From City Hall, we went to my dad's sister's house in Uptown.

We were out of our house at least for a year. My daddy never talked about any financial difficulties. I do remember the SBA helping with loans. They did most of the rebuilding themselves with the help of local black contractors. If my parents had the same sense of loss that a lot of people in New Orleans felt after Katrina, they didn't articulate it. We have one picture of us with me as a baby sitting in a chair and my two brothers standing next to me. All else was lost. I actually don't remember a sense of woe is me. I just went to school and did what I had to do. My family believed that faith meant you deal with trials and tribulations by being functional.

Betsy was the first time I really recognized that we were living in a precarious situation. I didn't even think about it before then. You could actually walk on the levee back then. Sammy the Snake Man would catch snakes and play with them. But they told us they fixed the levee.

I do think the levees were blown in '65. When I was listening to the shortwave radio, I do recall it was almost a discussion as to exactly what was going on. And the way the water came was extremely, extremely fast. It went from like nothing happening and everybody going to sleep to terror, rushing water, and climbing on the roof within a half hour easy. It's not impossible in our concept that they would allow us to sink or swim.

For the tremendous tragedy and widespread devastation that Betsy was and the uncounted people who died, there wasn't much discussion at all about Betsy. You never had it in a history class. People just dealt with it. A lot of families moved out, houses were abandoned. I feel that a lot of the discord, the crime, and the drugs in the Ninth Ward were because of the damage that was done to the housing and the stress to the families. All my friends that got into really bad trouble did so in their twenties. The Ninth Ward went from a fairly stable environment to a very unstable environment after Betsy, and it never recaptured the sense of purpose that it had before.

Segregated South

I remember reading one book called *Jean-Christophe* by Romain Rolland. I was a teenager, and it was a two-thousand-page book, an autobiography of this French kid who was battling hardships. I identified with the hero in his epic struggle.

My first time going to school with white kids was when I went to Cornell University. The paradox of the segregated south was that some of the schools were pretty good, not because the schools were equal in terms of physical plants or in terms of books. They weren't. But the teachers were dedicated. Those were the best minds and they really, really, really believed that teaching was the most important thing they could do for those kids. We were taught everything right. Everybody got up and went to school.

I went first to Joseph A. Hardin Elementary School, which is right down the street. Then I went to a public school named Rivers Frederick, and subsequent to that my parents sent us to what was considered the best school for black men at that time, St. Augustine. It was run by the Josephite fathers and brothers. The Josephites don't belong to any parish, and they were specifically chartered to help African Americans and Indians. The history of the Josephites was one of activism and dedication. Some of the Josephites actually came from the Northeast. Most of the fathers were white. They felt we could become leaders and change the environment. When I was in high school, they sent me to Atlantic City for a national debate contest. I was on the first team of all blacks who competed against whites on television. People in the neighborhood would say, "There's the Prep Quiz Bowl boy."

I could not have competed at St. Aug, if it were not for my black public education. Many of the Catholic, Creole kids from St. Aug came from parochial schools. They were from the Seventh Ward, so to some extent, I was an outlier even within St. Aug. I was black, Protestant, Lower Ninth Ward. There was a separation of perceptions of class and status based on skin color and white admixture. I come from a militant, pioneering, military family, so it's not like they could make me feel bad about being who I was. We didn't feel second to anybody.

Cornell University

I went to Cornell in the fall of '68. (A lot of the St. Aug boys actually went to Ivy League schools and became judges, lawyers, and politicians.) I had a full academic scholarship called the Telluride Scholarship. So you grow up in the corner of a corner, and then you go all the way up to Ithaca, New York. My parents put me on a bus. I caught a bus to a plane, and then went from JFK to the Port Authority, where I caught a Greyhound bus.

Our education back home gave us a sense that the sky was the limit. In fact, the funny thing about it was that although we came from the segregated South, when we entered the larger world, whether it be the military or Ivy League schools, we weren't impressed. We knew that what we were doing was as good as what anybody else was doing.

I was walking past Willard Straight Hall in October '68 in the snow with my militant friends, and we were trying to decide what to do with this education thing, the "white man's education" we called it, because we didn't really know whether we wanted to continue. I said, "I think I'll be a historian and tell the story of our people." And they said, "We got enough people talking. You're young. Do something that's going to be of service to people." So I said, "I'll be a doctor." Of the six dedicated militants, four became physicians. It absolutely always was my 150 percent unadulterated goal to come back to the Ninth Ward and serve my community.

We thought there was going to be a revolution at that time. You can look at things different ways. We got into "trouble" or we disturbed the system. I got credit for two semesters, because the takeover of Willard Straight Hall was in April of '69. I chose to come home because what had happened closed what I felt was a chapter in my life.

Tactical Education

I went back to LSU in New Orleans, where I got a B.A. in biology. It was a state school. There was a rule that state schools had to accept everybody. I was so smart that my grades were fine. Xavier University wouldn't even accept me. Why would they want to invite a militant?

But at that time I had already made the decision that this was going to be tactical education. "Tactical education" is when you go from point A to point B to learn something in contrast to "vital education," which is what your parents and your community teach you.

I worked my way all the way through school. I don't remember my parents giving me anything once I went away to Cornell. I was a mailman one summer. I was a carpenter's helper. I bused tables at the Desire Oyster Bar on Bourbon Street, just whatever it took.

We were blackballed after Cornell, but when it came time for med school, you had to take a test called the MCAT (Medical College Admission Test). My history is that I can do any standardized test they put in front of me. I was a National Achievement Scholar. There must have been a way of knowing that I was what they called a "Negro," so I got letters from all of the top schools in the United States: Harvard, St. Louis University, and the University of Chicago. "Dear Keith, You would be a great student. Please apply." And then I would start getting rejections.

You had to have letters of recommendation. Marion Mann, an old army general, who was then Dean of the School of Medicine at Howard University, a historically black university on the radical fringe, was doing a swing through the South interviewing students for medical school. I asked him if I could come for an interview. He allowed me to come up to his hotel room at the Royal Sonesta. He said, "You are obviously a great student. Your MCAT's are the highest that we've seen. You know you have good grades. But have you ever seen your recommendations?" My recommendations said that I was antisocial, and I should not be allowed to be a physician. They made me sound like I was a mass murderer. He said, "Son, I'll give you a chance."

I went to Howard from 1972 to '76. At Howard, I did loans and I got a grant because I was a student from Louisiana. I did work-study too. I was an audiovisual aid. It was my job to set up the equipment. I would clean up the blackboards and have fresh chalk ready. I did very well at Howard. I was a member of Alpha Omega Alpha Honor Society. Then I came back to the US Public Health Service on State Street in New Orleans for a year. The government had my record, but they would take anybody. I wore the uniform. It was a tactic to pay back my time and loans. It was in New Orleans, so I was able to come home.

When it got time to get a residency, I went to LSU Charity, where I did two years of Internal Medicine, then two more years of Cardiology. My last year, I requested from Howard that I be allowed to come back and finish there because they had a modern hospital. I wanted the best techniques, and I was getting depressed watching the abominable conditions at Charity's ER.

"Hardest Working Man in Medicine"

From 1981 to '83, I was with Medical Associates, the largest black professional group in New Orleans, and one of the largest in the nation at the time. They were on Louisiana Street, and they were doing very well as black doctors at a hospital called Flint-Goodridge. When they brought me in, I was the first black board-certified cardiologist in New Orleans.

But the plot from Willard Straight Hall in the fall of '68 all the way to '83 was to come back to service the Ninth Ward. Fifteen years of planning and sacrificing. My wife Daphne and I founded Heartbeats Life Center in '83 on Poland Avenue right by the St. Claude Bridge. We took out a two hundred thousand dollar loan at 14 percent from a local bank that demanded a key man life insurance clause. I told the banker, "You're ripping me off. That's usury." But no other bank wanted to give me money. There was nothing to say to a bank that I was somebody who would be anybody. We made a down payment. My wife and I had saved up a lot through extremely frugal living. Daphne sewed. We ate beans and rice, and grew vegetables. My wife worked as a registered nurse. We paid the loan back within the decade.

You could walk in and get a full work-up at Heartbeats. The type of stuff that I was doing with the seven hundred thousand dollar nuclear machine, the echocardiogram, the board-certified cardiologists, and full-time techs—nobody else was even thinking about that kind of equipment and investment in a "poor" or "ethnic" area.

Heartbeats was placed on Poland Avenue because I wanted it to be available to serve the community from which I sprang. But the catchment area for Heartbeats was quite wide. I had seventy-six hundred active patients. Some people would drive in from Mississippi to see us, because they felt they weren't getting the type of care they needed. We had people of different economic strata, even though it was located right on the street in a working-class neighborhood. Professionals, dentists, attorneys, and judges also would come in. My staff and I purposefully never treated any patient differently based on their socioeconomic status.

I have patients who came to see me for years. One of my patients, Scoop Jones the Founder of *Louisiana Weekly*, was a writer for the *Black Data News Weekly,* a small New Orleans paper. He came in for a regular office visit. My daddy smoked and drank a little bit, but he wasn't a smoker or a drinker. Scoop was the typical New Orleans player. "Your blood pressure is doing ok, you aren't having any chest pains—everything's fine." This grown man in his seventies starts crying crocodile tears. He's saying, "Why did God have to take Val? Vallery was a good man. I've been smoking, I drink. I run the streets." I had multitudes of patients who would come in and before you go to the doctor-patient thing, they would talk about how my mama was a great woman. Or they'd talk about my grandfather Copelin.

Working with patients energized me a lot. It's a labor of love. I'd get up between 3:00 a.m. and 6:00 a.m. and start exercising. I'd be at work by 8:30, 9:00 every day. I would bring things to patients' homes. I would go to the pharmacy. I stayed in my office on Poland Avenue until 10:00, 11:00 at night. I'd put in ten to fifteen hours a day as a routine. As a physician, I used to work like my daddy worked as a laborer. In a way, it's in honor of his work ethic. My office was open to the last day. I remember I worked a full day on the Friday before the hurricane.

The American Dream: From Tupelo
Street to Lake Forest Estates

After I married and returned to New Orleans in '76, we bought our first home at 2427 Tupelo Street only a block-and-a-half from St. Maurice for ten thousand dollars. It was my big brother Kenneth who convinced me to purchase the home, although I was a working poor medical student with no real income. He assured me that the only way to become a homeowner was to buy a home. It was a small box: a small front room, a bedroom next to it, a small kitchen and another bedroom. The floor was just slats on cinderblocks, so you know if you took a hammer and punched it through the floor, you'd go down to nothing. But it was our first home.

As I became a physician, I accumulated a little bit more means, but I wanted to stay in the Ninth Ward, so we went from that four-room box across St. Claude and bought this two-story house at 1027 Flood Street for forty-seven thousand dollars. We were "Big Time" then. It was viewed as a step up to go across St. Claude to the histori-cally white area now known as "Holy Cross" next to the river. It was a very nice house right around the corner from Fats Domino. This one had wooden siding, a front porch, and a big yard.

Then one time I took my family to Astro World in Houston. When we came back, my heart hurt really badly, because they had kicked in the door and taken everything that could move. And the Ninth Ward is so connected that whoever stole all that stuff is somebody who knew who I was. A couple of my hardcore friends came by to say, "It wasn't me, man."

In the late '80s or early '90s, there was this new area being built up called New Orleans East where you could get a really nice house, and you could have a little lake in the back and a pool. You could build these big brick houses. It was like the American dream. We had a sun-splashed back of the house with a lot of glass in Lake Forest Estates, which at first was a white neighborhood. Within the last ten to fifteen years, it's become a multiethnic community, mainly black, but with Asians and a few whites, like the lady who sold a lot of houses in Lake Forest, Joan Blackwell, who wasn't orig-inally from Louisiana. It remained a very prominent neighborhood with black judges, lawyers, and doctors living on my street. New Orleans East gave you the convenience of Atlanta, Houston, Dallas, but it still had the feel of New Orleans because a lot of the people who lived out there are New Orleanians you knew from childhood.

By moving out of the Lower Ninth Ward, I was trying to make my family as com-fortable and happy as I could. I used to go back to the Ninth Ward all the time. I had patients there. My transcriptionist lived on Renee Street in the Lower Ninth Ward.

Katrina

In 2005 I'm still a news hound. I fearlessly questioned the need to flee as Hurricane Katrina approached August 28, 2005. I was acutely aware of these storms, even though I never ran from a storm, because I always felt that I could figure out what was going on with the storm and ride it out. On Saturday night it was obvious that the storm was coming. So my wife, her mother, and our two daughters left Sunday morning at 6:00 a.m. for Jackson, Mississippi.

My oldest son, Kamau, and I drove into the Bywater to get some plywood and nails from Ken, and to see what he was going to do. My brother was literally loading up his van with food and supplies as he prepared to flee New Orleans with his family. I always quietly looked to Kenneth to reassure me and give me a sense of stability as various small crises arose in my personal life. So I said, "My big brother is going, maybe I should re-evaluate." So 11:30 a.m. on Sunday, Kamau and I drove out across the Twin Spans to Jackson.

The storm moved in early Monday morning. You could see what was going on in New Orleans on TV. You could see the Superdome roof coming off. I got a call Tuesday morning from one of my friends here in Atlanta. He was congratulating New Orleans because the report from the cable news was that it was a glancing blow. I had a cold chill and was tearful, and I said, "No, that's not the way it works. The low-lying areas could be flooded, and a lot of those people live in some very, very shaky houses." My concerns were that shotgun houses would collapse or the roofs would come off of substandard buildings, because you could see what was happening to buildings as stable as the Hyatt and the Superdome.

I don't think they had to go dynamite anything. They're not that stupid any more. I think they underdeveloped the levees, so that if there were a storm, it would break itself like some machines have built into their mechanisms, so if they're working too hard, they'll break.

When the flooding started, I knew the typography of the flooding. I knew exactly how it felt to sit on top of a roof, and I knew that when they showed these heroic pictures of the Coast Guard helicopters plucking people from the roof, that was for public consumption only, because New Orleans is a pretty sizable city. Once it became obvious on Wednesday and Thursday that they didn't have the big Huey helicopters and the amphibious assault vehicles coming in picking up people, I knew people were just dying left and right.

I knew where my clinic was situated was bad. There was an electronic way you could go on the Internet and see your house. Both from having lived through one of these storms and having looked at those satellite pictures, I knew that the house and the office were basically gone. In two days, I had to wrap my head around everything that I knew no longer existing.

With the power out in Jackson, it was hot and sticky, so I decided we should go east because there were too many people going west. I knew people here in Atlanta at the ABC, a nonprofit organization dedicated to eliminating cardiovascular health disparities in people of color. I had served as chairman of the board from 1990 to '94.

We were on Interstate 20 between Jackson and Atlanta. So we had two cars, my wife, her eighty-four-year-old mother, three adult children, and whatever we had is what we had. I had one hundred dollars in my wallet, because my father told me to always keep cash for an emergency. We're on the cell phone trying to figure out where to go. All of the 504 numbers were down. My youngest daughter, Aminisha, had a cell phone with a 917 area code, because she used to live in New York. I called several people in Atlanta. One was Dr. Mark Walker, a surgeon and a former classmate of mine at Cornell in '68. He actively found a place where we could stay. My wife talked to her sister in Albuquerque, New Mexico, who had a son who lived outside of Atlanta. That's where we ended up staying for several weeks. My youngest son, Jua, was evacuated to a shelter in southwest Louisiana. He later joined us in Atlanta. That is just how arbitrary and difficult the decision of where to go and where to live was.

"All the Things That I Defined as Normal Are Gone"

I spent most of my life in the Lower Ninth Ward. You go back and the house that your daddy physically built and the first and second houses my wife and I owned—all of that is either heavily damaged or gone. I drove and looked at all of the different places: my brothers' houses, my grandfather's old house, my aunties' houses, and my cousins' houses. None of them are ok. There's almost nothing in most of the Lower Ninth Ward that's substantially salvageable. Even those houses that still stand, like my grandfather Copelin's house, would need a complete overhaul to be viable. One of the homes slated to be bull-dozed in the Lower Ninth Ward is 1027 Tupelo Street, my first family home my wife and I bought together. It's totally collapsed now. I felt a sense of loss when I saw it on the list of homes to be bulldozed. In my neighborhood in the East, everything is dead.

You're not mourning the loss of your '76 Buick. You're mourning the loss of friends and colleagues who may have died or been crippled, the everyday common things like the store you went to, the church you attend, and the gas station in New Orleans East that has been leveled. You mourn the loss of the city, your sense of your neighborhood.

The practice, from a psychological point of view, is dead. The outside of Heartbeats Life Center is ok, but internally it's been severely damaged. We had at Heartbeats almost a million dollars worth of equipment that represented a personal investment. My medical equipment was insured but it's been quite a struggle with my insurance company. They say that the flood caused the damage, the hurricane didn't, and flood is capped at two hundred and fifty thousand dollars. I think I'm going to get it because I've been fighting through the system. But if they had capped it at two hundred and fifty thousand dollars, then I would have lost money both on my house and the office. I had worked so hard to be solvent, financially, and now I had all this equipment thrown out on the street. There are 120,000 people in New Orleans as of January 2006. My practice had 7600 active charts. I saw 3000 people a year, so I must have had 10 to 15,000 patients myself which is like 10 percent of the people who are there now. I could go to another town. But I'm from New Orleans and the Ninth Ward, not from out there somewhere in space.

I've seen my patients on television being interviewed. They're scattered everywhere. All the bodies buried amidst the rubble of collapsed houses and the bodies washed away that are not accounted for, I just know that tons of them are my patients.

A lot of my depression around Hurricane Katrina was the fact that Heartbeats wasn't just a job, something I did, but it was what I had focused my whole entire adult life to do. It almost felt like I was paying back the Ferdinands, the Copelins, my grandma, my mother, and my father for all the things they had done to make this little boy successful. Then the community and that life work was taken away from me in two days. Now what's your whole purpose?

Conclusion: "I Wake up Every Morning Trying to Decide What to Do"

In the immediate aftermath of Katrina, people were calling, because I'd done research and lectures, and been on panels around the country. I decided to try and channel a lot of that goodwill into trying to help the people who only had whatever they put in a plastic bag as they were fleeing for their lives. I created the HOPE Initiative out of air. My thing has been garnering resources and helping people, because I have to get my own head together. It was an effort to channel the money differently from some of these big organizations, where a lot of the money would just go to make their organization bigger. This is a fund separate from the general fund of the ABC, and it's managed by me. I get no salary for it. We just did direct aid to evacuees, usually through a non-profit like a church or a community group. Now we're into a mode of trying to identify some needs that we may have overlooked in terms of medications, health status, and long-term issues.

After New Orleans was evacuated, martial law remained in force even after the flood waters receded while specialists assessed the damage to the infrastructure. The safety and health of the citizens was the stated reason for the delayed access to belongings that might have still been salvageable.[2]

The first time we went back home, you couldn't get in. My wife wants to see the house. I say, "Daphne, we can't get back in." She said, "But you're Dr. Ferdinand." I said, "These people don't know that. That's old." So we roll in on the Highway US 11 bridge, a little bridge that goes across Lake Pontchartrain. Local New Orleans police have a road block, and they're turning people around, because you only can go in if you have a pass or if you're a reporter. We roll up to the joint. A black policeman taps on the window real aggressively. I roll down the window, and I said, "Dr. Ferdinand, Louisiana State Board of Medical Examiners." He said, "Go in Doc." He didn't even look at my papers. As we drove away, I said, "I am somebody! I may not have a house, but I am somebody! I may be unemployed, but I am somebody!"

I just now, July 10, 2006, got a private check for my house from Allstate. I don't know what I'm going to do with it, but at least right now I have some of my equity back. If I fail to maintain my status, I can't go to Kalamu—he's a poor writer. Kenneth is a struggling businessman. Many black professionals are a banana peel from being under.

I'm not clear to what extent I want to go back and rebuild in an area that's been so decimated. First of all, I have to think about my family. My entire nuclear family was raised in New Orleans. At the same time, they all have choices. My eldest son's gone back to Japan.

For work, I come to the ABC, use the computer, do a few things, and make more contacts. Being at the ABC gives me a national platform. Universities, industries, and some of the medical education groups will pay you to speak. I already had those contacts, so that's what I do. And I help out at the Saint Thomas Community Health Clinic in order to try to fill that hole in my heart.

THIRTEEN

Charles W. Duplessis*

Born in 1951 in the Seventh Ward, Charles W. Duplessis moved to the Lower Ninth Ward after the death of his mother. Charles married Thirawer, a strong, stylish woman with deep Lower Nine roots, and served in Vietnam in the early 1970s. In 1976, he felt a calling to the ministry. A graduate of New Orleans Theological Seminary, Charles started Mount Nebo Baptist Church (Mount Nebo) in makeshift quarters in Metairie. In 1988, the congregation was able to purchase a rare two-story building and four adjacent lots in the Lower Ninth Ward. Both the church and Charles's home served as community gathering spots before the storm.

On February 15, 2008, Charles was interviewed at the home of a Mount Nebo church member, who welcomes the Duplessises into her home whenever they are in town.[1] Charles is a towering, lean man with a build reminiscent of Abraham Lincoln. He was casually dressed in a navy blue, short-sleeve, rayon shirt with dress slacks. His leather church shoes were polished. Unlike those who, however reluctantly, stay away from New Orleans and are reticent to talk about its future, Charles was uninterested in discussing the pre-Katrina past. No matter what the question, he tenaciously pulled the conversation back to present-day struggles and future visions.

Before the storm, the Duplessises led thirty-six people from New Orleans to Tuskegee, Alabama and experienced some of the least hospitable responses given to Katrina victims in the rural South. His home on Tennessee Street was two blocks from the levee break. Mount Nebo was in the immediate vicinity of the two-hundred-foot barge from the Ingram Barge Company that was either drawn through an existing hole or caused a rupture in the Industrial Canal floodwall, causing tremendous property damage as it moved through the Lower Ninth Ward.[2] From Tuskegee and Marrero, he continues to mobilize resources, protect the interests of the neighborhood, and model the principles of Martin Luther King Jr.'s beloved community. Charles Duplessis is an example of the original, I-am-here-to-stay, pioneering spirit.

I was born in 1951 on Marais Street in the Seventh Ward. We were renting a house at that time. My mom died when I was five years old. When I moved to the Lower Ninth Ward, we moved on Flood Street, and we had an outhouse. Eventually, my father remarried and bought the house on Tennessee Street.

I served in Vietnam. I was in the army. I was stationed at Cam Ranh Bay in Vietnam. I was there when I was eighteen, nineteen years old. I did one tour. My company was at an army base, and they were pulling out. When I arrived, we had over fifty thousand people, and in less than three weeks, we had less than a thousand. I was among that thousand left behind to close up warehouses and things like that. When they gave me that weapon and said, "Lock and load," I knew I was in danger: before that time, they would give me a locked box of ammunition, and tell me I couldn't open it unless I saw the enemy right there. If you went on security, they gave you a locked box with a tag, and you couldn't break it. If you broke it, they would give you an Article 15 because they didn't want you to have any bullets unless you saw the enemy. The enemy's not going to come and say, "Here I am." Nonetheless, we made it through.

I met my wife, Thirawer, on June 19, 1970. We haven't been apart since. We didn't know each other before that, even though she lived right around the corner from me. We met through her aunt, her daddy's sister. (Since the storm, her father and her aunt have died of heart attacks.) We've been married thirty-seven years next month.

We have three daughters, Angel, Violet, and Tangela, plus those who have adopted us. We have a son, Michael. He adopted us when he was sixteen, and an adopted daughter, Janet, who also was sixteen. He's going to be fifty in August. He's got children, and they've got children.

We had been living on Tennessee Street thirty something years. My dad died in '76. We moved in and took over the note and paid off the house. Our house was a wood-frame house. We had just remodeled it. It was a major renovation—leveling, siding, and we added on a screened-in front porch, and a deck on the rear of the house. We were getting our homeowners insurance and flood insurance that Monday because we had a builder's risk policy, and that would have covered the mortgage we took out to repair it.

God called me into the preaching ministry in '76. At the time, I was a deacon serving at Zion Hill Missionary Baptist Church. I would later serve as their interim pastor, until I was called by the Hill of Zion Baptist Church in '84, where I served as Senior Pastor. Upon leaving Hill of Zion, God led me to establish The Mount Nebo Bible Baptist Church in 1988. Mount Nebo especially focused on children in the neighborhood. We were at 1720 Flood Street almost seventeen years before the storm. We had more children than adults. I guess it was the yard. We didn't have it fenced in. I'd go there and I'd see them on the open land. Most times they'd run, and I'd say, "No, come here. There are rules. No cussing, no fighting, and no trash." Ninety-nine percent of the time the neighbors didn't have to call me to come around there. It was a big place, twenty-three thousand square feet of property. I didn't have to go and pick up trash normally. There wasn't any fighting for years. Eventually they started coming to the church.

Katrina

We weren't taking the storm seriously until Saturday evening. After we watched the weather on Saturday evening and heard the mayor's address, we mobilized our church members, our relatives, and our friends. My wife called our niece who lived in Tuskegee, Alabama, and enlisted her assistance in finding hotel rooms. Everything was booked in

Auburn, Opelika, and Montgomery, except at the Kellogg Conference Center affiliated with Tuskegee University. Among the relatives we called in New Orleans, there were people who didn't want to leave because they didn't have enough money to make the trip. My wife told them that we would pool resources. Now is not the time to be proud, she said. A member who lived across from the church said they were not going to leave, so I told her that she had a key to the church.

After forming a circle, joining hands, and praying, we left on Sunday morning at 10:00 a.m. with thirty-eight people in eleven cars. Ages ranged from two-weeks to seventy-six-years old. We had eighteen children and twenty adults. One car died in Mobile.

We attempted to fill up with gas whenever the gas gauge reached half empty, because we had three gas stations in Mississippi along Highway 90 refuse to sell us gas. One wouldn't accept cash, but we surprised them because several of us had credit cards. Another gas station attendant further down the road said they were saving their gas for their regular customers. A third gas station claimed to be closed at 6:00 p.m. on a Sunday evening. We arrived in Tuskegee between 2:00 and 2:30 a.m. The drive normally takes four hours. We only had clothes for three days. We stayed at the Kellogg Conference Center.

"The Churches Saved Us"

We had over seven hundred evacuees in the city of Tuskegee, and the Red Cross and FEMA wasn't going to come. We were advised to go to Auburn or Montgomery for service, but we had gone to Montgomery before to seek aid. When we had gone to the Red Cross in Montgomery, a guy with Red Cross came in the seating area, and he saw all of us watching the news. There were only minorities in the room at the time, except for one lady who looked like she could pass for white. (She was Creole.) He came in, looked at the television, looked at us, and said, "What you all doing, looking at soap operas?" And I knew what that meant. The Red Cross guy was saying, "If you need shelter, you need to go here, you need to go there." I said, "We don't need shelter, because we have shelter. What I need personally is formula for my grandbaby, and the other babies that were at the conference center," because we were cash strapped. All of our money went to support the people in our caravan who didn't have money. Gas prices were up. He said, "We can't help you with that." I said, "What kind of stupidity is that?" And then he went off. He said, "My daughter was in Florida and she didn't get no help."

Eventually the husband of the Creole lady said to her, "Go over there and talk to the lady," one of the Red Cross volunteers at the Center. And she went over there and played coquettishly with her hair and they gave her vouchers. They said, "You can go to a hotel. We'll give you a voucher. It will only cost you twenty dollars for a room and two meals at this place." Because she looked like she was white, she was treated differently. I didn't get upset about that. I didn't want to go in nobody's jail; we had too much going on. I would have gone to the hospital for sure, if I had allowed that incident to bother me. The Red Cross came and they gave us vouchers for 230 dollars to each family unit. They sent us to a church. That's where we found out that another church in Montgomery was calling and saying, "Are there any African Americans that we can help at your shelter?" It was Pastor Osby who said they weren't going to their national conference the weekend of Katrina as planned. How could they, "with so many hurting people," so they used the money instead to pay for the rooms at the conference center for our family for several days. They also brought three ice chests full of food. We shared the food with another family at the center and ate from those ice chests until we left.

Churches helped us across denominational lines when we got to Alabama. I mean there was Methodist, there was nondenominational, there was Catholic, there was Baptist, and even some Muslim mosques. It was black and white congregations.

Stress and Illness

My sisters and the majority of my siblings, except for my oldest brother, are in California. So they sent for us and we went up to Oakland, California around Halloween to preach. That's when I started having symptoms of the second blood clot. The first time I had a blood clot in my lung, one lung was 100 percent blocked, the other was 70 percent blocked, but I had no symptoms, no pain, no breathing problems, and no blood pressure problems. It was amazing, they said, that I lived. We went to the VA clinic in California. I told them about the first one. They did x-rays. My oxidation level was normal, my blood pressure wasn't up, I wasn't sweating, and there weren't chest pains. From a blood clot in the lung, there are certain signs you're supposed to have, but I wasn't having them. They just dismissed it as stress.

And even when we got back to Alabama, the doctors didn't believe me there either. But from my prior relationship with this doctor, he saw that something was wrong, and so he sent me to Montgomery, because the Tuskegee facility had closed certain services. I went to the Emergency Room of the VA Hospital in Montgomery. They saw me on Veterans Day, and they did not do the tests, because everybody was off. They narrowed it all down, but they didn't do the tests until that Monday or Tuesday. The next day after the tests, the doctor came in with his head hung down. I said, "You need to listen to your patients." And he sat down, and he listened to me. I said, "I could have died waiting for y'all." They put me on Coumadin.

As pastors, we hear so much. At a meeting of thirty-six pastors, we heard reports of deaths after deaths after Katrina. In talking with Dr. Kevin Stephens, the New Orleans City Health Director, the second wave of Katrina-related deaths is just not reported, because some people don't want statistics out there because it looks bad for a so-called nation. So we've had a lot of people die, not only because of killings, but because of stress: heart attacks, strokes, and blood clots. Post-Katrina there has been an extraordinary amount of people, who came out of New Orleans, and who are dying at a far greater rate than I would call normal.

Rebuilding the Lower Ninth Ward

In the Lower Ninth Ward, as you know, there was Katrina and Rita. The powers that be reflooded it during Rita to keep the other areas from flooding. The water stood there for about three weeks. So we weren't able to get in there until November, and by that time it was so bad.

The waters and the wind pushed our house off of the foundation. The front of the house moved straight back to where the back of the house would have been, and was sitting on the deck. The Mennonites have offered to do the labor for our house, if we can get the material. Our biggest struggle is trying to get the funds together. When the insurance companies came back, they only gave us enough to pay off one of the mortgages. We only paid off part of it. We didn't get any money from the insurance at all, because it went directly to the finance company. You get to a point where you just want to get it over with.

We haven't gotten Road Home money yet. With the Road Home, we applied online when it first started. We had our initial interview on January 25, 2006. They sent us, what they call a "Gold Letter." We signed the section that said, "Option One," saying that we were going to go back and rebuild, and we sent it in. I knew the process was going to take a long time, so I waited for three months. I called and they said, "We're going through closing." I said, "That sounds great." "Closing" means you've been approved and they're going to give you a grant. It's a cyberspace thing. We're *still* in closing. I've been calling them and calling them. They never asked us for anything else, except when they lost the papers, we had to go and resign the papers. Over the months, they've never asked us to send anything else in, because we'd brought in everything: the driver's license, the credit card, the insurance papers, tax papers, whatever they needed. It wasn't easy because everything had been lost, but we did most of it online or by phone. We don't have computer skills, but we knew people who did, and they were able to help us. You've got to go through almost a year-and-a-half of the so-called Road Home; well I call it the Roadblock Home, because it has blocked us. Louisiana's not the only corrupt state.

Our house was like the house of the neighborhood; it was called "The Kool Aid House." Even before we did the pre-Katrina renovation on our house, people would flock to our house and we just couldn't understand it. Why would they come in? We had something to offer, ourselves. And if you offer yourself, somebody's going to come. That's why the Mennonites extended themselves because they usually do eleven hundred or twelve hundred square foot houses. We said, "That was not our house." People say, "You take what somebody gives you." I'm not just going to take something because you offer it to me. It has to fit my needs. Now I think about it. Is that being unkind? But no, there are limits you set for yourself. You appreciate the kindness, but we believe that God is going to provide. So the Mennonites decided they were going to build a house the size that we had previously. There's a church out of Flossmoor, Illinois that has agreed to help us rebuild the church through Churches Supporting Churches (CSC).

I also work with CSC. One of the things CSC is trying to do is you have five areas. One is in the Lower Ninth Ward, and I'm the area coordinator for that. We've carved out sixteen-square-block zones in those areas to rebuild in. It's centered around the churches in each zone. We're trying to help the CSC churches we're partnering with and then reach out to the other churches that are not a part of CSC and say, "What do we want to do in this sixteen-block area?" We're trying to get our parishioners back. It's not just about the people of the church but it's about the beloved community.

"Cutting Grass"

Originally there was a moratorium on rebuilding in certain areas. It didn't turn out the way they wanted, I don't think, because they wanted the mayor to not develop the Lower Ninth Ward at all. When the city issued permits to finally say you can rebuild your home, the Army Corps of Engineers started knocking down people's homes without the homeowners' knowledge, even though the latter had a permit in hand to redo their property.

According to Mtumishi St. Julien, executive director of the Finance Authority of New Orleans, "Mortgage companies can go to court and file a Writ of Seizure and Sale asking the court to sell a property to satisfy the debt owed for failure to pay on a health, housing, or grass lien. Elimination of blight," he continues, "is another public purpose which can result in the use of expropriation. The definition of blight is that the property must be vacant and the property is a public nuisance. Upon due proof of blight, the court can order the title expropriated from the individual to the New Orleans Redevelopment Authority."

We met with some of the ministers and community groups. So that's how we got started with some of the groups coming together, like the Lower Ninth Ward Neighborhood Coalition and CARE (Community Awareness Revitalization and Enhancement Corporation). We wrote a lot of petitions and resolutions to City Hall. That slowed that bulldozing process down. We've been cutting grass in the neighborhood on the weekends. The city had begun to put a lien on the property if your grass is not cut, and if you put enough liens, you can take the property. There are a lot of elderly people who owned their own homes who can't get back to New Orleans every two weeks. We're also trying to keep down the idea that this is an area that's not coming back and you can come and dump your stuff in our area. So we started on October 13, 2007. We had over a hundred and some people. The mayor was down there, Ed Blakely, the city's Recovery Czar, was down there, and the sheriff had "prisoners," cutting grass with us. That was to get it on their radar. That was right before the elections. Sometimes we get students in. One time it was just me and my wife. Some people have seen it and picked up on it.

Failure to pay ad valorem taxes can lead to a sheriff's sale that would allow the buyer to gain a tax interest in the property by paying the taxes. However, the original owner would have eighteen to thirty-six months to make good his tax arrears and five years from the date of sale to challenge the legitimacy of the legal process.[3]

Louisiana's got some laws that are ignorant, but they're the best laws in the land now, and you just can't come in and grab somebody's land. Believe me they're trying to figure out something in this legislative session or in the future so they can take the land without the struggle of going through years and years of me having the opportunity to come back and get the people's property. We're going to be vigilant. We're working with Policy Link and other organizations that do the research, and then we can get the information in order to stop that from happening.

My vision for Mount Nebo is that it be even more an outreach place, a community place. Not only that but also interacting with all of the churches that want to interact, so we can have a further-reaching ministry. Instead of duplicating ministries, the churches will support one another. We're trying to align with the Make It Right program that Brad Pitt started, and with what Edward Blakely and others are trying to do in the Lower Ninth Ward.

Conclusion

If you make it through two blood clots, Katrina, Rita, and Betsy, and a lot of other things in between, you're not afraid of much. I believe this is what God has prepared us to do as a couple and as a family. None of the children or their family members who were in our church died in the storm. When we talk with our members, some of them want to come home. I say, "If you're doing better, you stay. You have to make up your mind. It's not about Mount Nebo per se, it's about you and what's best for you and your family."

We don't have any income from the church now, because people are scattered and struggling. They can't help us because they're trying to rebuild their own lives. All we had was my disability check. I still pastor the church by God's grace, through CSC and other friends and family. Normally we only have service once a month, here in our member's home in Marrero, because of going back and forth between Alabama and Louisiana.

The vision is to get New Orleans rebuilt in a just way. Fifty-nine percent of the people in the Lower Ninth Ward owned their properties. Some people have taken the Lower Ninth Ward and put it where it ought not to be. People had a mind to worship, they had a mind to work, and they had a mind to own and try to maintain their own property. But some people didn't and so there was blight. Well there's blight through-out the city. But the idea is to build a striving community of businesses and people who want to be there, not that you can keep anybody out, but to get a community like there was before. Whatever influx of immigrants, we ought to make room for whoever needs to come. That's what America is about. But it ought to be in a just way.

God has sustained us by faith. We're rebuilding the house and trying to rebuild the church. If it wasn't for the churches, I know we wouldn't be where we are and so many other people we've talked to say the same thing. The government's inept. In a major city in America, because of all the bureaucracy and the red tape, you can't get services to people.

Concerned Americans can pray. If you can come, you need to come and see. You can support CSC or any group you choose to. Some people have expertise that the com-munity may need that has nothing to do with coming out of your pocket. You could come down and help somebody set up a nursery, teach in a nursery, or hold what we used to call backyard Bible studies for a week or so. If you've got computer skills, teach people computer skills. If you're an outsider trying to help, communicate with people, instead of saying, "You're going to do it this way." Sit down in a reasonable way and then look at that person in the community from a historical perspective. It's bringing minds together to do a greater work and make this city a better place. That's what we're going to do, by the grace of God, even if nobody helps us.

Willie Pitford

Raised on Marengo Street in uptown New Orleans, Willie Pitford is one generation removed from the rural Mississippi upbringing of his parents, who taught him the art and science of surviving a hurricane. He learned the elevator business from the ground floor up. Finally tiring of discrimination within the company he worked for, Willie started his own elevator business. In an unusual twist for this book's narrators, his mother-in-law had the financial means to stake his business. A father of two, Willie lives with his wife, Eva, on Lake Bullard, an upper-middle-class development in New Orleans East. Along with his brother-in-law, Joe McNeil, and son-in-law, Nakia Hodges, a New Orleans Emergency Medical Technician, Willie rode out the storm at home.[1]

On Memorial Day in 2006, fifty-three-year-old Willie Pitford narrated his account from his large two-story home complete with a pool and a separate two-story game room.[2] In the front yard was a FEMA trailer, a white Hummer, and two SUVs. His dress was casual: denim shorts and a t-shirt. His manner was modest and self-effacing.

This narrative is an example of men with means who chose not to leave: risk-takers who wanted to protect their property. Willie, Joe, and Nakia rescued between one hundred and twenty and one hundred and fifty people of all ethnicities in the first five days after the storm. Not only did they eat and drink in fine style throughout, but they also shared their food with those they saved. When an official rescue boat finally arrived, these three black heroes of Katrina were accused of vandalizing Willie's home. Within months of the storm, Willie had his business back up and running, and was rebuilding his home from a used RV he purchased shortly after he was evacuated from the city. For Willie, a diehard New Orleanian, it is impossible to think of the city not coming back.

My father, Albert Pitford, was born in Richland, Mississippi, but he spent most of his life here in New Orleans. He and my mama, Ionia, got married, and they just came and stayed here. I guess he came for work. Longshoremen made good money. Uptown, that's where I grew up, on Marengo Street. We was renting a big two-story double.

When I graduated from high school, the Urban League was coming to the schools to talk to seniors about getting a trade. So one of my friends, he got into electricians school, and my real best friend, he went to be a plumber. I went to Southern in Baton Rouge to play football. I didn't wind up playing football, I just wound up partying up, but I went to my classes. I went back the next semester, because I said, "I'm going to try it again." I still didn't do it, and I couldn't find no job or nothing up there. So I said, "Let me get a job," so I came home.

A guy called me one day, "You want to go to work?" I say, "yeah! Where?" He say, "Otis Elevator. "I said, "OK. It's a job." I went, and I just got good. I worked thirteen years for Otis. Once I got there, the guys started telling me, "Go to electronics school and you'll move up." I went to electronics school. It took me out of construction. I went out there and saw the elevators. I was doing service work, and within a year or so, they put me in with an adjustor. He does all the fine tuning. Once I got to work with him, I would watch everything he do, ask him questions, and he would tell me. He was real good. Within the next two years, they made me a mechanic, which usually takes four years. They had me do a lot of service work.

Then they put me on my own route and gave me Charity Hospital. Nobody wanted to do it, so I said, "It's fine with me." It was kind of rough. They ran me around, but eight, nine years, I made it. They were going to give me a break, let somebody else go over there. They gave my route to a new guy. About a year later, they called to complain. The other guy wasn't keeping the elevators going and stuff. They needed me back over there. They said, "They're going to be putting new elevators in there in about another year." So when they started putting new elevators in there, I was there for it, and I learned. That's when you learn about microprocessors and all that. I learned it all. I can do it all in the elevator world.

Not much later on, I started my own business. I had been thinking about doing it anyway. One day the maintenance men decided to go on strike. "Ok, I'll go with you all." Then I went back over to work to pick something up, the construction guys told me, "You can't go in there." I would never have done nothing like that to nobody. Me and the other guy filed charges against them through the union, and then they really made me mad. All the guys in the union running it were white guys, and all these construction guys were white. The time for the union meeting in the past was always at 7:00. So me and the other guy, we go there for 7:00. They say, "We had the meeting already. We voted." And I said, "You said 7:00." That kind of really made me mad. We started to appeal, because the other guy was black too. We went to EEOC (Equal Employment Opportunity Commission) and they told us we couldn't do it. I just got fed up with the union and everything. Not much later on, I started my own business. (I'm still a union company in order to get the guys that work for the unions.)

Then me and four other black guys started an elevator business. One guy, at first he didn't want to work in the projects. I said, "Man, that's work." I went over there working on an elevator one day, and he come over there and was trying to tell me how to fix it. I said, "You work in the office. Don't try to tell me how to fix all this." He was trying to do something wrong, to get over or something like that. I said, "No, I ain't for that. We're going to do it right." So I just left and went to start my own business. It was me and one of the other guys, but what helped really was my mother-in-law. She came up with the money, and then she really became like the president of the company, because if you own more than 10 percent of the company, you can't work in the field. And I had to work in the field. We've been in business now for ten years.

After I got married, me and my wife moved across the river, and stayed over there nineteen years in Westwego and Avondale in Jefferson Parish on the West Bank. We've been here in Orleans Parish like fourteen years now. We built this house. Jefferson Parish was nice. I kept my house in good shape over there. I had a pool and stuff. I wanted to raise my children up right, and I wanted them to have stuff.

I moved over here because at one time there was only the Huey P. Long Bridge, a four-lane bridge on Airline Highway, and the bridge had so much traffic. I was trying to get my wife to move out to the country like Covington. I could have gotten four to six acres of land. She said, "That's too far." I said, "I'm not going nowhere unless I'm on some water." So her and her mama, they got together and they went to looking, looking, looking. The realtor brought them to Bayou St. John over by City Park. We looked at one house that was like four hundred thousand dollars and it needed a lot of work, because it was kind of old fashioned. So then they came out here and found Lake Bullard, and built the house from scratch with Alan McKendall of McKendall Construction Company.

I love New Orleans. I like the night life. It's open twenty-four hours a day. I like the French Quarter. I love seafood. Even when I lived in Jefferson Parish, I still felt like I was part of New Orleans. I was over here every day. I wouldn't want to stay nowhere else in the world.

Katrina

My father always let us know: "Always keep an axe and a hatchet. If a hurricane came, make sure you had water, canned goods, and stuff like that." We always filled the bathtub up and put water in it. I grew up doing it. My father really schooled me on hurricanes.

So after I got married, when hurricane season came, I started preparing my children. You can ask them right now what you got to have for a hurricane, and they're going to say, "You got to have an axe in case you got to go through the roof."

I'm the one who wanted to stay. My son-in-law (son) decided he was going to stay with me. My brother-in-law (brother) said, "I'll stay too." So it was on. I probably would have stayed by myself. But we had everything. I had a big old axe and stuff. My brother, he brought a chain saw. We had ice, because I kept ice in my freezer. I had water. I filled all my tubs up. I had big old garbage cans. I filled them up, and I put a garbage can in my bathroom for flushing the toilets. Everything I had that I could put water in, I put water in. We were even able to have baths. We had a powerboat and we had a paddleboat. I took all the food out of the freezer. I was ready to stay there till the water go down.

The night before the flood, in the neighborhood, all around the lake, we were all flashing lights just to let you know, we here. We wanted to know where people was.

Saving Lives

I didn't go out all the time. Sometimes I stayed home and cooked. We fed everybody we rescued and gave them water. After I cooked and after my brother and son come back from their first run of the day, they'd get ready to go back out and I'd go out there with them.

That first day we saw two firemen and we asked them, "Is it alright to be out here?" I told them who I was, and stuff. "I already got a boat, man." A lot of people was worried. I didn't want to be out there and somebody come get you and take you where you didn't want to go. When I first went out with them, me and Alan McKendall, we went out on the paddleboat. Alan and I rescued white lawyers on the corner down here. They were hollering, "help, help." We said to them, "come on down." So they came on down and brought their stuff. And they said, "Where are we going to go?" I said. "It's dry on Chef."

We were in my paddleboat and we decided we needed both boats, mine and Alan's. So as we were transferring one of the lawyers to Alan's boat, the guy started falling in the water. So we pulled the boats back together and pulled the lawyer into the boat. Then we paddled them on out there, and they were so glad and happy that we got them out.

On the way back, we heard people hollering down that way. We said, "Ain't no sense us going way back there. Let's get these people right here with the paddle boat," so we went on over there. They had some pit bulls, and one of them was clinging on the fence, but I was too scared to go mess with him. So there were two guys. They were up in the attic, so we told them to come down, but they were scared of the water. The water wasn't that deep right there. I say, "It's about three feet. You can walk in it." So we finally talked them out. They got in the boat. As we were leaving out, there was another guy on the roof waving a flag. With the first two guys on my boat, Alan went and got the other guy. Alan was going back home, till we heard some more people hollering. So he went and picked up somebody else. I went back there and got them, and came back to Chef. We got our exercise that day.

When we went out in the powerboat, we went over in the other direction. One lady had her head there just out of the hole in the attic, and my son-in-law saw her, so we went over there. She was a black lady, and she was green from dehydration. She was maybe in her late forties. And so my son jumped up there with that chain saw. I say, "Wait till the lady moves her head. Tell the lady to get back." So she got back. And he cut her out. (We cut a lot of people out of their roofs.) She said she didn't think she would have lasted another half an hour because she was just gone. The attic was hot. The lady said she stayed because she had to work that day.

Another guy we got out fixed appliances and stuff like that. He said, "Man, you all ain't got to worry about nothing. I'm going to fix all you all's appliances." "Man," I said, "You aren't going to have to worry about fixing nothing. We're going to all need new appliances." We brought him on out there, and we picked up some other people on the way.

The next day, my in-laws went to Lakeland Hospital, down the street on Bullard. They're getting people out of the hospital and bringing the people to Chef. When we were loading the boat up with more people, one guy heard Alan's voice out there. He said, "Alan that you out there?" And Alan say, "Yeah." He said, "My mama, she's eighty-nine years old. She's at home," he say. "I don't know if she's dead or alive. Can you all bring me to go see?" Alan looked at my son. My son said, "Come on man, let's go." We brought them people, a boatload full, to Chef first. And then we came back and went to the house. Everybody was praying. He jumped down in there in the water, the water was about four feet and he was a short guy, and opened the door. He walked in there. "Mama! Mama!" She said something, and he turned around with tears in his eyes. And my son-in-law jumped down there to go help. They were in there so long. They finally came out with them, and her hands and her feet were all white. They said she was in the bed. She said she had done went in the attic. She stayed there till she couldn't take it no more. She came down to get her some water or something, and she couldn't make it back up there. She laid down on the bed, and the bed floated. Her hands and her feet must have been hanging in the water, because they were all messed up: dried, white-looking, and all that.

So we picked her up, and got her on the boat, and her feet were hurting, but she was glad. She was telling us on the way, "You all don't have to worry about nothing." We said, "Miss, we don't want no money, no nothing!" She was talking about she was hungry and thirsty, so we said, "Come on, let's go by my house." So we carried her in and fed her. I think that day we had some sausage and some bread. My brother-in-law was there, and I told him to give her some food. Then he saw how the lady was, and he hurried to get the stuff. He was rolling the food up in some aluminum foil. She said, "You all, I need to eat *now*!" So we opened it up for her. We said, "We can't bring her to Chef," because she couldn't walk. So we brought her to Lakeland Hospital, because they were air-lifting sick people up and out, but the rest of them they were leaving there.

The next day we went back that morning, because they were saying they didn't have no food and stuff. I say, "People have to eat. Let's cook it all up and just bring it down there and bring some water." So we brought food and water down there, but when we got down there, most everybody was gone. They had come and taken just about everybody out.

"Why You All in That House?"

Seems like we were out there three days without anybody, but it was just us and two firemen. So we just up there chilling in the back, and my son-in-law was in the bed in his room. And we heard him talking. So we go see. He's hollering, "We ain't going nowhere!" I went out there. "Who are you talking to?" They got some white reporters and stuff on the boat, see, filming and stuff. And they're asking questions. "This your house?" I said, "Yes, this is my house." "You sure this your house? Why you all in that house? Why you all still here? You all don't need to be here. You all need to be gone." I said, "Look man. We've been here three days rescuing people, and you want to come around here and tell us that we can't be here." But he kept talking like we looting the house, like we ain't supposed to be here. So I say, "Look, if you want, I could show you my ID. If you want, you could come through the door and see my pictures on the wall." I heard them on the radio, "We got looters at **** Bullard." I got so mad. Alan McKendall eventually came to the window and shouted, "Look out here! I built or own all this for nine city blocks. Get off my property!"

So then they kind of went, but they came back later on with some other people. So when the other people came back, I believe it was Wildlife and Fishery, I told them, "Look man, I'll show you my ID. This is my house! If you want to come inside, I got pictures of me on the wall. You think I came in here and put my pictures on the wall?" He said, "No, we'll believe you," so he came on around and he was just looking at stuff. I said, "You can come on in if you want." He said, "No, no. That's alright." I said, "We've been out here helping people. We ain't no looters!"

We was talking about how we was going to stay, but then our wives kept calling and stuff. We had phone communication the whole time. It would come in and out.

Alan had some kind of worldwide phone and we would call and talk to our wives. They kept telling us, "Why don't you all leave?" We said, "Why we going to leave? We out here saving people's lives. There's nothing wrong with us." We weren't in danger. We never did really get to the canned goods, because I was trying to cook the meat first before the ice ran out.

In retrospect, I wish I'd stayed until the water went down, because the National Guard did enter my house after we left and broke out a window upstairs. They also kicked in my back door and the door to my game room. When we came home, some things were missing.

Reclaiming Existence

We left on Saturday. Someone gave us a boat ride to Chef, where we were picked up in a truck. We worried we were going to be dropped off at the Superdome but instead were taken to Louis Armstrong International Airport. We started to stand in line but were told we would have been flown to San Antonio, too far away from New Orleans. We left the airport on foot toward Baton Rouge along Airline Highway. Alan called his wife. She was in Opelousas, LA. Someone picked us up and drove us all the way to Gonzales, where Alan's wife met us. She drove us to the Baton Rouge airport, where we rented a car. We drove to the different evacuation points of our wives. My nuclear family stayed together. I had a brother in Clinton County, Illinois, who got us private rooms with single beds and separate baths at a military base.

So we started looking for an RV. We couldn't find nothing. Finally, the day before we were going to head out, I called these people up north. They say they had one. It sounded real good. So I said, "You all, let's go up here and go see this RV." It was an '87. All the rest we looked in was burnt up. You opened that door on this one, and man, it was perfect. Perfect! We brought it down the highway. He said, "Come on, you drive." So I drove it. So I said, "We're going to buy it." But I had trouble getting all my money out of the bank. He wanted ten thousand dollars for it, so I think I had like about seven. So he just said, "No problem." I said, "You're going to get your money. Trust me." So he believed me. We went back and loaded up the RV. My brother didn't want us to leave. He wouldn't even help us load up.

I didn't have any trouble getting back in to my house in the East, because I called somebody, and I told them I was working. So they told me to fax something. They faxed me a permit. I'm in. And once I went to City Hall, they gave me an ID card.

I called my pastor and he said, "I'm not coming back." So he offered us his place, and we rented that from him. We were over there six months. But in the mean time, for people coming to town, we had the RV parked out in front of our pastor's house. They could come stay in the RV. After that, I took the RV and I posted it by my office in the Broadmoor area. I stayed over there some nights. And then I brought it out here to my house in the East, and I stayed here. The best night of sleep after the storm was the first night in that RV next to my house. I got my money's worth. After we started getting stuff together, my wife started coming on weekends, staying with me, and cleaning up stuff. That RV really came in handy. It still comes in handy, because right now, I'm using it as my office. I was handling up without FEMA.

Conclusion

I know we ain't going to take a vacation this year. I want to finish here. All my neighbors swear they're going to come back. My neighbors to my right really don't know what they want to do, but everybody else is coming back.

Me and my cousin talked the other day. We were blessed. We ain't lost nobody. But I felt pity for all the people who did lose somebody. They are still finding people.

Every time I think about it, and I ride around this city, ooh, it just gets me. My heart aches to see all this destruction. I'm going to do whatever I can to get it back.

FIFTEEN

Mack Slan Jr.*

At the time of his interview on April 24, 2008, fifty-nine-year-old Mack Slan Jr. had been living in New Orleans for over twenty-one years. But his relationship with New Orleans began much earlier in life. Like many New Orleanians, he grew up in rural Mississippi. The technical education he received in New Orleans from 1967 to '68 allowed him to trade in his military draft card from the Marines for a Navy berth. After discharge from the Navy in 1972, Mack successfully pursued a career in business in San Diego. When Mack and Bernadine, his wife, came back to the South in 1986, they chose New Orleans, for its economic viability and its proximity to their childhood homes. He worked in a number of jobs, ranging from the human resources department at Avondale Shipyard to the position of deputy director for a jobs training program, before striking out on his own as a contractor in 2003.

This interview took place in Mack's state-of-the-art kitchen in his completely renovated three-bedroom home in the Edgelake neighborhood of New Orleans East.[1] Either his cell phone or his house phone rang approximately every five minutes. A short, substantial man, Mack was dressed in neatly pressed black dress slacks and a loosely flowing, short-sleeved, linen button-down dress shirt. His answers were as polished as the marble tabletop around which the conversation took place.

Mack's political and cultural sensibilities reflect the values of the culturally conservative, business-oriented black middle class of his generation. In late September of 2005, Mack made national news by leading a 74-car caravan of New Orleans East homeowners back into the city, despite threats of arrest for civil disobedience from the chief of police. He speaks about political organizing for black homeowners predominantly in eastern New Orleans, which, for tactical reasons, temporarily included Gentilly and the Lower Ninth Ward. Of this book's twenty-seven narrators, Mack had the most financial gain from the aftermath of Katrina.

I was born December 29, 1949, and raised in a town called Woodville in Wilkinson County, Mississippi. Woodville is a small, rural community. The whole county had approximately nine thousand people, made up of quite a few blacks.

My father, Mack Slan Sr., worked at an all white funeral home in Woodville. He did a lot of things from working with the embalming process to general maintenance of the place, because he basically was the only worker besides the owner there. My father was brought up during that time when blacks had to succumb to a lot of things, being demeaned and things of that nature. Although he took some abuse, he basically was in charge of himself, and he learned a lot.

The big income bringer in the family was my mom, Amy, a cosmetologist. She always had a business at home. She was one of the best. She would travel to the different conferences and shows that they have and come back with different styles. I traveled with her sometimes.

My grandfather, Dave, on my father's side, was a barber. Everybody on somebody's side of the family took up a profession. I remember Sarah Holmes, my grandmother on my mother's side, very well. She died in '88 at 102. She was the type of woman that took care of us when my mother wasn't available. As a young baby, I had thyroid problems. My grandmother carried me up and down the highway to the doctors in New Orleans. So she and I developed this great, great bond. When I was a sophomore in high school, I started living with my grandmother.

Woodville had no industry. In order to work you would either have to work at some of the stores around, or some people made a living farming. A lot of the people went to the fields, like the potato fields and the bean fields, for harvest. They got paid by the crates or the Kroger sacks of goods they would pick. Then you had St. Francisville, which is about twenty-five miles south of Woodville, where they had a potato plant. A lot of people got employment there. You had the mat-casting field where they designed those big concrete pillows for the levees. Then you had Natchez, where you had a saw mill. Baton Rouge was fifty-four miles away. So a lot of people got employment in those places.

To me work was necessary, if you wanted anything. I started working at Martin's Drugstore. I went there at 4:00 after school and I stayed there until 6:30 in the evening. I used to deliver medicine on bicycle. Then I'd work on Saturday from 8:00 to 8:00, and I got paid twelve dollars a week. I had more money than any other kid in my neighborhood.

I used to attend civil rights meetings for the NAACP, the premier organization for black people to join back then. We'd have our meetings in churches out in the woods. They were announced in church. We'd go to meetings out there in the woods, secluded, and everybody's wondering if somebody's going to come through the woods and just shoot up the building. I remember one time up at the Methodist church, they had a meeting there, and they shot a guy at the meeting. Too many men threw their hands down and said, "I'm going to let the Lord take care of me," and people killed them. We would always have one or two guys out there protecting us with some type of gun. The media portrayed Malcolm X and others as being outlaws. We were always taught to do things decently and in order. We were raised Baptist. I consider myself to be a moderate activist. I could be a really radical individual, if I thought about all the things that I've experienced, but I'm not. Some things are just not worth the fight.

We all had strong family ties. The ones that lived in Louisiana and other places would come to Woodville in the summertime, or we would come to uptown New Orleans. When I came to the city, I was amazed at the tall buildings, the movement of everything, whether it be a street car or bus. People were always on the move. When my aunt, Lucille Anderson, first moved into the Calliope projects in the '60s, she had furniture, a bathroom upstairs, and hardwood floors. I'm saying, "This is nice." Then I'd go back home, and our house was a modest house. When that house was first built, the restroom was not attached to the house.

The city helped shape me. I knew I couldn't stay in the country, because what was I going to become? A field hand or a worker in a potato factory? My life was going to take me to some heights much higher than that. Once I got out of school, I went to electronic engineering school here in New Orleans. I did a course for a year-and-a-half.

In '69 I was drafted into the Marine Corps, but I went down and joined the Navy before I had to report, and never looked back. Especially black marines were being trained how to use a gun and go to Vietnam and die. I knew that. I was lucky I had some technical background, some schooling, and I was looking to excel. An M-16 in my hand wasn't the ticket.

I went to boot camp in San Diego. I was transferred to Middleton, Tennessee. Then I went to Virginia for six months. When the squadron was recommissioned, they said, "Mack, you've got your choice to go anywhere you want." I went to the West Coast, and I got on a squadron there and worked on an EX-16 computer system.

In 1971, I was shipped out to Vietnam. My squadron was aboard the USS Oriskany. We did a seven-month tour. I'd been in situations where the racial tensions got high, and fights broke out when we were socializing and drinking. But when we got back in the field, places like Vietnam, that didn't happen. When we got back to the states, I said adios to the military.

I don't regret that Navy experience. It helped discipline me. It helped shape me even more. I always knew how to be a follower, but it gave me a leadership role, because I was the second in command of the squadron, and then I won honor-man privileges at the end of our boot camp. That's a privilege that's given to those that do exceptionally well, and show a lot of leadership capabilities.

When I got out of the military, it was a time when companies were opening the doors for minorities, and I was ready to make some money. In 1972, I would have accepted no less than $3.50 an hour, and the company I started with started me at $3.54 an hour and a month later they raised me to $3.72. I became a manager in that company, and then the director of training. I left that company and went to another company as a manager of all their retail products for microprocessor based computer systems.

I went on to get my degrees. I have a degree in business and a degree in electronic engineering. I joined Kappa Alpha Si fraternity. I went from electronics to business because I knew that business could always be used with electronics. I learned a long time ago in business that the biggest deals would be pulled off with those individuals who knew if they had to call on me, they could depend on me. San Diego was a career builder for me.

I loved the community, the climate, the outdoors, and the culture. They weren't as family oriented as people back south. Most of them were striving to get somewhere. There was also less racial tension in San Diego. San Diego had its own black community too, and the affluent ran in a certain circle. I was always part of that fun-loving circle.

We moved back here in New Orleans in '86 after my company in San Diego had dissolved. The guy who bought the company wanted to move it to Illinois, so they gave us buy-out packages. I took the package and said, "Let's go back south."

I became an entrepreneur. My wife got her license to do cosmetology work. And I opened up a beauty salon and a beauty supply business, but I didn't do my homework. The market was really slow. So I went to work for Avondale Shipyards, and that's when the reality of the South hit me, that it really hadn't changed all that much. Then I went to work for the governor's office as a consultant, and then I ended up with New Orleans Jobs

Initiatives, where I became the deputy director, and that was a comprehensive training and employment program that was funded by the Annie E. Casey Foundation, the charitable arm of UPS. I worked there for eight years until their funds ran out.[2] Two years before the storm, I moved on into this phase of my life of being an entrepreneur and a contractor.

When we moved back to New Orleans, we moved to Frenchman's Wharf in the East, and we stayed there for a year. After that, we moved on Apple Street in Hollygrove, where went through a fire in 1990. We lost everything. We didn't have no renter's insurance. We never thought that a fire would come through and destroy everything, but it made us think about the importance of having insurance and the importance of friends, because people came to our aid that we never thought would show up.

Then we bought out in Edgelake. It was one of the first established communities in the East. I didn't buy in Uptown, because I don't like shotgun houses. Uptown you had hardly any off-street parking. I wanted something more modern with more space. The East reminded me more of California for the landscape. It was a well-structured house, and I loved the layout.

The first thing that caught my eye was that this house had plenty of room to park a boat next to the house. All the other houses he was showing me were right next to each other with no space for a boat. I'm a sportsman, I love to fish. I fish Delacroix Islands and Shell Beach. I normally go with a friend of mine, Willie. Most of the time we go out for speckled trout and red fish. The attraction of fishing is relaxation and anticipation of that big catch and just thrill of bringing them in.

It's an integrated community, but it was established by white people. The old guy that lived next door when we first moved in here was a retired fire chief. He was white. This was his daughter and his son-in-law's house that he built for them. He was a very talkative old man. I used to look out my window to see if he was outside. He'd almost cross that line and use the "n" word. I said, "I'll just let it be." His son-in-law is a police officer. He's got the same mentality.

Before the storm, I didn't know too many of the neighbors. I knew the ones over here and behind me and right next door. One of my neighbors works as a supervisor for Entergy. He and I are really close. We share fish. I'll give him trout, and he'll give me perch. He's from Waterproof. My neighbor over here, David, is an entrepreneur. He's got his own AC refrigeration company. He's from Liberty, Mississippi. We work together. Butch, the guy that lives in the back, is an entrepreneur. He delivers antique furniture and stuff all over the country. He's from Lafayette. If we go out of town, we all let each other know we're going out of town.

"Katrina Became a Healing Piece for Me"

I've always been the type of person that makes, I think, fairly good, rational decisions. I'm not going to stick around when a major storm is coming. Everybody else was probably wondering about their properties and stuff like this.

My wife and I went to my mom's house in Woodville. We were there for like three or four days and the storm had actually touched that far. It was hot. No air, no power. We found out they had power in Baton Rouge at my cousin's, and there was twenty-seven of us, so we all went to Baton Rouge and brought the ones from Mississippi too.

I have one cousin, Charles, we've been not only cousins but friends as couples because he and his wife Sylvia don't have any kids and we don't have any kids. They've been married about thirty-eight years. We've been married thirty-six. They told us, "You've got to come stay with us." They've got this large house there in Baton Rouge, so we had like the East Wing and they had the West Wing. So we moved in there and stayed there until we moved back to the city.

After Katrina, San Diego called on us like crazy. They wanted to send money, but my wife and I are sort of peculiar people. We know who we are. I could have received thousands of dollars after Katrina from my best friends, but I didn't need it.

I didn't know what the rebuilding costs would be, but I wasn't worried about whether we could rebuild or not. I started saying, "OK, you weigh your options. You're going to get x amount of dollars from your insurance". One year prior to the storm, my mortgage company had informed my insurance company that I should raise my insurance at least up to the value of the house, because the flood insurance that I was carrying was from the purchase of the house. Well I purchased the house at $110,000. So we boosted it up to $207,000. That money came right down the pipeline. I owed, I think $98,000 on the house, so all the rest of it was money to repair. Then I had my homeowners' insurance because my roof blew away. Homeowner's wanted to pay from the ceiling up. But I said, "All my furniture and everything, cabinets and everything, was wasted from water that came in through the roof." So they had to pay.

I paid this mortgage off in 2005 after the storm. You don't know how your blessings are going to come, but with my insurance monies and everything else, it covered the house and the repairs. How much more can you take off your shoulders? I'm at a point in life now that I can retire, and I don't have a mortgage payment tied to my retirement.

Shortly after the storm, I began to think where I am now could be a blessing. There was a lot of devastation here, and I saw the potential of uniting both my contractor background and my managerial experience. In November, I had companies calling me to go to work, so I started up a couple of projects here, and they were very good projects. Lucrative! For one, my crew cleaned up the Iberville Housing Development.

Plus we were doing the community organizing in Baton Rouge. At first, we started Eastern New Orleans United and Whole, because nobody was giving us any information about our homes. In September 2005, Tangey Wall and I met at a Shoney's, and we said, we need to get together and organize where we can have a meeting and get our community leaders to come out and share with us what's happening, because they were supposed to be able to get in contact with people on the emergency management side. They had more information, we thought, than they were sharing, but we didn't know where they were. So I went to Reverend Dennis Hebert of True Light Baptist Church and asked him could we use his multipurpose center to have a meeting for all Orleans residents.

In late September, the city was letting selected people into their homes and getting their possessions, and we weren't allowed back into the city, and we didn't know what was happening to our properties. It seemed like the people being allowed in were given special treatment because of their status and networking. I could have easily, probably, got in here and went back to give people some information, but they needed to come in here for themselves. The chief of police defied that. He told us it wasn't safe, and if we did come, we'd be arrested. We said, "We're coming." We decided that we would get a large contingency of people and just form a caravan. We publicized that we were coming home, with or without your blessing. We had at least forty or fifty cars. We had a plan, and the plan was we knew that they had these checkpoints, and if they were going to stop us, they were going to have to stop us at the Read Boulevard exit coming in, and then they were going to have to tell us to turn around, so we'd have to come up under the bridge and go back. Our plan was to turn up under the bridge with everybody just going in different directions. In the ninth hour, we got a call that said the police captain of the seventh precinct was going to meet us at the theater up here, and that he was going to let us go in to see our homes. We got that information when we were on the highway already. We drove up, and they had a press conference and the whole works.

From that time on, the organization grew in strength. We got a lot of coverage from the *New York Times*, and it helped expose some difficult times for people who were hurting because of a disaster.[3] Then the so-called leaders of the community started coming to our weekly meetings and voicing their opinions. One time we had the mayor to come up. We opened sort of a one-stop where information could be channeled to them. If people had different concerns, we put them in touch with different agencies. We continued meeting for about a year. Elmgrove Baptist Church also housed us for a spell.

In January 2006, my wife and I moved back in the FEMA trailer here and stayed in that trailer for about six months. I don't know how people do it. It's alright for a picnic or a camping trip, but to live in? It gave us a place to lay our heads down and sleep, a place to cook, take a shower, and sit down and relax, but when you're accustomed to something different, you learn what cramped really means.

Conclusion

I'm a little bit shocked at where we are now as a city, because we are not too much further than we were before, especially in eastern New Orleans. And if it had not been for the homeowners coming back to do their own things, no progress would have been made. There is no economic growth being put on the table. The essential services that we need are not here yet. We have no hospitals at all. Little individual things are beginning to pop up. You look at other business entities that we had to go out and fight for like Walgreens. Three years later is a long time to say you haven't infused this community enough that you can see some real changes.

For me personally, Katrina took some burdens off my shoulders going into retirement. Between me and my wife, we're all right. I'm vested in three companies and we've got some liquidable assets, we've got some fixed assets, and we've got some retirements in place. We've already calculated all that. My wife is an operations manager for Tulane University's Barnes and Noble Bookstore. She's been there now for twenty-one years. If we were sixty-two right now, we'd be smiling.

We used to have a two-story house that was across the street that was section eight, and they would always have some pretty rowdy tenants there, but the storm took care of that. The whole building's gone. I won't say any community is barred of crime, but I haven't heard of any major crime committed around here.

In ten years, I see me and my wife in a big, coach type motor home traveling the country. There will be more fishing. That's inevitable. I find myself still knowing that there's going to be community work, and there's going to be kind of keeping an eye on your family members, your relatives, and close friends to make certain that they're still striving to be the best that they can be. I plan to be somewhere, so that, if they need me, they can call to ask questions.

Top (l–r): Aline St. Julien, Leaughan Salaam, and Kiesha Salaam; Narvalee Copelin
Bottom (l–r): Narvalee's father, Noah Copelin; Leatrice Joy Reed and Edward Roberts

Top (l–r): Leatrice Roberts's son, Edward Reed Roberts Sr.; her grandson, Edward Reed Roberts Jr. *Bottom (l–r)*: Irvin Porter; Porter's mother-in-law, Polly Morrison, his oldest daughter, Gail Porter, and his mother, Emmaline Porter

Top (l–r): Leonard and Dorothy Smith in front of the Zulu Club; Pete Stevenson
Bottom (l–r): Parnell Herbert speaking at Congo Square; Cynthia Banks

Top (l–r): Cynthia Banks's father, Richard Gullette Sr., with Cynthia's brother, Richard Gullette Jr. and his wife; Denise Roubion-Johnson's mother, Marlene Roubion
Bottom (l–r): Denise Roubion-Johnson; Kalamu ya Salaam, Keith C. Ferdinand, their aunt, Neomi Foy, and their father, Vallery Ferdinand Jr.

Top (l–r): Keith C. Ferdinand and Kalamu ya Salaam's mother, Inola Copelin Ferdinand; Vallery Ferdinand, II in his military uniform
Bottom (l–r): Charles and Thirawer Duplessis; Willie Pitford with his family, standing left to right: Ka-Ron Pitford, Willie, Aisha Pitford Hodges, and Nakia Hodges; Eva Pitford sitting in front

Top (l–r): Willie Pitford's parents, Albert and Iona; Mack and Bernadine Slan
Bottom (l–r): Rochelle Smith; her mother, Lillian Smith

Top (l–r): Eleanor Thornton and her daughter, Chelsea Thornton; Jermol Stimson with his mother's hand behind him
Bottom (l–r): Kevin Owens and his wife, Elise Ramsey; Demetrius White rowing two of his neighbors to visit their mother following the storm

Top (l–r): Senta Eastern graduating; her son, Korey Lewis
Bottom (l–r): Her daughter, Alana Eastern; Yolanda Seals

Top (l–r): Yolanda Seals's father, Freddie Seals Sr.; Aldon Cotton
Bottom (l–r): Ben W. Cooper and Noella R. Cooper, Leslie Lawrence's grandparents, holding her Aunt Penny and mother, Gaynell; Leslie dancing with her aunt, Carmen Morial

Top (l–r): Le Ella Lee standing with her mother, Velbert Stampley, and baby brother, Mikey Stampley; B.J. Willis
Bottom (l–r): B.J.'s friend, Chad Charles; Toussaint Webster in Geneva, Switzerland

THIRTY SOMETHING

SIXTEEN

Rochelle Smith*

Thirty-nine-year-old Rochelle Smith is a devout Christian and dedicated LPN. She often left the hospital where she worked with a migraine from working the jobs no one else wanted or from taking on her patients' problems. In her childhood, she shuttled between uptown New Orleans, where she lived with her mother, and Baton Rouge, where she lived with her uncle and aunt. A mother of two almost grown children, Rochelle was renting the house she had dreamed of buying in the heart of Gentilly at the time of the storm. She recalled her life before Katrina as comfortable, well organized, and filled with work, friends, family, and church. Rochelle chose to ride out the storm at an uptown hospital five blocks from St. Charles Avenue.

On January 10, 2006, Rochelle, a trim, vivacious, middle-aged woman of moderate height, brought her eighteen-year-old daughter to the interview room set up by Senta Eastern and Renae Stephens, coordinators of the Houston Urban League's Hurricane Katrina/Rita Relief Program.[1] Renae positioned herself at a nearby desk out of Rochelle's line of vision but close enough to provide emotional support for both mother and daughter as they spoke about the memories that still haunted them four-and-a-half months after the tragedy. Rochelle narrated their experiences interspersed with tears and laughter, while her daughter sat in silence.

More than the loss of worldly goods by the floodwaters and black mold, Rochelle's family was traumatized by the week of fearful nights spent abandoned in front of the convention center.[2] She and other staff and patients were directed there after they were deposited on

St. Charles Avenue by a white fisherman, to whom Rochelle dedicates this chapter. She is one more example of hardworking, health care providers in black New Orleans who were trauma- tized by loss and violence, and who strive to regain normalcy in settings not of their choosing.

I was born in 1966 in New Orleans, Louisiana. I'm an Uptown girl. Growing up in the '60s, God, it was a good time. I remember a lot of Sam Cook music I heard. I'm the youngest of three children. I have a mother, Lillian Smith, who has suffered from mental illness since I was born, but only when she endures crisis does it throw her back into her episodes. She's on medication the majority of the time. My mom raised us by herself. When she would get sick, we would go live in Baton Rouge with my uncle and my auntie. She'd come right back out of the hospital, get back on her feet, and get us.

My mom has cooked at a lot of major restaurants in the French Quarter. She would be considered a chef without a degree. She wasn't well paid, but she would bring home what she cooked a lot of times. Food would get low, but we would never be hun- gry. Fried chicken, collard greens, cabbage, sweet potatoes, and baked potatoes. I never liked black-eyed peas but my mom cooked them. A lot of stuff my aunt grew in the garden. She sent it to my mom.

Before the storm, I lived in Gentilly on Elysian Fields Avenue in a very old house that had just been renovated on one side. It was a very unique, beautiful house with arches, cathedral ceilings, and a stairwell. I was renting.

I'm a licensed practical nurse. I have worked excessively! I was a nursing assistant for five years, and after that I decided to go to school for nursing. Charity had a waiting list that was until 2006 or something, so I decided to become an LPN first, because I needed to take care of my children. I got paid decently, but I was exploited. I was a black nurse who worked for a basically white company, and I got a lot of jobs where I would be the only black nurse on the floor. Some areas some people don't go in. And I didn't care. I went to make money. I didn't care what they thought about me or how they felt about me. I would go home with a migraine in the evening, but I made it. And I paid my bills.

I was drawn to nursing by a calling from God. I'm a giving person. I love taking care of people. I believe that you haven't fulfilled your destiny until you help some- body. I have a friend. She had started a nonprofit where you give out clothes and pray for people. She had a sister that was battered and murdered by her husband. A lot of Thursdays, we would just feed and clothe people. We were counseling people. It was on North Claiborne Avenue.

I grew up Baptist, but I go to the Church of God in Christ now. We had a very compassionate pastor. He teaches us about buying property and investing. He teaches us how to be business-minded.

I love Uptown. I love the streetcars, I love walking in Audubon Park every day. I love the people that I met, and I conversed with every morning around the track. It was the same people walking every day. I love how some Saturday mornings or eve- nings, a girlfriend and I would make our journey through the French Market, just to look. I used to take my kids when they were small to see the clowns, get free balloons, and drink their nonalcoholic daiquiris. What I liked least about living in New Orleans was the crime, the poverty, the corrupt politicians, and the fact that everybody thinks that every black woman from New Orleans does voodoo.

Memorial Baptist Hospital

I heard that Hurricane Katrina was approaching on the Monday before the storm, when it was still in Florida. When they started evacuating, I was at the French Riviera Spa. A little old lady and I were discussing it as we walked on the track. We both said at the same time, "Here we go again."

I didn't know what to do. I was both fearful and undecided. I was upset because Luther Vandross had died, and nobody made a big deal about it.

When the hurricane hit, we were in the Oncology Unit of Memorial Baptist Hospital (Baptist), located on Napoleon Avenue four or five blocks from St. Charles Avenue. That's where my sister works. She's a desk clerk. She worked through hurricanes, because you can work around the clock and that's her chance to make more money. I hated it. The whole time I was there, I hated it, but I didn't want to leave my sister. I took my daughter and my mom. I attempted to bring my twenty-three-year-old son, but he went his own way.

At Baptist it was OK at first. Then, after a while, the administrative people that ran the hospital kind of started downsizing on how they fed us. The meal went from a plate to a cup. We had electricity that Sunday and that Monday. The generators came on that Monday and lasted maybe a couple of hours, and I think water got to them, so we stayed there Monday night and Tuesday night with no lights. They put us out Wednesday, August 31. They told everybody that they had to leave. They told us that they had buses that would meet us on St. Charles Avenue and take us to Houston. When we got there, no buses came for us.

There were some fishermen from northern Louisiana who came with their boats. They took us out, and they boated us to St. Charles Avenue. The doctors went to another place. We could see our boats going one way and their boats going west toward Metairie in Jefferson Parish. Lot of whites went the opposite way. When we got to St. Charles Avenue, a police chief, supervisor or somebody, told us he was not going to leave us because martial law had been declared and that as of 6:00 or 7:00 that night, we had to be off the street.

Technically, martial law had not been declared, since no such provision exists in Louisiana law. However, the state of emergency declared by Governor Kathleen Blanco on the eve of Katrina had the same effect.[3]

So he got a whole bunch of his police officers, and they got some people, and they hotwired whatever they could hotwire. Sick and dying people had been left on the curb. No doctors, no nurses, no nothing. They started putting us in vehicles, the handicapped, the elderly, and the sick first. I guess they took them to St. Charles General Hospital, a small community hospital a block from Touro Infirmary. I think they were still open.

There was a lady out there that was actively dying. I talked to her husband, a dialysis patient. "How long she's been like this?" He said, "She's just normally like that. She doesn't respond. She lays there, and breathes heavily, and doesn't open her eyes." So I said, "If that's the way she normally is, then OK." But I knew she was actively dying. They did take her and put her on a little zoo bus that they found and took her to the hospital.

At that point, I wasn't working as a nurse any more. You weren't worried about somebody else's family member. At that point you were worried about your own family and what you were going to do and where you wanted to go. The police officers kicked open a Rite Aid, and people were going in the Rite Aid to get stuff. They said, "Help yourself."

The first two truckloads out were all white and foreigners that were loaded up. I don't remember seeing them after that again. The whites and foreigners just kind of ran to the truck. They just kind of ran to the truck like they were scared to death, and got on the first two trucks, and went out. They didn't go toward the convention center.

People were panicking. I told my family, "We're just going to stay here. We're going to get a ride." We just kind of stayed to the back and waited. We ended up on a tool shed truck with about fifty other people. There were no white patients left out there. What I saw were black patients. They had started loading us. We were taken to the convention center. The police escorted us, and the truck was returned.

Convention Center

When we got there, we went inside and got chairs, and we sat down on the curb. We didn't stay there long. A fight broke out over water. There were people who had been there two days before we got there. Stress levels were high. We politely got our chairs and moved on till we were sitting underneath the Mississippi River Bridge. Later on that night we had to pick up our chairs and leave again because somebody else was fighting.

Every night, same thing. Somebody fighting and the military running up on us. They looked like SWAT teams. Some dark blue, some black uniforms. They were carrying guns with radars, the little light beam, on it. They just panicked. They were afraid, we were afraid. There were other people in the city that were committing crimes and then just running there where we were and hiding among us. Folks whose homes weren't overtaken by the water or folks that just didn't care. It was mostly teenagers, younger people committing the crimes. There were a lot of well-known drug dealers inside in the convention center. I knew from the symptoms—folks hyperventilating, people throwing up—that it was heroin. Not everybody, but there was a group of them there that was dealing drugs right there and keeping up the chaos. Stealing vehicles, doing whatever to go make their connections, and come back. Someone would come up in a vehicle, run inside the convention center, because there are exhibit halls inside. They'd go in there, and they'd hide. They'd bring the heat where we were, and the military would come and put the guns on us, as if we had committed the crime. They wouldn't say anything, they'd just put the guns on us and shine the light, like they're looking for someone, but it didn't make any sense to me. If you're looking for somebody, you go in and you find who you're looking for.

Every day they were terrorizing the crowd. Oh God, it happened twice a day some times. It happened maybe once in the middle of the night. The first two nights it was ok, because you had light. But between 12:00 and 3:00 in the morning, it was chaos. Everybody had to jump up and just run. You don't know what you were running from, you just knew that something was going on. I would grab my mom, scream for my daughter and my family, and just run. You couldn't sleep. We sat up on chairs on the curb. We didn't sleep for four days.

I listened to my transistor radio day and night till my family screamed, "Turn it off." That was Thursday night when the mayor came on the radio and was screaming and begging for help. There was one reporter, Garland Robinette, a renowned newscaster and decorated war hero, employed by WWL—AM 870, and WWL-FM 105.3. I thank God for him, because he stayed on eight clear channel stations day and night, him and a guy called Bob. He just cried out for our help. He broke down and cried on the radio. Actually it seemed like three days before anyone realized we were at the convention center. A police officer had his wife call in and say that we were there. There were thousands of people for blocks there.

At first everybody was kind of just, it's me and my family. But that Thursday (September 1) when people realized that nobody was going to come help us, everybody started sticking together. I met a friend that just emptied his pocket and gave me all the money that was in his pocket. One girl had some kind of old fashioned charger that she charged her radio on to help us call for help. One family fed and clothed us. Had it not been for those people, I don't know if we would have made it. I had two, maybe three friends there that I had been knowing for years. Everybody else who helped was strangers. Everybody that was out there was black. I remember an old friend of mine who I went to junior high school with was there with her four children.

That Thursday CNN came and the *New York Times* came. And we all were just running up to them and telling them, "Please tell somebody we're here. Please just go tell somebody what's happening." And they started broadcasting live. I remember sitting back. This has got to be the end of the world. That Wednesday we saw three trailer truckloads of food come in. It wasn't released to us till that Friday, after President Bush had come through in an armored car. We had to stand in line for a bag of dehydrated food and a bottle of water.

Once the military came that Friday, you couldn't go anywhere, you couldn't do anything. You had to do what they told you to do. We no longer had rights. There was one guy that was killed right in front of us. He tried to get the attention of the police or whatever. He had to be in his thirties. He didn't say anything. He just ran out in the crowd, leaped up on the car, and came up with a huge pair of scissors. He went down like he was going to break the glass, and there was a sharp shooter attached to the car. He just shot him. Whoever was driving the car went mad and just started driving all up on the sidewalk where we were. We all just went down to the ground. I tell you God was with us, because I don't know how he missed us. He straightened up the car, ran over the guy's body, and then kept going. I can still hear the sounds which his broken body made. Nobody came back until very early the next morning. When they came back, a whole bunch of troops were standing there with guns in our faces. They stood in front of us, laughing in our faces. One of them pulled the cover off the young man and spit on him. They left him like that in front of us as a warning: "You get out of line, you die too." That was Friday. When we got on the buses on Saturday, we left him lying there.

September 2 was a horrible night. They had a lady who had a nervous breakdown and her family had to physically knock her out and make her be quiet to keep the military from killing her. Nineteen-year-old people were telling me they were having chest pains. They had a girl there that went into labor. The babies cried all night. I remember a young lady screaming at her baby: "You got to shut up, you got to be quiet." I was like, "poor baby." The baby was hungry. The elderly began to collapse and die all around us from dehydration and heat exhaustion.

We got up on Saturday morning and we moved. We moved way further down closer to the end of Canal Street. It was chaos. I remember when we moved down, there was a lady laying on the concrete like she was asleep, but she was dead. There was a man not too far from us, sitting up in the chair. He died sitting up in the chair. They covered him up with a blanket. We kept looking back at the yellow blanket to see if somebody had moved him.

The presence of the military didn't comfort us. If anything, we felt that they came there to kill us. They were just threatening us. One day somebody came and told us, "The military says, 'If you clean up the sidewalk and shovel all this filth, they will let you go and let the buses in.'" They were sending the buses across the river to the West Bank to another area that wasn't even hard hit like we were, to people who had lights and water. That night I counted 132 buses. They sent buses away. This girl's brother got one call-in on the phone and said, "My job has given me a bus. I'm coming from Houston to pick you all up." She told us to get ready. They told the guy when he got there and approached the center to stand down or they would shoot to kill. By that time, every street that we wanted to go down had military personnel.

I remember a little girl had a seizure across from us. The military said they were not there to help us. They were there to maintain order. And no one would communicate with us. Nobody talked to us until our Louisiana reserves came in. They laughed with us, talked with us, walked with us, and explained things to us.

The evacuation started that Saturday (September 3) at 3:00 p.m. There were no priorities in getting people out. You ran for your life to get on the bus. They made you line up in two long lines on both sides of the street. People were passing out from heat exhaustion. I remember when they got to us, they said, "There are no more buses." Everybody started freaking out. And then some guy said, "Wait, here's some more buses." And four buses came.

"The Kindness of Strangers"

We didn't find out until we were en route that we were going to Murray, Kentucky. Murray, Kentucky sent four buses for us. I remember just getting on the bus, and I remember the people. Finally in a safe place, I did let my guard down. Everybody was passed out. The bus driver was a gentle man with a soft and kind smile. I remember us stopping in Memphis, and the driver went into a McDonalds and ordered meals for all of us. When he told the manager that the food was for people from New Orleans, the guy gave it to us at no charge. That's when we realized that people were doing things to help us.

But we didn't realize how many people were helping until we got to Murray, Kentucky at 5:00 that Sunday morning and the whole community was there. The doctors, the nurses, everybody was there. They were white. There was only one black church in the community. One pastor and his wife sent to Louisville, Kentucky and brought hampers of brand new clothes for all the kids that were going to school. And then there were two churches that came from Paducah, an hour away, and they started helping us, bringing us hair products and stuff.

We stayed in a camp for children called Camp Wild. Rich kids go there for the summer. It was our first experience. There were maybe eighteen of us in one cabin.

We asked to be moved, and then it was just us and another family. So we shared a cabin together. It was like we were living outdoors, but it was better than sitting on the sidewalk, even though there were snakes. We stayed there two weeks until the governor got tired of the people that were taking care of us complaining. He came and saw for himself, and then a friend that he knew in Congress sent the Red Cross and FEMA from Cincinnati. That's when we started getting help.

They fed us three meals a day till we got fat. People sent so much stuff. They sent a school bus full of clothes. Everyday there was somebody coming to get us, bringing us places, doing things for us. We couldn't say we needed anything without it being done. That's when I realized that God was with us, and He hadn't forsaken us.

It felt so weird after going through what we felt like was a racially inflicted crisis to have an all-white community do that for you, but it was such people that befriended me. One girl took me to her house and said, "You need some privacy." Her name was Sherry. I met her at the dentist's office. She could see that I was just frustrated. There were 154 of us at that camp: too many people. She went into her closet and gave me things for my daughter, some stuff like nail polish that didn't even matter anymore. She said, "Go take your shower and have some time by yourself." She doesn't know how much I appreciated that.

My son ended up in New Orleans as long as we were. I think he went to the Superdome, but he left, and I think that's when he went and stayed at the school. They picked him up off the roof of the school and brought him to Metairie on I-10 near Causeway and left him there for two days until the buses came to bring him to Houston.

A couple sent for my son because he was freaking out because he was by himself. They flew him to Nashville. We drove and picked him up. When he got off the plane they put money in his hand and clothes on his back. They brought him to the camp. And once we all were together, we decided that we would come to Texas, because my brother was in Dallas. The camp flew me and my daughter to Houston. My sister got a U-Haul, because the people in Murray, Kentucky gave us so much clothes.

"Thirteen New Grey Strands of Hair"

We were going to Dallas, but we stopped in Houston. A friend of mine manages a hotel here, so we stopped there. When we got to Houston, we couldn't get any help from anybody. I called the Urban League in Washington or somewhere. I was screaming at people, and that's how I found Renae Stephens, who has been a blessing to me.[4]

The biggest issue was my daughter's senior year in high school, and I was frantic about finding a house so she can finish school. If I could get my daughter stable, she could finish school, and we could get some normalcy, because it still didn't feel normal. Now we've got furniture and enough money to maintain us through the day.

The people of Murray gave us clothing, but what they gave us wasn't church clothes. It wasn't something you could dress up and go to an interview in. Renae blessed me with that, because I wanted to find a job. I got a dress to find a job in and to go to church in.

I'm still calling FEMA. They put me off for four months now. They went and looked at my house a week ago. The first FEMA inspector lied and declared me ineligible. He told them he called me on Thanksgiving Day and all my numbers were inoperable. I have caller ID on the cell phone, on the TV, and on my home phone. So I had to call back and wait for two weeks—two weeks!—to come back eligible, and then they still told me I cannot receive anything from FEMA. I can receive housing assistance, but I cannot receive any financial monies for my personal property losses until SBA denies me. SBA has held my case up for four months.

On December 12, 2005, in McWaters v. FEMA, the district court stopped FEMA from requiring everyone to apply for an SBA loan before they could receive temporary housing assistance.[5]

Renae and staff have helped me look for work, but I don't feel that I can go and work the hospital floor right now. When I was normal I would go home with a migraine. I was taking Topamax twenty-five milligrams. So can you imagine me now? I have gotten thirteen new grey strands of hair from this ordeal. It's a lot working the floor. You take on other people's problems. I would be in an insane asylum if I went and took on somebody else's problems right now.

I'm trying to work from home where I can be there for my daughter to finish this last year, where I can be there for my mom through her crisis, and I can still be able to get some rest, because I feel like I need rest. I just got past waking up every morning exhausted. The first two weeks, I would cry myself to sleep. If I watched anything Katrina related on TV, I would cry. I'm better now. I still get emotional when I talk about it, but I don't burst out in tears.

Conclusion

We had to drive home just to remind ourselves that there is no home. During the holidays, we were so homesick. We drove over eight hundred miles roundtrip to go to church to be with our church family just to feel normal, to see everybody, and to know everybody's ok. Most of all our pastor, because he's like a father to us all. Just to hear his words and hear his voice.

New Orleans is the same way it was when we went there two months ago. There's no change. The only thing that is functional is part of the CBD, the French Quarter (their money maker), and Uptown where the attorneys, the doctors, and the old money are. There was no water there from the beginning. Gentilly is shut down. Before the storm, I had to wobble through traffic to get to my driveway, that's how much traffic there was in my area. Now there's a few people, maybe, every other block, working on somebody's home. There's no lights, no water.

In New Orleans, there is no place to stay. I would have to go back to my old rental house and try to clean it up. I was offered a trailer in Alabama or Mississippi. I'm in Houston! The people in New Orleans don't want the people to have trailers there. Can you imagine? Nobody wants to put us anywhere. Everybody's fighting, "We don't want them here, we don't want them there." I wish somebody would tell me, where do you go?

In Louisiana, fifty-six of sixty-four parishes did not welcome the trailers. Thirty-two parishes explicitly prohibited trailer villages. In the fall of 2005, Mayor Nagin and the New Orleans City Council clashed publicly over trailer villages in New Orleans.[6]

You know New Orleans has played home to everybody. When you come to New Orleans, you experience our hospitality. They've come. They left their trash. And it's up to us to go out, clean it up, and make it right. Now we need to be welcomed. Don't get me wrong, Houston has been kind to us, but Houston's not home, and there are some people that don't want us here already. Everywhere there's evacuees, locals are having a problem with them being there. That's cruel because there is no home for them to go to. West Bank (Algiers, Westwego, Harvey, and Avondale), Kenner, and Metairie are overloaded with people. They don't want us there. Baton Rouge didn't want us there. I know because my friends have told me. I had a friend in Oakland, California. He just went back home. They don't want them in Oakland either. Why? There's a group of people that causes some crime, but there are some excellent people. Hello! Everybody has their own issues. Obviously you don't want our issues, but what do we do? Nobody has a solution.

I don't know what would help me the most. Peace of mind? I don't know. Just to know that I have the financial means to help my child begin her new life, because she'll be eighteen and this is the beginning of her whole life. To know that I have a place to call home, and not a temporary eighteen-month home?

SEVENTEEN

Eleanor Thornton

Born and raised in Algiers, forty-two-year-old Eleanor Thornton enjoyed a childhood rich in spirituality and love. As an adult, she took enormous pride in her ability to work double and triple shifts for banquets and special events as a waitress at the convention center. Eleanor took great care preparing each event as if she were about to serve guests in her own home. Summer time in New Orleans was notoriously slow for the service industry, and just before Katrina hit she had spent the last of her savings on school clothing for her daughter, Chelsea.

Between January 4, and February 6, 2006, four interviews were recorded at a three-bedroom apartment in a gated complex in Smyrna, Georgia, that Eleanor shared with Dwayne Chapman, her fiancé, and Chelsea.[1] Eleanor, a short, athletic woman with a new hairstyle to lift her mood, was impeccably dressed in one of the mix-and-match outfits she purchased on sale after the storm.

Eleanor's ordeal began at a hurricane party at a shoddily constructed apartment complex along I-10 in New Orleans East. She eventually reached the convention center only after braving a long journey through flood waters up to (and sometimes over) her neck and then a long wait in the scorching August heat on an I-10 overpass, before she accepted a ride on the back of a Ryder truck. Unlike Rochelle Smith who witnessed events at the convention center from the outside, Eleanor's narrative takes us into the halls of the sprawling complex. From the convention center, she was bused to two different military bases in rural Arkansas and then to a church retreat near Hot Springs, before finally receiving the money for a Greyhound bus ticket to Atlanta, Georgia. Eleanor's experiences in Georgia are typical of the highs and lows of evacuee life experienced by the working poor of New Orleans.

The West Bank consists of Algiers and Jefferson Parish. I identify totally with Orleans Parish. I have nothing in common with Jefferson Parish, other than the fact that we go to the same shopping center at Oakwood. Except for the New Orleans Center, which is right down in the CBD by the Superdome, the major shopping centers are in Jefferson Parish.

I grew up in Algiers. Algiers is a smaller part of New Orleans. All of my family grew up in Algiers—my grandparents and my great-grandparents. The biggest employer of Algierians is the naval base. There is no other facility that employs no large amount of people from everywhere. Most people from Algiers work either on the East Bank or in Jefferson Parish. You didn't have as much violence on the West Bank as New Orleans proper, the East Bank. "The West Bank is the best bank," we always used to say. When I want to party, I usually go to the East Bank because everything is on the East Bank. I couldn't hang on the West Bank, because you just see the same people over and over and over. Ain't no difference, ain't no flavor, oh no! I go across the river where you see all kinds of people.

When blacks started becoming homeowners and business owners on the East Bank and the West Bank, affluent whites started moving out to Kenner, Metairie, or Slidell, but they kept their businesses in New Orleans. It used to be more Caucasians than blacks in Algiers, but Algiers was predominantly black the whole time I grew up. We know they was prejudiced, but there's still "good morning," "Hey, how you doing?" "Where your mama? How your mama? How your grandma?" You're going in their stores and all. And then as we started getting older, they know you just like they know their own children.

My mama was a nurse aide, then she started working in a kitchen in Algiers at Touro. It was a nursing home. She worked in barrooms as a bar maid. She was prissy and put on her heels. Most of my mama's friends, the ones I know, they all, even if they had a regular job in the daytime, their second job was in a barroom. My mom always had two jobs too.

I had four brothers, and I have two sisters. My mama wanted to go places and do stuff, but you know she didn't have enough money, so my mom used to give bus rides to Houston, to AstroWorld or Six Flags. She would sell tickets and give suppers to pay for her buses. A supper is when you cook dinner and you sell a plate from out your house. So that's how we all started being in the kitchen, because we had to help. We'd wrap up the forks, wrap up the cake, do everything, because she cooking. We had a lot of families go with five, six, and seven children. My mama knew they might not can't pay for all five or six of the children, and theyself. So she'll tell them, "Pay for yourself and however many children you can pay for." If at the last minute they say, "Bert I ain't got." She'd say, "Come on. Bring them!" If they didn't have no food, she'd say, "Eat with us." She never made a profit, but everyone had a good time.

My mama had two rooms, a bedroom and a kitchen. We lived over a grocery store. Me and my little sister slept in the same bed for a long time. I don't think I knew to feel deprived. I know my mama did the best she could, so I never did get caught up with what somebody has, and what I don't have. I didn't feel that, because I felt a lot of love with my mama and my brothers. And she was fun. Even when she tired, we just all get in the bed with her, watch TV with her, and crack jokes. After my mama died, I just never got back right, period.

Growing up, I always had my own money because I went to the store for all the old people in the neighborhood. Back in the day, you could return that bottle for a nickel. That was our hustle, running to the store. I used to buy all my brothers and sisters something. But that was all a part of growing up with a single parent.

I've never lived in a house with my daddy, but I was a daddy's girl. My daddy's people came from Donaldsonville, Louisiana, and they were white. His daddy was white, his daddy was white, and my grandma's grandfather was white. My daddy was a presser. He actually passed for white to get jobs, because he was your color with blue eyes and straight hair like yours, not like mine. We fair-skinned, but we know off the bat we black. Our daddy's black, our mama's black. The color's just different. My daddy and I have the same skin color as Bush and his ma and pa, but I'm African American all day long. I am blacker than black, and proud of it.

My daddy had a little grey Nova. I wanted that car. He owned a house, my mama never did. Me and my two brothers inherited all his property—a double house in the front and a house in the back, and a whole other lot with nothing on it. We didn't have no insurance because there's nothing to have insurance on. The houses was raggedy. They need to be tore down.

I went to O. Perry Walker High School. I was a good student. I still love to read and write. My oldest brother, he used to say, "read this!" He used to sit down and do math with me.

"You Need Two Jobs to Be Comfortable"

I started working when I was fourteen, and my mama got a permit for me. Until I got to Georgia after Katrina, there never was a time when I just sat around. It wasn't always that I worked two jobs, but I had two jobs even when I graduated from high school. If I'm giving my mama money, then I still wanted money for myself too. It was a habit. Just like now, I got to pay bills with one job and I have a kid. I've had good jobs. I done worked for the School Board, worked in the mail room at NASA, and all that. I was cleaning the offices that I worked in for a second job. Since I don't have money sitting in the bank somewhere, and I didn't get my college degree, I've got to work like I work. The biggest regret of my life is that I didn't finish school.

My favorite job is being a server. I worked in the Superdome, in the New Orleans Arena, and I worked in restaurants. We'd go to conventions and do Tipitina's, an uptown Rhythm and Blues nightclub. It's with people, so it's not boring. You go to work, you're doing something. It's a rush, and the money's good.

You have waiters that make more than school teachers and policemen. I tried to be one of them. The secret is how you treat people and just working. Every time they have work, work. You have temp services for waiters. You have functions all over the city. You get off your job, ok, but they got a function, and you got work. And that's a whole other check. Sometimes I worked doubles and triples, two or three different jobs in a row on the same day. Your feet start hurting after a while. We always was starched and pressed. I'm telling you, we give you top of the line. You might work twelve, fourteen hours three or four days in a row. Servers make money when you get into the right places, and if they have work. Most people have two jobs when they do that kind of business. An average week is four to five hundred dollars.

I'm not going to give nobody no glass that I wouldn't drink out of. I'm going to wipe every last glass. Management get pissed off about that, because it's all about speed and making money in the industry. I'm ok with speed, but I'm not going to give them no dirty utensils, cups, plates, or stuff that just come out of the dishwasher. I'm going to wipe it, because I don't want to touch it, let alone eat off it. Why would I serve that to somebody?

It's seasonal, but you prepare for that. You put money aside. In New Orleans, in the summertime there is hardly no work because it's hot and you don't have that many functions. There's nothing going on in August. They're saying we poor because we got stuck in New Orleans. Sh**! So I'm poor! We ain't never had enough, even if we're working, because the money's not consistent. But we were resourceful. We knew how to make money, and we knew how to spend it. I wasn't stressed. I could always have used more money, and I don't mind working for it. There are people who are worse off than me. We had Medicaid.[2]

My last dollar, if you need it more, it ain't doing me no good just to hold on to it. Why wouldn't I give it to you? I do give a lot to people on the street. I feel if I tell them no, I'll be punished for it. Your riches is your spirit. Everything else is lagniappe, an unexpected bonus.

New Orleans

Before the storm, I had a car, but it died. In New Orleans, public transportation was better than driving a car. I could park at home for free, and spend a $1.50 to get in and read a book. You hop on a bus and you can go anywhere.

The Sixth Ward, Tremé, was my favorite place in New Orleans. I liked Louis Armstrong Park on North Rampart Street, the people, the atmosphere of it. You at Congo Square. You're looking at all the chairs, all the cans, and all the strong black people. It's the electricity of it. That's the heart of New Orleans. That's historic New Orleans.

What was going on in New Orleans was awful, as far as the murder rate. I never felt like I was in danger because if you don't bother them, they don't bother you. Now if you going to walk around places and act like you lost, crazy, or you don't belong, like you too uppity, then you're going to have problems. But if you just go ahead on, "Hey, how you doing? What's up?" You alright. I go anywhere I want any time of the night. I love the Iberville Housing Development. That's my friends who I worked with and who I partied with. I'd walk through the Iberville anytime of the night. I ain't never had no problems. Leave them people alone, and they going to leave you alone. Don't go in there with your ass on your shoulder, because they going to knock it off. They're going to speak, and if you don't speak, just walk past them, then you going to feel something coming down on your head because they going to get you. There are rules.

The last ten years or so, it just started getting worse and worse. The violence was out of control, because you could be at the bus stop, walking to work, and it just happen. I wasn't used to New Orleans being like that. There's a lot of people I see our age who didn't have the parents that we had. They basically raised themselves. All of them not bad. It's not like they want to be selling drugs or robbing. But they done been to jail for one reason or another, nobody don't want to hire them. They can't get a job, so where they going to go? To the street.

I met Dwayne Chapman, my boyfriend, four years ago today. Dwayne is the kind of man who'd lay his coat down for you. Chelsea, my daughter, was in the ROTC (Reserve Officers' Training Corps) in school. I actually encouraged Chelsea to go into the military because they'll help with your schooling. I let Chelsea go to Landry in Algiers. Then Landry wasn't up to par so I sent her to John F. Kennedy Sr. High School, a magnet school located on the edge of City Park.

Katrina: Landfall and the Flood

I had heard about Katrina. Probably the whole week I knew it was coming, but by that Thursday I knew it was going to be serious. They still was saying that it might turn, so I wasn't panicking. We never evacuated for a hurricane before. Our house in Algiers was always a house that everybody came to. I have family in Plaquemines Parish. They always came to the city, and we just rode out the hurricane by us. After the hurricane over, they go back to the country. In fact, we had hurricane parties, where kin folk or neighbors all pile up together with their children. You're going to cook two or three meals before the electricity go out. But then you're going to have all your canned food and your tuna fish meat, your sandwich meat, enough bread, water, and a lot of beer. You need the beer—that's the party.

There was a total of four people with us: me, Dwayne, little Kevin, and big Kevin. It was Kevin Marshall's apartment in the East. Nobody out of the four of us could swim. We were laughing at the ones evacuating at that point. They'll be back in a couple more days, have spent their money, and nothing happened.

During the actual hurricane, the first thing that happened was the building shaking. We were on the third floor. I'm like, "Lord have mercy," because the wind's coming from all around the building. You heard glass crashing across the street. So where you looking across the street, you're looking straight through their house. Then you got people trying to run to the front of their house, because the back of their house gone. That there alone went on for hours.

That's where you really started hearing people as the buildings collapsed. To hear everything being tore up and people screaming, "help me." You can't do nothing. The houses are falling into other houses. As tore up as the apartment building was, we could have been killed in that alone. If they'd have had five or ten more minutes of that hard wind, our building was going to collapse. Then when the water came, everybody start screaming, "You alright?" At this point, Dwayne and me is sitting on the steps. "Do you think we're going to die?" I said, "God, I can't go like this. I don't want to leave my child like this." I said, "This is not part of that natural disaster." The hurricane was over.

The most we could of did was like when some young boys, because their family in trouble, tried to swim to where we was from across the street, not realizing how high that water was and how deep. The current was pushing them, and they almost drowned. We grabbed two as they went past us. We used to know them to say good morning and good evening to them, but not to be in their company. But I know they was working people. You see them going to work, bringing their children to school, stuff like that.

We saw one man go under twice. And one of the girl members of the family kept on just hollering his name. She had to swim and bring him back up. She was saying, "Come on, you got to hurry." She was telling him to swim, but he couldn't swim. He went under again, and he died.

The next day we was counting the steps. As the steps clear again, then we knew that the water was dropping. And once it got down to like two or three steps, then we knew that it was like about six-and-a-half feet deep. We started seeing helicopters picking up people, but they was picking up old people. So we stuck it out until it got dark on Tuesday, the second day, so we couldn't leave then. We said, "They ain't coming back, so they must have made it somewhere." We didn't know how far we had to go for dry land. Dwayne wanted to go out on a scouting mission. I said, "You not coming back and getting me. We leaving. If we going to die, we're going die together or we going to survive together, no matter what."

"From Crowder to Chef through the Water"

I'm five foot eight inches. Coming out of the apartment complex, the water was so high we couldn't just stand up and walk. First we had to hop a fence outside the complex. We held the iron fence all the way around the complex. Trees were down. I grab a tree full of red ants. The red ants started biting me up. I finally got around the tree, but I'm blistering. I'm having a real hard time even holding on to the fence. I started trembling. Dwayne started pulling me. He was saying, "Come on, Eleanor. Come on! You got to hold on!" When we got to the end of the fence, we had to hop another iron fence that blocked off the apartment from the street. I had on heavy tennis shoes. I'm already bit up, so I couldn't get my leg over the fence. I kept going backward. They done hopped the fence, but I'm still on the other side. I told them, "Just leave me. I can't get over this fence." Dwayne said, "You're going to get over it." So even after I got one leg over the fence, my other leg got caught. I had to be pulled across the fence. I'm bleeding, so I'm just going to be chop suey for everything in the water.

We walked from Crowder to Chef through the water. When we got to Chef, we're going to walk the interstate. We stayed to the side like what would be the shoulder lane because you couldn't see it, it was covered with water. You got people walking from Michoud, where the Vietnamese live. As long as we bounced, we could stay afloat. But the people who were short, like the girl with five children on their air mattress, were in trouble. We started pulling her and them children along with her husband, so nothing don't happen with this girl.

The National Guard flying overhead saw us. They started saying, "How many children?" We were saying, "five." He came down and he started saying he going to take the children and the mother. And she said, "No! My husband have to go too." He say, "No, your husband will find out where you're at." She said, "I'm not leaving my husband." So the man started trying to grab the children off the mattress. The propeller is blowing us. We're like, "You all about to flip all of us over. We know you all are trying to help, but this lady's not letting you all just take her and her children. They got to go as a family." We start pulling the children back. He say, "OK." They was as nice as they could be. I can't say nothing bad about them.

Now we done got to Chef. But we can't cross no more, because we know from living in New Orleans, there's a big dip. So we really don't know how high the water is, and we at a standstill. That's when the people who had boats, out of the kindness of their hearts, appeared. One man said, his wife asked him, "Where you going?" He said, "I'm going to New Orleans. I'm going to help them." All of these people with little fishing boats was white. When I tell you they had food and water on their boats! They was getting their personal stuff for us. They was 100 (percent) with us. They say, "We going to get all of you all." They kept making trips to get us. We was letting the old people and the people with children go before us. They was bringing us on the other side of Chef. It was dry over there. When we got on that overpass, they had to have about five to seven hundred of us.

Now after they did that, they had two military trucks. Those military trucks had food and water on them. When I got hungry, them MREs was good! After that, we just sitting on the overpass at the area leading up to the high rise near two hotels on Chef by the Downman exit. The military trucks took some of the people but nobody never said where they was going. It's getting dark. It was hot. Regular citizens start taking the Ryder trucks across from where we was and they start putting people on the back of the trucks. We was told that we was going to the convention center to be put on either a boat or a bus to go to Houston.

As we rode to the convention center, we were seeing people we know on the interstate walking like bums. It was like they done dropped bombs on New Orleans or like we committed a crime. Being black, just being poor—I guess that's our crime. Just being a regular working person's not good enough in the United States.

Convention Center

The first day we made it to the convention center, Wednesday, there was a body actually sitting on the neutral ground that they said had been sitting there the day before we got there. The legs on this lady were swollen as big as this chair we're sitting in.

When we got there, it was just a mad rush of people. You could see people starting to scrap for carpet to sleep on. We got there kind of earlier than a lot of other people on Wednesday. So you see people looting, but they're feeding each other. We was not stealing. We was feeding people. Now, they have some, of course, that are going to do wrong.

The police that did ride around, if you get close to them, they're pulling their guns out. They was shooting persons, asking questions later. That's how they killed that boy. He was asking about the buses. He's got a grandmother that's sick, so he walks across and jumps in between this chain of police cars. And when he done that, they jumped out, shot him, and they left him there. This is the person on TV you see laying in front of the convention center with the stream of blood running from his head to the curb. He had to be about nineteen years old.

The convention center was total chaos. That was devastation every day. First of all you're barely sleeping because you're scared for your own safety. There's no police, and before the military came, they was fighting with pipes. The little boy looked in my face, because I guess I wasn't moving fast enough for him. I'm really in shock that all this is going on. That little boy told me, when they come back through here, whoever in our way, going to get it, so you might as well move. They thought it was a game. It's funny to them at that point. I would say they were between the ages of fifteen and nineteen. You had bad ones, but you had good ones too. Everybody not drug dealers or hang on the street or not in school. Our children ain't no different than those little Bush children that got drunk and had fake IDs. You love yours? We love ours.

I can't say nothing bad about the military when they came into the convention center because it just wasn't nothing that they could do, but give you an MRE and some water, because they didn't know no information either. All they know was they was just sent there to secure things, to get order. That's what they told us. The government itself, what Bush did himself, is a different story. But them, they was 100. If they wouldn't have came, no telling what would have happened because it was getting worse and worse every day.

The military came in the daytime on Friday. They cut off the electricity. You can't see because it's pitch black around 8:30, 9:00. That's how they got order. You couldn't even be mad with them about that, because like they said, they had to protect themselves to see what was going on. The military knew some of those boys had guns, because they knew they stole guns from the Wal-Mart. I guess on Saturday is when we first started seeing buses. But the buses wasn't by us. Those buses was on Julia Street. On Sunday is when they started bringing the buses and when they started flying helicopters.

They treated us as badly as you could treat your fellow human beings. I don't care who you are, what color, where you're from, or what's your background. I wouldn't treat a dog like that. The United States blows up Baghdad, and they'll drop them food. (My brother is serving in Iraq.) But they didn't do it for us. They didn't bring food for the convention center until the fifth day after the storm. I would love to know how they sleep at night.

You have good white people that go to work. They lost too because they aren't the right financial status. But they got the whites out of there. They got them to Jefferson Parish, and immediately started taking care of them. That's why we didn't see them. We was left to fend for ourselves. But we know how to live with everything, and we know how to live with nothing. They going to get the Caucasians and put them into safety, and you going to bus us to a POW (Prisoner of War) camp. We was told we was going to Houston.

Arkansas

From the convention center, 250 of us went to Fort Smith. Some came from the Dome, some came from the convention center, and some came from Causeway. Some was flown in, some was bused in. We slept in a dorm: Mens, women, children, and everything.

We was in Arkansas for a month. We spent the last two weeks in Hot Springs. The minute we left the convention center, we was treated with love by human beings and the military. They did as much as they probably can do. I had on a pair of shorts and a t-shirt. We didn't get clothes until we got to Fort Smith, and they had a thrift store set up for us for alternative clothes. Those people was nice. This was a military base with regular volunteers that came in. The military people wasn't mean, they just didn't know how to take us.

They called it a registration camp. They told us at that time we were not citizens of the United States. We were listed as not existing. The only way we could get back into the United States is we had to get our names back into the data base to be a United States citizen.

The people coming from New Orleans looked traumatized. Austin Leslie of Chez Helene was one of the biggest chefs in New Orleans. You could see his movements was bad. Dwayne's phone was one of the only phones working, so he let everybody use it. When we finally got in touch with that man's children and wife, he wanted to see his wife. He said, "Where's my sweetie?" They ended up putting him on a bus to go to Texas. He died of a heart attack a few weeks later.[3]

The food in Fort Chaffy, our next stop, was good. On the military base, they had hot food for breakfast, lunch, and dinner. Still it felt like a concentration camp because we couldn't leave. We had no say so about going or being where we wanted to be. We all needed drawers, we all needed socks. The ones who had money and wanted to go shopping, couldn't because Arkansas didn't want us in their towns at that time. They told us that they wasn't letting us leave the base, because when Castro asked the United States to take Cubans, they didn't tell them that there was Cubans who had just been released from prison. They rioted in Fort Smith and Fort Chaffy. And they said they was not going to let that happen again.

They had so many young children that were separated from their parents because of the way people were bused out. They got splitted up on the buses. The bus could only fit forty-eight people. If your family is big, and they already got so many seats on that bus, then half your family went on this bus, and half your family went on another bus, but that don't mean those two buses are going to end up in the same place. They told us we was all going the same way, but we ended up in separate places. They had a little girl, sixteen years old, and she looked like Chelsea. I grabbed that little girl and I said, "Baby, don't move." Her little brother was on heroin, and he was pimping her out to other heroin users in the camp. He beat her up when she wouldn't have sex with somebody. She said, "I don't have nobody." The little girl's mama was bused to Texas.

The church people was uncomfortable. You talking about all these people done been bused from New Orleans, their lives in total turmoil. They'd have been looting and now they're in our space. That's how the media portrayed us. So that's how them people received us. Until you start talking and then they say, "Everybody is not like that."

At Hot Springs, the barracks had bunk beds. The retreat was like a little dormitory room, but you still in a room with strangers. It was like a twilight zone for everybody involved. You want to take a bath, but there ain't no privacy.

I'd drink my coffee in the morning, but I wasn't really eating. They say what I had done been through had me sick. We were really trying to get our bearings to see what was going on and see how to get to Georgia because they wasn't offering no transportation out of there. So we really didn't know. We figured if we get our FEMA checks, we can pay for our own self to get out of there because we knew we wasn't staying. They kept on saying, "FEMA people coming, FEMA people coming," but nobody never came.

When we left Arkansas, we caught the Greyhound bus to Georgia because Dwayne's children was here. Chelsea didn't like Texas. The people at the church retreat either paid for the bus tickets, or they did it through Red Cross. We didn't have a crying cent.

Reclaiming Existence

We got Red Cross money two days after we got to Georgia. Hell yeah we were happy to get here! Just to see people period. Black people! I was ready to get somewhere where I could walk out the door, walk freely and go how I'm used to being at forty-one. We went to the Microtel Hotel first. When we told Dwayne's son Keenan we was coming to Georgia, he had booked a room for us through FEMA. Everybody in the Microtel was from New Orleans.

The first thing we bought when we went to Wal-Mart was the hot plate, the pots, and some food. (I love Wally World. Wal-Mart is for poor people.) We used food stamps from Arkansas and Salvation Army vouchers. Dwayne's voucher was for sixty dollars and mine was forty dollars, because I only had me and Chelsea. He had his two sons. The hot plate was twenty dollars, and we bought pots, cooking spoons, and plastic containers to put the food in. Then we went to the grocery store.

They hadn't had home cooked food since they left New Orleans. They was just feeding their childrens fast foods till we went in there with a hot plate. Dwayne was trying to bring some of New Orleans to whoever was there. They're smelling food up and down the hallway. And they came. People was offering us their little food stamp cards that the state had done gave them to buy food, or they'd buy the food today, as long as Dwayne cooks it.

I helped people get through to Red Cross and FEMA. I had like thirty, forty people in that room at one time. I'm the only fool that stayed on the phone twelve hours for everybody to talk. I would just say, "Since it's taking us eight to ten hours to get through, can we all just go through you?"[4] They let us do that till they got tired. When we found out something, we let everybody know. Us being from where we from, you frowned upon to be mean, because it's not natural. It also helped to pass the time, because otherwise I would have been going crazy. The rooms were small, but you ain't got your people. I moved out of the Microtel in October.

When we hit Atlanta, people was willing to help you at that time. Now America is getting back to normal. It's what you got, and what you going to get for yourself now. A church adopted us and helped us get this apartment. It's a white church. It's Apostolic. They stay on the Word. They always reaching out to youth. They have student-to-student programs there, and all kinds of Bible classes. They have to have over ten thousand parishioners. We was the only people from Katrina in the church. The church sent all their resources to Alabama. They didn't know we was in there. We started going to a Builder's Bible class. That's how they found out that we was from New Orleans. Then they really started trying to help us. They gave us furnishings. A lot of the furniture in this apartment is from them. This table is from someone whose mother had recently passed. They did it out of the kindness of their heart. We bought our beds. They gave us the sofas and those nice stands. We bought the TVs and our bathroom stuff. They gave us sheets. It made our house complete. Them Christians was like, "Know that God ain't going to leave you." They invited you to their homes. I went to a Christmas party: no singing, but food. This in an area where them people had limos with the chauffeurs parked outside all the way back there.

A member of the church owns this apartment complex. They paid two months rent and the deposit. All they did was criminal background checks on both of us. They didn't do credit checks. We can't afford to live here. It's 879 dollars a month. I like it because it's spacious with a balcony. It's quiet. The people in this building, you see them coming and going to go to work. They come home, go inside, and that's it. I like that.

Dwayne hates it. He was raised in a neighborhood where you could borrow sugar from next door. These ain't the same kind of people. He misses most the family atmosphere, the knocking on the door, just sitting on the porch. Everybody's just sitting around talking, enjoying each other's conversation. Here we don't even know our neighbors across the hall, underneath, and around the corner. We're learning that Louisiana is something special.

Finding Work

If they hadn't helped us, I would have returned to work sooner. Instead, I spent November and December getting my life back together. I knew for a fact that by January, I was going to have a job. For one thing, it's boring sitting around. Then I have to buy a graduation ring, I have to pay for graduation next year, so that means I've got to be on my game.

Borrowing a car yesterday helped me tremendously. Dwayne will just go and catch a bus and go everywhere—to the store, anywhere we need to go. I don't have a problem with catching public transportation, but then I realized by the time I get to where I'm going, you're talking hours. I say, "There's got to be a better situation." When I got in the car, I got lost. I said, "I ain't going to get nervous," because see I used to get nervous about getting lost in strange places. I was lost *all day*! Eventually, I rolled down on streets I know. It was a pretty day to be lost. I was passing a lot of restaurants, where you see signs, "Help wanted." I say, "All you need is a car."

I ain't worried about nothing now. That Bible in the bathroom is open and marked to the verse that says not to worry. I was scared. I said, "If I can get my rent, I could make it, because I'm employable." That's the only way you're going to make it. Because other than that, you're going to end up in a worse situation than what you was. Katrina not going to be nothing compared to being homeless in a whole other state where you not from.

The most frustrating moment I've had with FEMA is for them to be derogatory. Like, "Was FEMA paying your rent before you moved to where you at?" I want to tell them, "You're just little hired help. I could do your job far better and with more compassion." But I don't.

Conclusion

Chelsea alright, but she don't like it here. In New Orleans, Chelsea could go where she want. They could hop on the bus, go do this, go do that. She would see her friends at school, and still hang out with them after school. Here it's different from where you run with your little friends on the weekends. Here they're bused to and from school. Chelsea signed up for tutoring because they going to bus them to tutor them. From this situation, I want Chelsea to learn to be strong. No matter what obstacles you go through, fight through them. Be self-sufficient.

Katrina is making Dwayne and me stronger as a couple. If we plan to be with each other, we was going to go through something anyway. If we got through that and we survived, there ain't nothing too much going to knock us down short of death.

Canal Street flooded. Bourbon Street didn't. But white people don't go on Canal Street. Only blacks be on Canal Street. White people go in the French Quarter, and the French Quarter didn't flood. Everywhere we be, everywhere we walk, they flooded. It was like they had somebody funneling that water right exactly where they wanted it to go. And they want us to come back to work. Just call us "Foo Foo the Clown," one, two, three to a million.

I thought my ancestors, and Martin Luther King Jr. died so you and I could be treated like equals. You got people looking at us like we're the lowest people on the totem pole. You think that you can just dehumanize people to that extent, and we going to just lay down. We not! We ain't laying down, and we going to shout it from the rooftops: "You all tried to kill us." You preserved what you wanted to preserve, but what you preserved is not a damn bit important. If we're not there, there is no New Orleans. That's our city. That's our heritage. I don't care how many casinos and restaurants you build. If you all want your money like that, then get it. But get it the best way you can, because we won't be serving you anymore. And we won't be as nice as we was, because you don't deserve it.

You know something? We got a lot to fight for. We have our lives and our children. And that's our city. We going to survive this. Life's never been easy. A lot of us came up in single households, so we know hard to be relative to what's going on today. They ain't going to break our spirit.

EIGHTEEN

Kevin Owens

Raised in the Calliope Project, forty-one-year-old Kevin Owens, the only survivor of three siblings, prides himself on being a hardworking man like his father, a Vietnam veteran. Before the storm, he and his wife, Elise Ramsey, were renting a townhouse with a back yard in New Orleans East. His wife fulfilled the role of the neighborhood mom, and there was always a constant stream of guests in and out of their house. Kevin worked as a maintenance man in the B.W. Cooper Housing Development.

On December 9, 2005, Kevin, a short, muscular, extensively tattooed man with kind, expressive eyes, took off work from the University of Alabama in Birmingham to come to the Birmingham Urban League to tell his story. He was wearing work boots, blue jeans, a button-down flannel shirt, and a baseball cap. In the tightly scheduled interview blocks lined up by a conscientious Urban League staff, there wasn't nearly enough time to hear all that Kevin had to say. A follow-up interview took place in Kevin's two-bedroom apartment in Hoover, Alabama, a suburb of Birmingham.[1] Normally a quiet man, Kevin spoke on tape for a total of three hours because he felt driven to document the treatment of those he felt did not deserve to be criminalized in the wake of Katrina.

Kevin describes the experience of many working-class blacks who sought refuge in the shelter of last resort, the Superdome.[2] He arrived after he and his family abandoned an apartment in the B.W. Cooper homes, comprised of sturdy, three-story, brick buildings. After being at the Dome for a few days Kevin, a diabetic, was taken behind the guarded barricades and curtains of the main floor after he collapsed from low blood sugar. There, he observed the scenes he relates in his narrative. Within weeks of being evacuated from the Superdome to Birmingham via the Astrodome, Kevin had successfully negotiated an apartment lease in an upscale community that was initially hostile to displaced New Orleanians. He started a fulltime job on the University of Alabama's Birmingham campus where his wife's son, Carldell "Squeaky" Johnson, was a starting point guard. His reflections are recollections of black New Orleans from a man with a working-class background.

I was born in Charity Hospital in New Orleans, Louisiana in 1964. I was raised in the Calliope Housing Project. It was a rough area. My mom and them had some good values. They tried their best to move out of that area, which they did.

Right before the hurricane, I wasn't living there, but I was working there as a maintenance technician. I can just basically look at something and tell you what's wrong with it. If it's your door, your toilet, your floor, whatever. If it's broken, I can fix it or replace it.

We was living in eastern New Orleans before Hurricane Katrina came in, and we got about ten feet of water inside our home. Our house was like the typical American home. It was three bedrooms, two floors. It was really a townhouse, but it was a nice home. We had a yard with patio furniture and in the summer time, we put a little swimming pool there for the grandkids and some of the neighbor children. My wife was really like the neighborhood mother. She used to come out early in the morning and made sure the kids got on the school buses safe.

We knew all of our neighbors in our subdivision right off Downman and Chef. We was able, thank God, to be blessed to feed a lot of people. We felt good about that. Our home was comfortable. We was blessed to have a lot of material things, and I always looked at it like if you're in a situation to give, then you're above need. The Lord always made a way for us to be in a situation to give. We kept our house clean, and we kept company. We kept an open door.

Katrina

I stayed focused on the news days before, and we had a chance of getting out. We had transportation: a pretty reliable '99 Chevy Malibu and my stepson's truck. He had to stay because he was working for the Sewerage and Water Board, and they wouldn't let him go.

My wife just didn't believe that we was going to get hit like that. I was conscious of our levee system, but by staying focused on the news all day, I knew that we wasn't going to be missed. But I couldn't leave her. We wind up leaving our home in the East and going to her brother's house in the B.W. Cooper on the third floor. It's brick. She felt safe.

What happened was we got to her brother's home and Katrina came through. I was up through the whole thing. I was looking out the window. We watched the whole thing. They had a leak in the house that pretty much kept us woke too. The wind was pretty scary. But the building itself withstood it real good. They had some windows blown out.

When Katrina came and left, we started clapping. We was like: "We made it!" I looked out the window and I said, "Lord, look at this here. We got to roll up our sleeves and start cleaning up." A lot of the old oak trees in New Orleans is throughout the city, including the B.W. Cooper Housing. The extent of the damage was such that a lot of them just was uprooted to the point of uprooting the sidewalks, and so there was no way we was going to drive out for a while.

I had a little storm tracker radio that you wind up, and I kept winding it. I'm listening, and I'm looking at our automobile, and it was being submerged every minute that went by. I'm saying, "Something is wrong, because this water should be receding, but it's coming in faster." That's when I started cranking to find out what was wrong. I heard the man on the radio say, "If you're held up anywhere in New Orleans, you need to try your best to make it to the Superdome or the convention center, because the levees have been breached and we can't stop the water."

I turned around to them and I told them, "We need to get out of this house, because even though we're on the third floor, we're going to be trapped in the city." We was running out of food and water. Electricity had went out about 4:00 that morning.

The Superdome: "The Scariest Part of My Ordeal"

The Superdome was about half a mile away. We wasn't really trying to carry anything, because we left a lot of stuff in our house in the East, but we had a few little clothes and a few little light things like a toothbrush and your driver's license. So we had to walk through the water to about our waist. At the time, there was just a lot of stuff from the automobiles like gas, oil, and stuff like that in the water. We saw some dead bodies as we were walking, especially as we went up on the I-10 bridge. Once we walked up the I-10 bridge, we had to walk back down at the Claiborne exit to the Superdome and go under the bridge and walk through the water again to get up in the Superdome. We left about 10:00, and we got there about 11:00 or 12:00.

We thought we was going to a shelter, but it was more of a prison. Outside the Superdome, we was searched by the military. It was like a regular little pat down. I felt like, "OK we're being searched and they might be looking for weapons or whatever." I didn't feel violated by the search. Soon as we got in there, I knew something wasn't right. We saw the water rising up by the Superdome and they wasn't trying to get us out. There wasn't no effort to try and get us out, but they constantly was bringing more people into the Superdome.

You could count the white people in the rest of the Superdome on one hand. I say 99 percent was black and poor. I know everybody else feels like we weren't in a major city in America, our country.

When we got there, I'd say about fifteen or twenty thousand people were already there that had went there earlier before the storm hit. They were stationed around on the benches and in the bleachers and sleeping in front of the restrooms and in that hall way. With every passing minute, you could see the whole situation was deteriorating. It was hard for us to find a place to lay down and get some rest, because the football field itself was wet, and they were saying, "Can't nobody sleep on the football field." There weren't any cots.

Everybody was just coming outside on the ramps, because being inside during the daytime especially was unbearable. You couldn't stay in there.

When we first got there on Monday afternoon, they was issuing the Army rationed food and water. The next day it stopped. Everything stopped. They said somebody stole the food.

I was just looking at my wife and her niece and grand-niece. My wife is a very strong woman, and she kept saying, "Baby, I'm going to just take my life. I can't take this." I was saying, "Lord, I got to get them out of here." That's what made me so mad. I didn't care if the people shot me or nothing because that was so wrong.

Every day they bringing more and more people, and we know the conditions. We're glad to see some of them because we know them. But we're trying to warn them not to come in.

When we first got to the Superdome, my wife adopted some kids that had a kid. Both of the parents were sixteen years old. And they had a kid that was physically challenged. She fell in love with them because she's that type of person. They sat next to us in the Superdome, and they was telling us that they couldn't find their family, so they clung to us.

Everybody was trying to hold each other up. If it wasn't for just everybody trying to pull together and save each other, then a lot of more of us would have died. We all we had. It was like whatever we was able to do for the elderly, we did it. Whatever we was able to do for the women, we did it, and the children. We took some clean t-shirts someone had, and tore them and made some pampers for all the babies that needed it. It was little stuff like that. People who needed medical attention, we were trying our best to get it for them. The good that we wanted to do, it was just so limited, because of the conditions, but it wasn't like everybody turned on each other, was violent or none of that toward each other. We were just praying that one day soon, we'll be out of here. We starving to death, and we hadn't touched no water in days.

I got sick, because I'm a diabetic, and I didn't eat for so long. We was going back and forward from the military to the doctors who was in there and telling them, "Look, I need something to eat. I'm going to pass out or go in a diabetic shock." So they was just sending us backward and forward for hours and hours. Finally, I passed out. That was Thursday. Some people and my wife carried me to the infirmary, which was located in the arena where the Hornets play. I had come to enough to understand what the doctor was asking my wife, "What's wrong with him?" She said, "He's a diabetic. He hasn't eaten anything." So he tested my blood sugar. My blood sugar level was like fifty or fifty-one. He said, "Take him down the hall where the people at, and get him some food and some water."

We went down this hallway in the arena, and there was this room that was being guarded, and they had cases and cases of food and cases and cases of water. So I got a case of food and a case of water. There's about six rations of food in one case. On my way out, I had asked them, "Why you all not feeding us? I thought you all said that somebody stole the food?" They said, "We giving it to the elderlies," which wasn't true, because we was with the elderly.

So on our way back with the food, the military didn't feel safe for me, so when I was about to cross back over the barricades to what's called the general population, they stopped me. They said, "Where you going with that food and water?" I said, "I'm going back over here where I've been." They said, "You going to have to stay over here with us, because you going to get mugged for food." I said, "Why is it like that? Why don't you feed them?"

We sat there all day and all night too. That's when I looked back. We was leaning against the glass, and I'm looking and they had a curtain open, and I seen all these little white children and white people sitting in a suite. They was eating good food. They wasn't eating the army rations. They had fans. The power was out, but they must have had a generator. They were with their families. I think they had one black person in there, and the rest was white.

I witnessed a black man literally try to take a little girl out of her mother's arms. That was right outside the Superdome on the ramp. Several people knew him before Katrina, and they said he was mentally ill. He should have been in a mental institution, but he was roaming the streets like that. I didn't see a little girl get raped and killed, but I was there when they took her out. She was ten. She was an African American.

I witnessed the gentleman confess to doing it, and he was a white man in his late fifties. He looked raggedy. He looked like he probably had, and I'm saying probably because I don't know the man, but he probably had a mental problem before Katrina and that Superdome thing probably pushed him over. To me she's a forgotten child. But when she was reported, I wanted it to be known that it was a white man who done it regardless of his mental state, because so many black eyes had been placed on us, and, it was like I didn't want no black eyes to be placed on anyone else, but I just didn't want the world to continue to look at us like animals, because we're not. You got people that act rude and bad, but that's with every race. A lot of young girls in the Superdome was afraid, especially going to the bathroom, because there were reports of rapes. I think it was blown out of proportion. That's my feelings. But I do think some incidents is true. To be honest with you, it's probably three or four cases.

New Orleans don't have gangs. A lot of people from the same housing projects do stick together. Now if you want to classify that as a gang, then go ahead. But it wasn't much of that in the Superdome. In the Superdome, it was the people against the military. It wasn't like the people against each other. I had family in the convention center, and their stories were more violent than what I saw. They said a lot of people from different projects wound up together in the convention center, and that's when the violence probably started in there.

I think it was mostly the United States military that was patrolling everything. They carried M-16s on their shoulders and were pointing them at people. Everybody was in a military uniform, but we didn't see nobody in charge. No announcements were being made. I said, "You all need to make an announcement to let us know what you all plans are, because we don't see nobody running around in charge of you all. Who your captain is?" I thought they could tell us why we hadn't left yet. We've been in there three days. We're not eating. I didn't have my crank radio anymore, so I'm cut off from the outside world.

There was an incident in the Superdome where they were pointing their weapons at the crowd. I had asked one of the army personnel, "Sir, can I ask you questions?" He said, "yeah." I said, "What are your orders? To kill us or something?" He say, "I can't talk about it." I say, "You're pointing your gun at us, man. We're not criminals! We supposed to be in a shelter."

But you know what really got me? When I went to that bathroom for the last time and almost passed out. It broke me because I wasn't thinking about myself. I was thinking about the women, the children, the elderly who had to come to these same conditions. It was extremely dark. I done worked in some nasty jobs in my life, and that was one I wouldn't have wanted to work in. Nothing was being done about it. It was hot during that time of the year, and then in the Superdome with all them people with no kind of ventilation, no air. I say, "Now if I'm about to pass out, and I'm a man, how the women that are with us going to come use the bathroom?" I snapped. You can only take so much. This was Thursday night. So I said a lot of foul things to the military personnel, the ones over the barricades pointing guns at us.

A black man in charge said, "Why don't you settle down and let's talk?" I said, "Man, we're in a prison here. Your people point your guns at us like if we make a wrong move towards them or something, we're going to be shot. We not no prisoners, man. You need to show me one piece of paper where I committed a crime. None of us in here ain't committed no crime, man. We're in here trying to save our life. If they is out there committing crimes, man, you go deal with that. We here to get away from this."

That night around 3:00, a young man on the football field, where they had at least twenty-five thousand people sleeping all over, got up and walked up to a military personnel, took his M-16 from the young military man, and shot him. They shot him four times in front of everybody. I don't know if that young man died or not, but the thing that stayed on my mind about the whole incident was he couldn't take no more. He broke. That incident to me shouldn't never of happened. It was just one big, old, desperate situation for everybody. Military personnel compounded the problem.

Leaving New Orleans

The way the military handled the evacuation was extremely bad and confusing for everybody. First they made the announcement: "Everybody back away, we're leaving from such and such gate." Obviously, everybody was with their family and making sure they got the little stuff they did carry and was able to hold on to for them days. Everybody is sticking together and we're slowly marching to the barricade. Now you got all these thousands and thousands of people on this ramp that's going up, and you got them packed in like sardines.

When everybody gets packed in real tight, then they make another announcement: "OK we don't want nothing but women and children." That made everything worse because all families were together. So that's how me and my wife got separated for two days. If you wanted it like this, you should have said it up front. Barricades could have been put in line. "I want women and children in this line, elderly in that line, and all men in the other line." Let's keep it like that in order but everybody going to the same place, so nobody panicked about losing your family members, or being separated from them. When they caused the confusion, by everybody being packed in, now you got children, little kids, and women holding on to their husbands, their boyfriends, and their fathers. They're saying, "No! You're not going to separate us from each other because this is all I have in the world left."

Then they made a statement, "Nobody leaving. Come across this barricade, you will be shot." I'm like, "Hold up! Why would you make that statement that you going to shoot us, man? We're not criminals!" When they made that statement, a lot of us men turned around and said, "Let us make our way to the back and let our women, children, and elderly people get out."

So then they had women and their little children saying, "You know what? I don't care what you're asking. I'm not leaving my daddy." Or, "I'm not leaving my husband because we came through all this here together, we came through all that flood water together, and we staying together." If it wasn't for our women and our children not leaving us, I think we would have been murdered, to be honest with you. It was like a feeling I had, because it was like, "We going to provoke them to excite a riot." I went to telling all the men, "Look, whatever you do, don't start a riot because they already pointing guns at us for no reason and we're unarmed."

Five or six buses left already, my wife being on one of them. About an hour or so into that process, they make another statement: "This is what we want. All families, make sure you stick together." So being that I listened to the first order, now I'm separated from my wife and nieces. And it's not just me, it's a lot of us men. Now my wife's gone to Houston and if I stay back here for another day or so, Houston may stop accepting people. Where I'm going to end up at? This is already in my mind, because plans change so rapidly. They said, "Everybody in here going to the Astrodome. Don't nobody worry."

On our way to the Astrodome on Saturday evening, we had just hit Texas, the bus driver came on the intercom, and he changed the whole plan again. He said, "I have another announcement to make. This bus's destination is now San Antonio." So I jump up and I say, "Sir, I really don't care where you take this bus, but you're taking me to Houston. This is the end of the straw. You can keep rolling after Houston if you want, but drop me off and I'll walk to the Astrodome to be with my family." He say, "I can't do that."

We had to stop at a rest stop, and they had a lot of Texas law enforcement officers out there. I hurried up and jumped off the bus, and I said, "Who in charge? I need to talk to the person that's in charge." So they pointed at the gentlemen and I said, "Sir, can I please speak to you?" He said, "Yeah, what's the problem?" I said, "The problem is I've been separated from my family due to no fault of my own or Katrina, because we stuck together through it all. We got in the Superdome and because I'm trying to comply with military orders, I've been separated from my family, but they all reassured us that we was all going to be together in the Astrodome. My family's been there now for going on two days." I said, "This man here just made an announcement saying he's taking me to San Antonio. Sir, that's wrong. I don't have nobody in San Antonio, and the way I feel right now, I cannot be separated from my family no more. I don't believe none of us could." He was a nice man. He went to the bus driver and he told him, "Take this bus to the Astrodome, and if you have to, I'll escort you. But all these people's families are in the Astrodome, and that's where they should be." We had some sandwiches and water on the bus. Most of us was just breathing a sigh of relief to be out of the Superdome.

Reclaiming Existence

When we got there, the Astrodome still was taking evacuees. When I got there, it was dark. I got reunited with my family, and I went to seeing people who had been in the Superdome, paying attention to the looks on their faces. They was lost, but they was relieved because they was being fed, they was able to take a shower, and they was able to just sit back and think about where they needed to start rebuilding their lives. It was obvious to all of us that it was going to be a long time, if ever, before we was going back home. Just the shock of that was hard.

At the Astrodome, our oldest boy, Lonnie Johnson, was basically there to meet us. Our son said, "Pack your things. I'm here to take you all out of here." The children my wife adopted start crying with their little baby. But I said, "Don't worry you all. We're not leaving you." Our destination was Birmingham, by our younger son, "Squeaky."[3] At that time, he was a senior starting guard at the University of Alabama. So we called Squeaky and told him, "We adopted these children. They have a kid that's handicapped. They need help more than we do. We want to bring them with us." Fortunately, the girl's uncle recognized my wife from seeing her with the little girl one time, and he said, "Wasn't you with my niece? When you see her, please give her my number." So when they got reunited with their family, we left the Astrodome.

A Houston pastor who knew the assistant coach of the University of Alabama at Birmingham (UAB) basketball team, T.J. Cleveland, put Kevin and Elise on a bus for Birmingham.[4]

Arriving at Birmingham, we came by this lady, Miss Linda Cox, a good friend of Squeaky. Her and her two daughters, Stephanie and Sylvia, they opened their arms and their house to us. I just went to looking for a job the next day, and we was just trying to get around and put our lives together from that point on. We lost our car in New Orleans, but they was bringing us around, taking us everywhere we had to go.

We was trying to get in this apartment complex. It was upper class. It was hard for us to get in, because we black and we from New Orleans. Most everyone else is white, and they're almost all from Birmingham. At first the manager of the complex told us there wasn't nothing available. Then it was something else. I said, "What if we get a voucher from FEMA?" They said, "We don't take vouchers, we don't take Section 8." I've never been on Section 8 in my life!

In the fall of 2005, the National Fair Housing Alliance conducted sixty-five test cases to see whether African Americans were at a disadvantage compared to whites when trying to rent apartments in Florida, Alabama, Georgia, Texas, and Tennessee. They found discrimination against blacks in sixty-six percent of the cases.[5]

So I'm like, "What do I need to get in?" So he came up with 2300 dollars altogether for the deposit and first month's rent. The rent is 765 dollars every month, which we pay. We wanted to just have somewhere nice to stay where we could feel comfortable.

From there, we've just been trying to get some normalcy in our lives by me working. I'm at UAB, the same school my son goes to. They act like they're not used to work. People from Louisiana, we've been working hard our whole lives. That's how I am, because that's how my father was. He was military strict. I get bored if I'm not working.

A few churches have helped us. We found them through some friends up here from New Orleans that are searching around too, and they're carless. That sectional we bought on our own, but this table was given to us by a church. Another church gave us a Wal-Mart card when we first came. It was two hundred dollars a piece, and the Urban League gave us a gas card.

When we was allowed back in the neighborhood after Katrina, our front and back door was bust completely wide open from the water. We had a fence around the house. The whole fence had floated away somewhere. It was a wood fence but it was embedded in concrete. To look at something that was embedded in concrete completely gone, where you can't much look down the block and see it. To have about an inch-and-a-half of mud throughout your downstairs. The smell was just unbelievable. Looking at every house in your neighborhood, doors bust open: It's like the whole city was dead, I'd say from Michoud all the way to Canal Street. I don't care if you lived in a poor neighborhood, a middle-class neighborhood, or a rich neighborhood, they all was devastated. We drove around and didn't see nobody.

The whole neighborhood right now is up for grabs. Whether it is going to be rebuilt or totally demolished, we don't know. It's a fact that they're trying to keep certain people out of New Orleans. When we went back to look at our house and assess the damages, there was no construction done. This was a month and a half after Katrina. There was still no electricity, no water, and the prices of apartments that we're looking for—we'd at least need a two-bedroom—a one bedroom is over eight hundred dollars everywhere in New Orleans. We're talking about apartment complexes that ain't worth five hundred dollars a month.

Conclusion

I would love to be part of the rebuilding process. I do home improvement—my neighborhood, with some hard work, could be saved. Hopefully more and more people will come back, and then New Orleans will get its smell and its energy back. But I don't know what our politicians are doing down there. I feel like they're not for the people. They're for themselves and their friends and their families. They were corrupt before Katrina.

I believe we're deliberately being kept away. The main example is that there's nowhere to stay for the people who want to go back and start to help rebuild the city. Every time I was able to get to the Internet, everywhere I looked for a two-bedroom, it's over one thousand three hundred dollars a month. Most black people who come out of New Orleans can't afford that. So people will get their little money from FEMA, and say, "I'm going to try to make my life with family here," even though we don't want to. You got to look at what you can afford.

I feel like I been robbed of everything that I love. Even if it was going to my corner store—I had my people there that I hung with. I miss being able to cook the way I want to cook. They say people from New Orleans cook from the heart. Like one time when we first got here, we drove around for five hours looking for the type of meat we need: pickled rib tips and DD smoked sausage. We went to Bessemer, which is thirty miles away. We went everywhere. I was like, "I'm tired. Forget it." So it's a culture shock.

Every day I go through my routine of working, but every day it come out of my mouth, "I want to go home." I miss seeing familiar faces and having friends call and say, "What you all doing? I'm coming over." I miss the everyday things: sitting in my back yard and just enjoying the weather, family, and friends. I miss seeing my wife putting neighborhood children on the bus in the morning. I miss seeing their faces when we barbecue and tell them to come get the hamburgers. We fed the whole neighborhood, but that was so enjoyable. Our whole neighborhood was our comfort zone. We don't have that same sense of closeness, where we at.

I'm a grown man and I'm cried out. I'm not stupid. I opened my eyes up and I seen what you was doing to me. I can ask my God to show me what's happening in front of me. I believe with all my heart the military was going to kill us. If we had been the animals that they were portraying us to be, then they would have opened fire on us. There ain't no other explanation to continuously pointing your gun at unarmed, innocent people trying to escape a flood that killed us. We were following orders.

I understand the media was portraying all the bad that was going on, like the young men shooting at the helicopters and people looting. I didn't condone that but I understood it. Before the news cameras start rolling, they weren't pulling anybody off their roofs that was begging for help. I mean the wind from the helicopter blades was literally knocking some people off, and they were drowning. That's when the shooting started, because it was like, "We going to take pictures of you all, but we not going to save you." I believe if there'd been a faster response to rescuing people, then a lot of that wouldn't have happened. I feel like my government should have and could have handled it better. I can't believe that they weren't equipped to do that given the way Americans always respond to other people's disasters that we have nothing to do with.

So many people just didn't deserve to be treated like that by our military service, the people that we trust. I've paid taxes for many years. My father did and my grandfather did. So many people you might look down on as being poor, but they was tax paying citizens. My father fought in the military. He did two tours in Vietnam, and he's disabled because of it. Everybody there wasn't no gun-toting criminals. In fact, the majority of the people there wasn't. And to be classified as something that I'm not, it's just wrong. It's like somebody decided that black people's lives in New Orleans weren't worth saving, our culture weren't worth preserving, that we just weren't worth the money, the time, and the effort. Who has the right to decide that? Katrina was an act of nature, I realize that, but it's like someone took advantage of that to get rid of what they might look at as a problem. And that was wrong because my parents always taught me you deal with everybody individually, because no two people are the same.

To me, somebody needs to be held accountable for their actions, because we go to war behind trying to save another whole country, and what's going to be done about the injustice that was done to our own people? I hope that the Lord will allow me to live to see some justice be done behind the slow response, us not getting out of there like we should have, and a lot of deaths. People died senseless. You know if a man walk up to a man on the streets and shoot him in his head, he should be dealt with. If you deliberately commit a crime against humanity, then you should be dealt with and I don't care who you is.

You might look at New Orleans people as being violent or whatever. Believe me, people from New Orleans would not be a problem to you, if you don't be a problem to them. We fun-loving people. We ain't going to take your stuff. We are Americans, and we human beings, and we didn't deserve for our government to neglect us and look over us and let us die like that.

NINETEEN

Jermol Stinson

Born in 1969, Jermol Stinson joined the Merchant Marines at the age of eighteen and traveled around the world. While home on leave four years later, Jermol was a victim of violence that left him paralyzed from the neck down. In New Orleans, friends and family provided round-the-clock care at home, even before there was money to pay anyone. One young St. Augustine student, Anthony, did his homework on the floor of Jermol's room each night. After several bouts with severe depression, he began working as an inspirational speaker at the Bridge City Center for Youth in Jefferson Parish. A practicing Muslim, Jermol was recovering from an infection at Kindred Hospital in Uptown immediately before Katrina. He was trapped in New Orleans during and after the storm, and during the storm's aftermath he wandered in his mind throughout the city, as he actively worried about his family and friends. Once he was stabilized in Dallas, he entered his first nursing home.

On January 15, 2006, this interview was conducted during a lull in the stream of visitors in and out of Jermol's Dallas hospital room.[1] He was in the hospital for surgery required to address the bedsores he acquired in the nursing home. A white towel wrapped around his head like a turban, Jermol's eyes revealed much emotion as he talked. His frequent smile revealed three or four gold-capped teeth.

Jermol's story at times is a love song to New Orleans. His love for the city is sensual and represents the intermingling of deep family roots and a rich sense of history of place. Hurricane Katrina and the subsequent flood seriously damaged or disrupted the sources of Jermol's independence, something most disabled people value highly and struggle to secure. He lost his expensive, adaptive equipment and an intimate familiarity with his city of residence, but more importantly, the storm scattered his extended network of friends and family, who enabled both Jermol and his mother, Cynthia Banks, to live full lives. The constant stream of visitors in and out of his hospital room is an indication of his charisma, essential to restoring his stability so that he can draft a new roadmap for his life.

I was born in 1969. My biological father was never really in the picture. Even to this day, if I call him, we just have generic conversations. The only dad that I really knew was my brothers' and my sister's dad. I can still remember the way he walked with his hush puppies on. We were always made to know how important we were and how important it was that we be men.

I grew up in New Orleans East in the area called the Gap, where my grandmother lives. When I was very young, we would come down every summer from New York and spend time with my grandmother. My grandmother's house was full of love—full of hugs, kisses, kids, grandkids, and a lot of aunts. Good food all the time. It was a beautiful childhood.

Jermol's maternal grandparents built a house in the Plum Orchard subdivision near St. Mary's school. It was one of the first communities in New Orleans East. They moved from the Desire housing project, where Jermol's mother was raised, to Plum Orchard in the late 1960s.

We moved to New Orleans from Long Island, New York when I was thirteen, just making fourteen. I consider myself to be from New Orleans because I love New Orleans so much and it was where my whole family was. When we moved to New Orleans, it was a brand new breath of fresh air for me. My first girlfriend, first going to middle school, first going to high school was in New Orleans. In the mid-80's, we moved to Lake Carmel, a middle-class subdivision in New Orleans East built around a small lake. There were always a lot of kids from the neighborhood in my house. We're still friends until this day. We played football and raced from one streetlight to the next streetlight. It was part of growing up in New Orleans with the gold teeth and the whole nine. First and Claiborne. Everyone go get their gold teeth there.

I went to Abramson High School, that's also in New Orleans East. Later I changed over to Joseph S. Clark High School, a prominent New Orleans public high school serving the Sixth and Seventh Wards. That was a new experience because it was in the hustle and bustle of things. I made it known that I was there.

My grandfather Stinson lived in Uptown off of Louisiana and Freret Street. The uptown area was so different from the East. He repaired refrigerators and ran a bar and lounge on Washington Avenue. I would catch the bus up there and spend the day with him, learn about him and his life and what it was like. He accepted me, showed me things, and taught me how to cook my first pot of neckbones. I just knew my son and I would have been thick as thieves.

I worked at Jax's Brewery when it was Jax's Brewery. I was like seventeen. I wanted some pocket change, to be able to do some nice things when Christmastime came around.

I traveled around the world and worked on a cargo ship. We delivered goods, picked goods up. I was also involved with Desert Storm, bringing soldiers and equipment over for Desert Storm, and waving goodbye to them.

"I Wouldn't Want to Be a Quadriplegic Anyplace Else"

I was injured in 1992. I was twenty-two. Even during the accident, I always felt safe in New Orleans. As far as I'm concerned, I look at it like the enemy was assigned to me that day to take my life, but he didn't have God's permission.

At first, all I thought about was myself, what I lost, what I wouldn't be able to do, how alone I was, and how I wouldn't be able to be a lover to my girlfriend, or a father to my son, and a big brother to my family. I wanted to die every minute that I was awake, every day. I couldn't breathe. I had to be suctioned, it seemed like every fifteen minutes, and that was a pain I will never forget. I couldn't be left alone. I had giant bed sores. I felt like my life was over.

My mother is a soldier and so is her sister Pat. They have everything except the combat fatigues. In the beginning, it was so hard, because my mother fought tooth-and-nail for Medicaid/Medicare. And I watched my mother suffer and felt that she didn't deserve what life was giving her as far as her son was concerned. She wouldn't let me go in the nursing home. I wanted to go in the nursing home, because I knew in the nursing home I wouldn't survive very long. But I would wake up and my mother would be sitting at my bedside. Whether I was at the hospital or at home, my mother fed me. She has stuck by me for thirteen years.

After I began to realize that the enemy took two arms and two legs away from me, but God blessed me with hundreds more arms and legs, I realized that I wanted to live. After I started wanting to live, I started to see the sun through the cracks of my windows. I would be like, "Open up my curtains!" No matter how my day went, I would always look for one good thing out of a day, and think about that thing at the end of the day.

In the beginning, I had a male aide that would come and take care of me seven days a week. He would suction me, give me a bed bath, shave me, and change the dressings on my wounds: I had two giant wounds on my hips, a wound on my sacral, and both of my Achilles tendons were exposed. It would take him like two to three hours of his own time.

My next-door neighbor, when he first moved there, was a respiratory therapist, and he wound up becoming a neurologist. I asked him to be my doctor. I was heavily addicted to Xanax, a sedative and a muscle relaxant. I asked him to help me get off the drugs, so I wouldn't be dependent. I could open up my window and give my doctor a report on how I was doing.

I have a friend by the name of Calvin Fletcher, and my brother, Dwyan, when this first happened to me, they went to C & A School to become certified nursing assistants, so they could take care of me. Calvin, I've been friends with him for almost twenty years. He brought me more awareness of what it means to be a Muslim. My brother, of course, is my number one. Guys that grew up in that neighborhood, I mean they were younger when I was growing up, but they would come and get me up, and take me out. You know, "What's going on?"

As a matter of fact, one of my best friends, Anthony, I met him when he was fifteen, he was going to St. Aug. I was maybe twenty-five. I was just coming out of my depression. Instead of this kid being out on the street playing and doing the things that a fifteen-year-old should do, chasing girls or whatever, he was in my room suctioning me, coming home from school, and doing his homework on the floor. "Are you alright J?" He grew up to be the kind of man that I expected that I would be. When I tell you that these guys are family? They're my brothers.

Back home I could go anywhere I wanted. We would go out to the movies, to plays, to dinner, and to parks. They would have this thing in the park every other Sunday during the summer, and we would go out and enjoy the cookouts. I mean you had to step over the people.

We hardly ever locked our front door. And we're talking about New Orleans. We're talking about the place where crime was outrageous. I mean the front door was left open. If someone wasn't there, someone else was on the way. The people in my neighborhood, I mean north, east, west, south. My neck was always hurting from waving with my neck. Somebody was riding past to talk to me, "How you doing?" I love the people of New Orleans. I miss my family. When I say my family I mean everybody. I wouldn't want to be a quadriplegic anywhere else, not even in New York, but in New Orleans, because the people there made themselves so available to me.

I would talk to young offenders at Bridge City in Jeff Parish. There would be a group of young kids that would come into the room, and I'd already be waiting for them. They'd be doing their thing, and being as thuggish as they possibly could, so that they could be hard. By the time I finished telling them my story, they would be crying like little girls. And I would be crying because I would explain to them in graphic detail about what had to be done just to take care of me, and how much my mother had given up, just because of someone else's recklessness, because they couldn't control themselves for whatever reason. It was real therapeutic for me.

"New Orleans Is My Girlfriend for Life"

I mean there wasn't an area of New Orleans that I didn't like. I loved the music and the nightlife. I loved the second line music. I loved the originality of just the heritage of jazz.

When I was away traveling the world, I missed everything about New Orleans. I missed the streets, the ditches, and the raggedy streets in the Ninth Ward. I missed the houses, the architecture of the houses, the shotgun houses. I missed the smell of New Orleans. I missed the St. Augustine grass and the stop light at A.P. Tureaud where St. Aug was. What wasn't there to miss about New Orleans? The food, the people, the music: New Orleans had everything.

As a young man, one thing that really struck me when I walked down streets like Esplanade Avenue with the brick streets and the brick sidewalks: I'd say to myself that somewhere in my ancestors' history, I'm taking the exact same footsteps that they took. I took great pride in the fact that I had a large family and that I was from New Orleans. When I went to Clark, as raggedy as the school was, it was family. And as worn down as the wooden steps were, I knew that someone in my family had put their foot in the exact same place as mine. That filled me with so much pride.

I feel a great separation from Africa, not that I know anyone in Africa that's my family. I think about how cut off we were from Africa: how our culture, our language, our God was taken away from us. New Orleans is like, "Well, this is where my family comes from."

New Orleans was a very, very rich and wealthy place, and when I say wealthy, I don't mean money. I mean family. Where else can you walk down the street and get red beans on Mardi Gras day, on Fat Tuesday. "How your mama and them doing?" You don't even have to know their mama. Not everyone was blood family, but we considered ourselves family. I mean like Charbonnet. This guy owns a funeral home. But he was family, because when it came time to bury our family, we knew that Charbonnet would take care of our family.

Charity Hospital is where I went when I was shot, and that's part of my family too. I went back to Charity after like eight years, and I went back to Intensive Care. And they said: "Jermol Stinson!" I could not believe that they remembered me. It really makes me want to cry, because unless you've experienced New Orleans, I can never explain to you what we really lost.

It's like anywhere you drop me off in New Orleans, I knew my way out. I even had an understanding with the mosquitoes. I love the ditches in New Orleans, the canals, and the smell early in the morning. In the summer time, I knew it was going to be hot about 9:30, so I knew I had to get to wherever I had to be. I really loved walking through the French Quarter at night, the things that you just can't make up. I love the water in the air, the humidity. Even in the nighttime, it was so heavy. It was home to me. Home!

I tell people from Texas, don't get mad, but I think that people from New Orleans are very, very good people. I think we're special. We say, "How you doing baby?" We really mean it. You walk down the street and see someone sweeping their porch off. You know the uptown area as beautiful as it was, and as ugly as it was, it was still home.

Both of my brothers went to St. Augustine. My brother Dwyan is next to me. I don't care where we are, everybody know Big Banks. And the police officers, they'll see us and they'll pull us over in my big, wheelchair-accessible, red van. "What's going on?" They'll say, "You didn't turn your blinker on." Then it'll turn out to be a St. Aug alumni. That's another part of New Orleans, man. Always, everywhere you go, you got to see someone that you know.

Katrina's Aftermath

My doc transferred me from Methodist Hospital in New Orleans East where I was undergoing tests to Kindred Hospital in Uptown a few weeks before the storm. I did not know how extensive the damage was, because where we were, it was dry. There was just a lot of down trees. My room was the only room that had a television and an air conditioner that was working. My mother, my aunt, and my little cousin were in that room, and I was friends with everyone.

When the food started running out, it got a little tight, and everyone was uptight, worried about their family. We were able to watch the news on the television. Everyone shared with everyone. I couldn't have been in a better place. Kindred did everything it could to make sure that my family, myself, and other patients were comfortable and taken care of. They didn't abandon us when they had every excuse to. I was actually one of the last patients to leave. And someone was there from Kindred until the time we left.

The National Guard was very concerned about me. They were very concerned about all of their patients. It was real tense because they had their weapons out. And it was a serious situation. But they took time to come, sit down, talk with me, make sure that I was ok, and that I was comfortable. I was treated real, real well by the National Guard.

I was told I was being evacuated to Houston. I ended up in Pasadena, Texas. When I got there, they were ready. The first half hour that I got there, I was bathed by Felicia, a nurse who worked past the end of her shift. I really needed it. I was fed and put in a clean bed. They gave me a cell phone to call my mom, and I had never really cried for years, but when I was on the phone with my mom, I'm not going to front, I cried like a girl because I knew that she was safe.

After Pasadena, we were evacuated to a Kindred in Dallas, and once again I needed a bath when I got there. The young lady who came in with her husband this evening, Teresa, was the person that gave me a bath. I slept for two days. She was the last person

I saw before I went to sleep and the first person when I woke up. She's my first friend that I made here in Dallas.

First Nursing Home

I'd never been in a nursing home before Katrina. The nursing home was every bad experience that I had ever heard about. It wasn't clean at all. One of the first times I walked down the long hall—the levels of piss that I smelled—I was like, "I can't believe this." It didn't make me cry. I think at times they tried to. There were like forty people to one nursing assistant. I didn't have a call light. I had to depend on my roommate, a blind guy, to pull the call light.

They didn't have an adequate bed for me. I got bedsores as a result of not being turned enough because of the bed. The bed that I'm on now in Kindred Hospital is pretty much like the bed I had at home, so I really didn't have to be turned as much at home. When I was at home, I was up and out more. At the nursing home, I was surrounded by people that had Alzheimer's, so that really didn't give me a reason to want to come out of my room. The food was ridiculous. My mother brought me something to eat every day. But in every bad instance, I've always been able to find a friend. I found two or three people there that took good care of me.

When I came back to Kindred for surgery because of the bedsores, my doctor let me know that I was depressed because of the nursing home. When I was in the nursing home, there were no phones in the rooms. You had to take what they gave you and be satisfied. I shucked and grinned a lot. But what I really missed was access to my family and my friends.

Conclusion

I think about all that I lost, all of the friends. When I think about the strangers that I saw on a regular basis, you know where I would just wave to them, or give them a chin check "hi." I'll never see them again. I don't even know what happened to them.

I made real families at Methodist Hospital in New Orleans, where I heard there was so much disaster. I knew *all* the nurses. I knew the cafeteria workers, all the doctors, the CEOs, and the ancillaries. I knew everyone right down to the janitors and the gardeners. They'd even come and speak with me. I really, really miss New Orleans! There is no other place like New Orleans. My community, my neighborhood, and my city is gone now.

I lost my computer in New Orleans. The state troopers had a one-time deal where they allotted me and a lot of other quadriplegics a grant where we were able to get equipment, and be able to get our equipment fixed. I lost my Dragon Dictate, where it allowed me access to the Internet, to turn on and off a television, my radio. It was all voice activated. It answered my phone and raised my head on my bed. It enabled me to turn the lights off and on, and even to answer my front door, even though it was always unlocked. I could be left alone. I was connected to the rest of the world with the Internet. I really miss that part of my independence.

If I had to explain to the president what displaced disabled New Orleanians need, what I would say is this. I think they need as much assistance as possible, because, if they were at home, they're probably in nursing homes now. And I promise you, it's hell!

One thing that I felt about New Orleans, and I was going to make a big issue about it the next time someone wanted to be elected mayor, or wanted to be elected state representative, was start your campaigns from these nursing homes. Not when they have everything clean, when everybody's hustling to make sure everything is just right. Go into these nursing homes and expose these places. People are suffering in these nursing homes. And you don't really have a voice if you're in a nursing home.

We need as much help as we can. If you're disabled on any level, you already need an extraordinary amount of help, and it costs an extraordinary amount of money. I think it's triple the price than if you're an able-bodied person. If I had to pay for the friendship that I had just to get up and go out in the backyard, I wouldn't have been able to afford it.

One day not long after I arrived in Dallas, I called a local mosque, and explained to them that I had lost everything. They understood that included the Koran. Before the day was over, the Imam made a delivery with his son. For the record, Allah is the best Knower of all things. There is no God but Allah. Jesus is the Messiah.

It will be a while before I'll go back to New Orleans. My greatest heart's desire is if I go back, I would like to go back with a degree in something dealing with the body, so that I'll always be in the loop knowing what's available for people not very much unlike myself.

I look forward to new relationships. I feel that where I am right now is where God wants me to be. I expect to discover a lot about Dallas. I guess the only thing I'm really afraid of is disappointing God at this point. But I miss my city.

I know going home to the new place my mom has fixed up will make a difference because there will be some normalcy. There'll be a routine between home health, my girlfriend, and my new friends.

I know for certain that my mom worries about what's going to go on with me. Should anything happen to her, what would happen with me, because my family is so spread out. Whereas, if we were in New Orleans, all of those answers were self-evident.

I know that she's struggling, trying to find work now, whereas she had her own business, and she was pretty much on the road to being self sufficient and independent. She is sixty-two, which is still young as far as I'm concerned. I see my mother aging very gracefully, but I see the worry in her eyes. I see her exhausted, and it saddens me once again because she doesn't deserve it. Like most children, they feel about their mothers that they deserve so much more. And I wish that I was able to give her more, to do more.

My faith and my mother kept me strong through this ordeal. My mother, my mother, my mother! She was my first teacher, my first nurse, my first spiritual guide to God. The first opportunity that I have, I will get "mama's boy" tattooed on my heart.

Demetrius White*

Forty-five-year-old Demetrius White had childhood homes all over the United States before he stopped moving with his mother, Martha, and settled with his grandmother Gussie in New Orleans at the age of thirteen. He claimed his birthplace, New Orleans, as home because of his enormous web of relatives who resided in Uptown before Katrina. When he was married and living in Chicago, he flew to New Orleans every other weekend. His job as a technician allowed him to pursue his passions, including music and fatherhood.

Demetrius survived the storm in his two-level house in Mid-City. Afterward, he spent six days rescuing people, including a night he slept on a stranger's car in Carrollton, despite his Uncle Joseph's irritated and unequivocal order to leave the city.

On January 12, 2006, Demetrius drove to Houston from Austin, where he had settled after the storm.[1] He wore his long hair in a loosely braided pigtail. Demetrius is a realist. Dressed in Levis and a 2006 Rose Bowl t-shirt, he proceeded to narrate a detailed, sometimes self-critical, often amusing, eyewitness account of Katrina.

By virtue of his almost daily five-mile ventures between Mid-City and Carrollton, Demetrius was well positioned to report on official rescue efforts by a variety of groups: the New Orleans Police Department, the Louisiana National Guard, Texas Wildlife Officers, and Louisiana State Troopers. After his last rescue, he was taken to a staging area in a dump truck.[2] Demetrius and several thousand other blacks had a front-row seat along I-10 near Causeway as whites were airlifted out across the interstate divide.

My mother is a gypsy, not Hungarian, but she moves a lot. She worked for the Federal Reserve Bank for most of my life. For each year I went to school, I probably went to a different school, even though I never spent more than a year at any school. My favorite place growing up was New Orleans, especially Uptown, the Carrollton area. After my mom's move from New York to Furman, Alabama, I came back to New Orleans on my own. It was home. All my family was there on my father's side. I have a lot of cousins in New Orleans. I was thirteen when I came back. I was trying to get back to my oldest brother, Larry, and my father, John. My father was an EEOC officer for the T.L. James Company for close to thirty years. I lived with my grandmother from whom I learned most of my life's lessons on Adams Street in Carrollton.

When I finished high school, I went into the Army. I was stationed at Fort Hood in Killeen, Texas, but I traveled extensively: Japan, Korea, Germany, and a couple of places in the United States. I saw active duty but no combat.

I started working when I was fifteen. I was a dishwasher for Loyola University. Over the past eight years, I've been a telephone cable technician and a musician. I played on tour with brass bands. I played with a couple of world renowned musicians like Freddie Hubbard, for one. I like music best, but I have a son, Cassius, and I had to do a lot of road traveling, which took time away from him, so I made a choice to stop touring to spend more time with my son.

Back in the '70s, the average Joe coming out of high school, even if he didn't finish, got a job with one of the oil companies or something. You were going to make a nice piece of money, but when the oil bust came, a lot of people lost their jobs. The only people that were left in New Orleans were true and diehard New Orleanians and professionals.

I got married and I moved to Chicago for six years in the '80s. I made more money in Chicago than I ever made in my life. But I just wasn't happy there. I'd wake up, go to work, come home, go to sleep, and dream about New Orleans. I was fortunate that my ex-wife worked for the airlines, so I was home twice a month at least. I'd hop a flight and stay the weekend.

When I moved back to New Orleans from Chicago, I bought a house in the Mid-City area one block off of Canal Street, on the corner of Iberville Street and North Olympia Street, which is beside the former P.G.T. Beauregard Junior High School. (Now it's Thurgood Marshall.) My house is actually a structure that was built in '28 or '29. It's two levels. I have an above-ground basement because there's a garage door where you can drive your car underneath the house. There's an area down there that I cordoned off, where I had a TV and sofa for when guys come over, watch the game, and cut up. Upstairs I had three bedrooms and a bath and a half. It was good enough for me. I'm a bachelor. A mattress on the floor would be fine for me.

Katrina

I was in Uptown, New Orleans for Betsy, and that was no big deal. I was five years old, and remember it well. Went outside and played in the water afterward. My brother taught me how to make a little boat from pieces of wood. If I'd been in the Ninth Ward, I might have been dead. I was five years old, so I knew what was happening in my own little six feet of space on this earth—that's the only thing I can speak of. I've heard stories of how it happened, why it happened. Given the time, things that were going on, and some of the stories I'm hearing now about what happened with this one, sure I believe it.

Day One (August 28). I heard that Hurricane Katrina was coming a week, two weeks before. I always take them seriously. My grandmother never left for hurricanes, not when I was around. So I never leave.

People say, "You didn't take it seriously, if you didn't leave." A hurricane was no big deal to me. So you get some wind, you get a little damage, you might get some shingles blown off. You may even get a little water. But it's never anything life-threatening.

I'm not going to go outside and play around in it. I'm going to batten down the hatches, and just ride it out.

Day Two (August 29). During Katrina, I heard things hitting the house, glass breaking. It wasn't anything like a California earthquake, so it was nothing frightening. It was nothing unusual as far as a hurricane is concerned.

The electricity went out at 3:00 in the morning. I had a telephone until that time also, because I actually called Allstate during the storm. The lady said, "How do you know you're going to incur any damage?" I said, "Ma'am, you want me to put the phone out the window and let you hear it?" She actually took the claim during the storm, and I got the claim number. I figured I might as well be one of the first ones in line.

After the storm, I went out to assess the damage. The shingles were taken off the top of the house; it's down to the tar paper. I noticed where the foundation is cracked in two spots, where it wasn't cracked like that before. I got back to the house and started doing things around the house, trying to straighten up stuff. I noticed that the water was just to the bottom of the molding on the SUV, so that had to be at least knee high. The water came in very quick. I had some very expensive things downstairs that I would have loved to have gotten out, because I knew that I had minimal flood insurance on my contents. My house actually got maybe five-and-a-half feet of water. I didn't get any flood waters upstairs.

I started worrying about my own safety six days after the flood. I'm not the kind that panics very easily. It's just my personality. I was trained at home to respect life more than I was to take life. At that time at my house, there was my brother, his wife, and his four children.

Mary is my neighbor. (She's white.) Her mother and father in their nineties lived half a block away right on Canal Street. She had no contact with them, so she was worried. So I waded around there to check on them. Waded back. Let her know that everything was fine.

I spent six days walking in that water. Most of the time I was wading, it was chest deep. It got to my neck, and I had to start swimming at Canal and Pierce, which is one block from my house going toward the downtown area. (I've been swimming all my life.)

My brother and I spent that day trying to gather food and information on what was going on, how to attack the situation. You would see the Coast Guard or a helicopter go over every now and then. We spent the night flashing flashlights, yelling for help. By my house being a second-level house, I'd be hanging out of the window, waving white t-shirts. They would never stop. I might have slept thirty to forty-five minutes that whole night.

Day Three (August 30). That's when my brother really started to panic. I said, "I'll be back. You all just sit tight here." I walked from my house all the way to Zimple and Hillary (Uptown), which is probably four-to-five miles. It took me three hours to walk and swim that far.

I had some older friends and relatives that live Uptown. I was worried about them, because I knew they didn't leave. I got there and I told them, "You all got to go." One of them was a diabetic. They fought me tooth and nail. I told them, "I'll be back tomorrow." About two blocks before their house is when the water went from hip deep to maybe ankle deep, so you still could drive a car. I had a radio at the house, so I knew that you could head out that way.

Day Four (August 31). I started walking back uptown to Zimple. I got to Carrollton and Walmsley Avenue near the Notre Dame Seminary, and I heard somebody yelling my name. There's a school there called Lafayette Elementary School, where people had broken into seeking refuge. I looked up and one of my cousins from that area was in the window yelling my name. So I waded over and went into the school. At that time there might have been thirty people there. Half of them were relatives. Everybody was scared. You had people constantly showing up in boats, walking or floating on little rafts and blown up air mattresses. I had one particular cousin, Helen, in her late seventies, whose house I went to before the storm and I begged her to leave. But when that water came up, she made her way to that school, which was about seven blocks from her house. I said, "I know there's a way out," because when I left that morning, they started bringing air-boats to I-10 and City Park, because you could hear them running up there.

They said, "They were picking up people on Carrollton and Claiborne earlier that day but they stopped." They added, "They left us. Could you take us?" I said, "I got to go uptown and check on some things. I'll be back." Sherry Ann and Cookie just broke out, like when you take a child to a babysitter for the first time. Sherry said, "Please don't leave me." And I assured her that I would be back the next morning by sunlight.

I got back up to Jackie's house, and she said, "Could you talk to Harold because he's refusing to leave?" I had to go through the whole discussion with him again, and his eyes started to water. He said, "Son, I just can't walk off and leave everything I worked for. I'm eighty years old. I'm not leaving all my stuff." "Unc," I said, "either you leave your stuff or you going to die. Uncle Frank is here and he needs insulin. If you stay here, you're killing him." I fussed with him for almost two hours. He finally agreed to leave that next day. I told my aunt Jackie, "If I don't come back, all you got to do is get to Tchoupitoulas Street. Go straight down. You can get all the way to the Mississippi River Bridge."

By this time, it's too late for me to make it back home now. I wound up spending the night on top of a car on the street. That was the only dry place I could get to. I had my two cell phones in ziplock bags. About 2:00 in the morning, the mosquitoes started biting me and I woke up. I took my phone out, turned it on, and dialed my Uncle Joe in St. Louis. I knew that he would be able to get in touch with people. I said, "Uncle Joe, this is Demetrius." He said, "Where are you?" I said, "I'm still in New Orleans." And he went off on me. "I want you out of there right now!" I said, "Just get in touch with everybody and let them know that I'm still ok. I'm trying to help Helen, Gail, and Jackie." He started going off again, "You can't save everybody."

Day Five (September 1). I went back in and I talked with about one hundred people in the school cafeteria. "I just wanted to let you all know there is a way out." I saw a National Guard truck drive past that school. They had to know that people were inside, because people were hanging out of the windows. But the National Guard was afraid also. I later found out that the guard came by and dropped some insulin from a heli-copter to the roof of the school.

That morning my cousin Sam showed up at the school. My cousin Sam had a little aluminum bass boat, and he had two to three people in it. He said, "I got everybody that I needed to get." I said, "Could I get that boat from you? I'm tired, I can't walk anymore." I took the boat and paddled with a shovel all the way back to my house.

I went through some pretty rough neighborhoods I was raised in. I went through a section of town called Hollygrove, across Carrollton Avenue opposite of Gert Town. I was paddling down Monroe Street. I got to Monroe and Olive Street, and I saw some guys pushing those plastic storage tubs. They were just stacked up with food and floating. And one guy said, "You not going to give nobody a lift in that boat?" I said, "No man, I got to go somewhere." They looked rough. The area of Hollygrove was called Ghost Town and it is known for being very rough. They were in their forties. As they got closer to me, I recognized one of them, a guy I had gone to school with. I said, "That you Terry?" He said, "Yeah, who is that?" I said, "This is Demetrius White." He said, "What's going on man?" I said, "I'm just trying to go try to help my brother get out of his house." He said, "Let him alone." I said, "You all be safe."

It awakened me to the fact that I didn't need to be in that boat. I paddled to the interstate. The interstate only had about six inches of water, so I got out of the boat, drug the boat across the interstate, lifted it up over the center divide, drug it over the other section, and put it back. There was a hole in the fence, so I was able to shove it in through the hole in the fence. I got back in and started paddling toward the house.

When I returned home, I said to my brother, "I saw airboats when I left yesterday. I'm going to take you guys up to the interstate where the airboats are and get you out of here first thing in the morning. You all are cramping my style." I'm still listening to the radio. We dissed the idea of the Superdome because we heard reports on the radio of things going crazy.

Days Six and Seven (September 2 and 3). That next morning, I told him, his kids, his wife, and my friend to get some stuff together. I took them over to the interstate, and they got on the airboats. I put my neighbor, Mary, in the boat, and I paddled her around the corner to see how her parents were doing. One of her sisters had spent the night with them on the eve of the storm, so they did have somebody there with them. She was glad to see them. I told her to keep the boat.

I went back to the house, but I was trying to devise a plan, set some kind of guidelines on what I was going to do next because I'd made these promises. I like people to say, "He said he's going to do this. I'm going to look for him to do that." I'm still worried about the people that I told I was coming back to the school. I wanted to get to the school, get them out, and then get myself out before sickness started setting in, because now the water's starting to smell, and the last time I went toward the school, I started finding bodies.

There's a scene in the movie "New Jack City" where they've taken over an apartment complex, and it's a big crack house, and it's just horrendous in there. This is how that school was. The stench was unbearable. Can you imagine a hundred, two hundred people, even animals, urinating and defecating in waste paper baskets?

Myself and my cousin rounded up eighteen people from the school, three of them were seventy-nine and above. "Could you please take us?" I said, "If you want to go, you're welcome to go. I'm not going to leave you." They said, "I don't know if I can make it." I said, "You're going to make it because I'm not going to let you kill me, and I'm not going to leave you. Don't worry. If we got to stop every twenty feet and let you rest, we're going to make it." All of them made it three miles in the water.

There was only one time that I was afraid, and this was the time. The reason I was afraid was because I had all of these people following me. All I needed was for those boats not to be coming there. I was busy listening for the airboats: you can hear airboats from a long way off. That day I never heard airboats. They'd stopped. As we got closer and closer, I walked faster than everybody else. When I got there, they had little small boats. The water had gotten too high for the airboats to come underneath the interstate, so they had little small bass boats with gas motors that were able to come underneath the interstate, pick the people up right there by Seminola's, and then take you all the way down the side of the golf course about a mile down Metairie Road. Then they'd put you in the back of a dump truck and take you underneath the interstate at Causeway and I-10. They put us out there.

The gentlemen said they were from Texas. They were supposed to be like Game Wardens, but the guys actually had badges. They treated us very well. If I ever had anything bad to say about people from Texas, I take it back. Then those guys from Texas and myself went back. We actually got another fifteen people out of houses and off of rooftops.

One thing I've actually had a problem with happened after the guys from Texas, who weren't supposed to be in the city after dark, left me. I was right on Bernadotte Street next to the interstate. I started paddling back toward my house. There was a guy yelling on the second floor. "Help! Help!" So I said, "You got food and water?" (I had gotten a case of water, so people that I couldn't do anything for, I'd give them water. I'd commandeered some bread and crackers and Vienna sausages, so I would throw people something to keep them going.) I said, "What's wrong?" He said, "It's my mother. I need to get her out of here." The lady was every bit of 450 pounds. I knew I couldn't get that lady out of there, and she was a diabetic, who had been out of insulin a day. She just sat there very calm. But I told the guy, "I can't put your mom in that little boat. That boat will turn over and she's going to drown." I said, "If I see a guard truck or some policemen, I'll let them know." It was hard for me to leave knowing that chances are I just watched that lady die, and didn't watch her die at the same time. I always wonder.

Through telepathic messages my uncle kept telling me, "Get out!" I knew I wasn't going to die, so I was definitely going to see him again. And I knew he was going to kill me.

I walked in the direction of where they were getting the people out down Metairie Road. I got on the Golf Course, jumped the fence where the water got shallower, and got on a dump truck. They took me underneath the interstate. I just wanted to get out of the water.

5,000 People and 9 Portable Toilets

Being underneath the interstate was horrendous. Some things happened that I don't never want to see again. For those people to say that they moved in a most expedient way, I didn't see it. I heard a lot of empty promises. I was there two days. The trash was ankle deep; the little kids were playing in garbage. You had five thousand people with nine port-a-lets. That kind of math don't work. They were mostly black with a small percent Asians. I saw one white lady, and she had a kid with her.

They had one old lady over eighty years old. This lady stood up leaning on her cane for over eight hours. I couldn't do it, and I'm only forty-five. They had nowhere for her to sit, and I guess she figured if she sat on the ground, she wasn't going to get back up. I asked her, "You don't want to sit down?" She said, "No, I'll be alright."

It was like a concentration camp. The Louisiana state troopers that were there weren't very friendly. There was plenty. National Guardsmen would come and deliver the MREs and water. The troopers really didn't walk around with openly displayed weapons, but the guardsmen had M-16s, 45s or whatever they carry now. They weren't walking around pointing them at people, but they had them. I understand it, because anything can happen out there. They distanced themselves from the situation.

I know a family that took off walking one morning. The guy had been talking about it all night. I think his father had died, and they were trying to make it to the funeral. If they stayed there, they weren't going to be able to make it. They sort of melted away, because they weren't letting you leave. I found out later they made it all the way to La Place. In La Place, they made a phone call and somebody came from Mississippi and picked them up.

I can attest to the fact that when they brought in groups of whites, they were gone. They actually never came to the same area where we were. We were on the river side of the interstate underneath the Causeway. The helicopters were landing on the lake side of the interstate. The National Guard trucks, they call them duce-and-a-half's and five tons. They have seats in the back. They ride personnel back there. You saw truckloads of whites come in, and they'd be helicoptered out. They were dressed like everybody else. They'd gone through things like us. People saw it and were saying things about it.

My cousin Helen, the older one in her late seventies, started hallucinating. I had taken her over to the little setup they had. It wasn't really a tent. He looked at her. "She's a little dehydrated. She needs to get some water in her, but she needs to get out of here."

So a couple of buses came. Families got separated because there was no order in how you were getting out. You've got this long stretch of a service road. Three buses would come: People would crowd the buses. So everybody would gravitate to this area. Three, four, five, six hours later, three more buses would come, but they'd go down there. Finally they had a bus, and they said they were taking the people that needed medical attention, but they could only bring one person with them. Helen asked me to go with her, and that's how I finally got out.

Overall Reflections

I can count the policemen on no fingers almost that I saw the whole time. The first day of the flood I saw one. That's the one that told me, "I don't know what's going on, I don't have any communications. What you heard?" He was on foot. As a matter of fact, he was coming out of the Lakeview area. I saw four policemen on a boat. My brother and I tried to wave them down, and they sped up heading down Canal Street. Pointed guns at us and kept on going. They were afraid that we were probably going to try to take the boat from them or something like that. I'm from the real world. I saw one more policeman on Carrollton Avenue. He asked me what I knew when I was coming from the school one day. I saw a National Guard truck going down Carrollton Avenue one time. It was just driving down the street. I didn't see it pick anybody up.

There were a lot of bad things that went on. But they sensationalize the bad things like people shooting at the helicopters. I saw on the televised Congressional Hearing they were saying, "We don't know about it." I heard it and I saw it. *Why* were they shooting at the helicopters? I didn't know a helicopter came down until I was underneath the interstate, already out. I never saw them stop for anybody. I spent six days shining flashlights until the batteries were gone. I actually had them slow down, hover for a second, look, and take off. It's the old saying: "Desperate times call for desperate measures." They didn't know anything else to do. They were shooting to try to get somebody's attention. They were trying to say, "Hey, help!"

Texas: "I've Never Been Treated This Well"

They brought me here to Houston first, but there was no room at the inn. So they took us to Greenville. Since I've been here in Texas, these people have done nothing but be the sweetest in the world to me. Anywhere I've ever been, and I've been around this world three times, I've never been treated this well. I'm the type of person that realizes you don't owe me anything.

They fed us. They took all my clothes because they said there were biohazards because I'd been in the water so long. They brought us clothes. I mean there was a gentleman. His name is Jerry Morris. I wear a real common size shoe for somebody my size. There were no shoes for me. This man went home and gave me a pair of his shoes. He took myself, a cousin, and another gentleman out to the store. Bought us whatever we needed—personal items, toothpaste, and a toothbrush. I left New Orleans with one of those little plastic squeezable lunch pails. I had a pair of underwear, a pair of socks, a t-shirt, three cans of Vienna sausages, and my wallet. I had about four hundred dollars in my pocket. The man said, "Your money doesn't spend here."

I called my aunt in Austin, and she found out I was there. She sent my cousin within hours to come get me.

Damage Assessment

I used to see people on TV when they get their house burned down or whatever, and they're crying. I used to say, "Just go buy another house!" I'll never say that again. I went to my house in October, the first time I was able to go back. I went into my basement. I'm going to tell you tears flowed. Fifteen years worth of house notes, and it's just trash. My brother was on the phone. The same word just came to me over and over again. "Damn, damn, damn!" My brother's a minister, and I still want to apologize to him. But for five minutes, I slung things around because I was so angry. I couldn't believe that somebody just took it from me all in one swat. I didn't have time to feel that the six days I was rescuing people.

A lot of people who need FEMA money and deserve it aren't getting it. I think I deserve it. I worked. I got the two thousand dollars and the twenty-three hundred dollars but I still have to pay a house note. But where God has blessed me, a place to stay is not my most serious worry. My most serious worry is trying to keep some money flowing so I can pay my house note.

I was able to save two out of three vehicles. I lost a van that I was doing my cable work out of. It got flooded. I got it once, I'll get it again. It's just going to take me a little time.

Allstate is trying to give me five thousand dollars. And I know I've got about eighty thousand dollars worth of damage on my house. If I sue my insurance company, do you know what an appeal would take me? I'm going to lose the house in the four or five years it takes to get your money from the insurance company. Contents, I only had three thousand dollars coverage. I'll take that loss. That's my stupidity. But pay me for what you owe me. My foundation should be covered. I want my house fixed.

Conclusion

I want to go back to New Orleans, but then again I don't want to go back. I guess this is the second time where I'm afraid with this whole situation. I'm afraid that New Orleans is not going to be New Orleans. What a lot of people don't realize is that New Orleans is not a place. It's the people. If all these people are gone, New Orleans is gone. So do I really want to be there?

It is an experience I don't ever want to experience again. People ask, "Will you leave for a hurricane now?" "Yes!" It's an inconvenience you don't even want to have to experience. I don't regret the risks I took rescuing people, because I know had I left those people, it would have haunted me the rest of my life. Just like the lady I'm telling you about. When I went back to New Orleans, I went back around there to see if there might have been some sign that they got out ok. I still don't know.

What gave me strength throughout this ordeal is the way my grandmother Gussie taught me right. My father's mother didn't know no strangers. She stretched out of her way to make sure you knew what was going on. I appreciate her to this day, even though she's dead and gone. I'd give up ten years of my life, if I could have her back for ten minutes.

Senta Eastern*

Born in Natchez, Mississippi, Senta Eastern was raised in a shotgun double in the Garden District. Her father was a teacher for the Orleans Parish School District. Her mother's professional job often took her away from the family for weeks at a time. Senta earned a master's in psychology from UNO and had her own consulting business. In August of 2005, Senta, a mother of two, had just finished decorating a new home in Mid-City. Before the storm, Senta drove from the city to Brookhaven, Mississippi.

A borrowed office at the Houston Urban League on January 11, 2006, was the setting for Senta's interview. She is a petite, gentle woman with long, flowing locks. Her compassion for the suffering of others was evident as she not only coordinated interviews with other Katrina survivors (one of her jobs at the Urban League), but she also encouraged, calmed, and empathized with them, never leaving their side as they retold their heart-wrenching stories. Senta was dressed in an elegant, black, secondhand outfit she had chosen with care.[1]

Formally educated, sophisticated, and sensitive to her surroundings, Senta and her stable, hardworking parents are emblematic of middle-class black life in New Orleans. Senta is reestablishing her family and professional life near Houston. The narrative of her separation from her children is similar to the experience of many parents during Katrina. Senta sustained herself through trauma and crisis with her faith and her ability to forgive.

My father was born in Natchez, Mississippi. His mother died when he was nine months old, so he alternated between his aunts in Natchez and New Orleans. In New Orleans, he stayed mostly with his aunt Lil, who made a really decent living working for the railroad from age thirteen. My mother was born in New Orleans, but my grandmother died when my mom was about ten years old. So my mom grew up on a farm in Mississippi with her grandparents. My mom would come from Mississippi to New Orleans to visit her aunt Gussie. My parents met at the age of thirteen. My father would see my mom across the fence and say to himself, "One day I'm gonna marry her." He did! They're still together after all these years. My mom and dad are wonderful parents.

I was born in Natchez, Mississippi, when my father was in the military fighting in the Vietnam War. My mother had returned to Natchez to stay with my father's aunt. We moved to New Orleans when I was eleven months old. As kids, we would visit my grandfather on the weekend. He woke up every morning at the crack of dawn when the rooster crowed to tend the crops and feed the animals. We used lanterns at night. Water was drawn from a well and the bathroom was an outhouse about thirty feet from the house.

When we were younger I can recall my father working three jobs, just to make ends meet. Whether it was raking leaves for New Orleans musical icon Al Hurt, digging swimming pools, or managing a fastfood restaurant, he was making it happen. He worked while he attended Xavier University. He ultimately became a New Orleans public school teacher.

My mother retired from BellSouth after thirty-two years of service. My mom also has a very strong work ethic. Her job took her away, on and off, for several years. That was distressing to me as a young child for different reasons. I was afraid of her flying on airplanes—I was thinking, "I might not see my mom again." Or, "I wish that she was here to comb my hair," but we managed. My father being an excellent cook of New Orleans' cuisine didn't hurt. When my mom was home, she cooked, cleaned, ironed our clothes, sewed, and supported our school functions. I remember being so proud of the lion costume that my mother made for me.

I grew up uptown, in the Garden District. We were on Adams Street. We lived in a very modest, no frills, double with a high concrete porch. The neighborhood was quite low- to upper-middle class, with the exception of the mansions on St. Charles Avenue about four blocks away. My relatives had a series of double houses. We lived on one side of the street, while they lived on the other side and my aunt lived down the street. The neighborhoods were safe. The neighborhood was very integrated. The mother of the former governor, Moon Landrieu, lived across the street. My two best friends, Sonya and Ionia, were white. Their father was a doctor, and they lived in a much larger house next door. All of the neighbors knew one another.

I have a brother who is eleven months older than I am. We were very close growing up. We went to Mater Dolorosa Elementary School together, participated in the same extracurricular activities, made the honor roll every year, and took band lessons from Ellis Marsalis.

When I was ten years old, my parents moved to New Orleans East to get away from all of the industrialism. The East was still undeveloped at that time. I was brokenhearted. I always said that I would move back, and I did. I returned to live uptown around 2002. Not long after that, uptown New Orleans was experiencing sudden regentrification for the second time, around 2003–2005. It caused a groundswell of political and school-based battles.

Early Influences

In the evenings, we would catch the bus to our relative's house on Leonidas Street when we couldn't go to my aunt Gussie's house. My aunt Gussie was a saint. She took care of all of her relatives and the kids in the neighborhood, and just everybody. She was a strong caretaker. I would travel to different states with her each summer. My most vivid memory of her was her praying on her knees at night, for what seemed like hours. When she passed away, literally, people came in on school buses from out of town.

It was important to my parents that we attend church. First Free Mission Baptist Church, where we all grew up, is now a historical landmark well over 125 years old.

It was founded by freed slaves. The architecture is beautiful. It has a steeple, stained glass windows, cathedral ceilings, and red carpet. As far back as I can remember, my mom has always been very committed to service in the church, a place where she has always been very well respected.

I developed a precocious relationship with the Lord at age seven. It didn't keep me from making mistakes or sinning just like anyone else. I was raised in a Baptist Church, but was attending a Catholic school, and so I had a mixture of the two, and it was kind of confusing for me, sorting through all of that religion. After I got baptized, instead of playing with the kids at recess, I would be found sitting in the church, because I wanted too. I would light a candle and just go and kneel and pray and try to get to know who God was. I was a very shy, somewhat introverted child. My faith became a very, very integral part of my existence.

"My Purpose"

I received an excellent education from McDonogh #35 High School and Xavier University. Unfortunately, it got to a point where I couldn't afford to stay at Xavier. I was a Biology/Pre-Med major while working full-time and trying to get my partying in and balancing my personal responsibilities. I ended up at SUNO. I love being analytical and Southern had an outstanding Psychology and Social Work Department. The partying ended and I put my head down, took nineteen hours, followed by twenty-one hours, and when I looked up, I had graduated with a B.A. degree in Psychology, with honors. Before I could really grasp what was happening, I think I was led into my purpose. I ended up at UNO in their graduate Counseling Program. I went through in two years straight. Once again, I graduated with honors. My objective was to harness as much information as possible to make a difference in people's lives.

There was a transitional period when I worked for several years counseling female and child victims of violence, but then I felt compelled to go to seminary school so that I could integrate biblical counseling with the skills I acquired from psychology. I was working toward a second master's degree in Christian Counseling. I was commuting to the Ivy League School of Theology located in South Carolina. Its main campus is located in the Northeast. People need hope and I wanted to give people the best possible results for a successful life.

My former mentor, who has passed away, Dr. Campbell, was a wonderful, exceptional person. He preached on every continent in the world except Antarctica. He was also a professor at Boston University, and he was the first African American to be canonized in London. Wherever he went outside of the country, he would bring me back a souvenir.

My current mentor and dear friend, Pastor Torin Sanders at the Sixth Baptist Church, was very instrumental in helping to pull me back into the counseling arena. Working alongside of Pastor Sanders and his wife, Dana, opened a lot of doors for me.

I was the Director of Community Relations for the Young Urban Professionals (YUP) in New Orleans. The goal of YUP is to create economic empowerment opportunities and social networking opportunities for African Americans. I worked with the AALP on healthcare and education and with the People's Institute for Survival and Beyond. Finally, I went into business for myself doing consulting work with churches.

Katrina

I'd just moved from Uptown to Mid-City. I was still unpacking. Two weeks after I'd unpacked my pots and started decorating, the hurricane hit.

Right before the mandatory evacuation I went to speak with Gerald, my son's dad. We both agreed that Korey would evacuate with him and his family to Texas. He could hang out with his cousins, whose parents had just purchased a new house in a ritzy subdivision located outside of Houston near Sugarland.

I evacuated with my daughter, my niece, and my mom and dad to Brookhaven, Mississippi to stay in my aunt Eloise's trailer. I followed them in my truck, because I insisted that I needed my truck to get my son to and from school and to his activities. I couldn't afford to be on my parents' schedule. When we got there, I urged my parents to keep going because I was afraid of the trauma that the kids would experience from the elements of the storm. The radius of the storm was two hundred miles wide, and we weren't far enough away. After retiring, my mom had gone back to work doing contract work. She was scheduled to be in Kentucky. So we got back on the road and drove straight to Kentucky. We slept on the floor in my mom's hotel room. By now the levees had broken and the whole world was watching the devastation in New Orleans. We cried every day, all day watching our people suffer in the Superdome and convention center. After about a week, we had to go back to the trailer because so many people who had lost everything were passing through in need of a place to stay, and we didn't want to lose the only housing we had.

Loss

Mid-City was the bottom of the soup bowl. All that water sat there. I had about eight feet of water. I remember standing in my living room, in that slimy dredge and funk, thinking, I *had* two wooden book shelves. I was standing on the other one. All the wood had rotted and the water and mold destroyed virtually everything I had. No family pictures survived except a set of pictures that my daughter had requested prior to the storm. I walked away with the set of tiny glass teacups from the top of the curio and the angel that sat on the mantlepiece. The teacups were a gift from Dr. Campbell, my mentor.

You don't just lose your belongings. You lose a sense of self and purpose, a sense of direction, a sense of safety, and definitely a sense of family, because your family is God only knows where, existing however they exist, if they survived at all. Since the storm, my aunt Eloise died. My parents are still in Mississippi renting the trailer with their savings. My father, a school teacher in Orleans Parish, was one year from retirement. He was fired along with the rest of the school teachers after providing over twenty-five years of service. He lost his only source of income. My parents' house in the East sustained a lot of damage; everything downstairs was completely destroyed. Whatever the water didn't get, the mold did. Upstairs there were some things that were salvageable, and that was a blessing.

The loss of my consulting business in New Orleans is a problem because I spent all this time building rapport and developing relationships in the community. Now I'm in an environment where I don't know anyone out here. I lost my support network, business network, and additional source of income.

I miss New Orleans: the hospitality, the food, the closeness, the driving proximity, and, most of all, the close knit relationships, my former church and my relatives. I miss its natural ambience: the scent of magnolia trees and the shade from one-hundred-plus-year-old oak trees. I miss the sense of security and safety that I had playing outside in the street, because the neighbors were watching out for you. I miss Street Car Po-Boy Sandwich Shop, the Clover Leaf Donut Factory, the Oak Street Snowball Shop, and the Adams Street corner store.

Reclaiming Existence

After communication was restored, a dear friend of mine would call regularly and check on me. He said that every time he called, I was crying or depressed, because I missed my son. I had no money for food, shelter, or transportation. He asked if I would like to catch a ride with him to Atlanta to apply for some assistance. In Atlanta, I found a recovery center and received some Red Cross assistance. After that I borrowed his car and got on the road and drove straight to Texas. Once I got here, I went to about twelve different hotels in one day. I took the only two-day vacancy that they had. It was three weeks since the evacuation, but I got to see my son.

After that, I stayed on the road going between Mississippi, Atlanta, Texas, and New Orleans. Sometimes I would be on the road for twelve hours straight. Finally, things started to come together in Atlanta. I was being offered housing from church members at the Greater Piney Grove Baptist Church in Atlanta. I picked out a school for my son and myself.

By the time I made it back to Houston, my son was enrolled in school. He had tested into gifted classes and was on the football team. Korey has always been very active. He earned a black belt, when he was six years old. He plays sports twelve months out of the year, while staying on the honor-roll. He was adjusting to his new environment, and I decided that I didn't want to disrupt his stability. So I made the decision to move to Houston instead of Atlanta.

FEMA hasn't been very cooperative, compassionate, consistent, or efficient. They're all over the place, and because they are, people's lives are all over the place. After having gone to New Orleans twice to find a FEMA representative to try to get someone to inspect my home, I wrote letters, I had them faxed, and I made call after call. I finally got the initial two thousand dollars. You almost have to exist as if FEMA doesn't exist.

In Houston, I also helped out the AALP by conducting research at the Urban League's office in Houston. At least there, I had access to free computers and telephones. Renae Stephens took notice of me and asked if I would be interested in applying for a job as a part-time intake specialist. We did the interview on the spot. I started working that following Monday. I requested to have the Urban League host a dialogue and interaction with the New Orleans ministers to help them get more quickly restored so that they could help restore others. Some of the pastors lost their homes, their belongings, and their congregations. That's really devastating.

I do put in quite a few hours at my job. Longer hours are going to have a big impact on my ability to see my son. If I don't leave work at 5:30, I don't get home until 7:00. I struggle with taking him out of his routine at his dad's to see him for a couple of hours at best.

Finding housing was much more frustrating. I went from hotel to hotel in Houston trying to find a vacancy, or someone who accepted FEMA. I ended up at the Hampton Inn, and they were so wonderful. The Hispanic lady at the front desk was one of the few people who actually checked the computer to even see if they had rooms.

I remember driving my truck all around Sugarland trying to find *anybody* who would accept FEMA rental assistance. There was an Asian lady who felt so comfortable with me that she put her arm around me on our way to seeing the vacancy. However, when we got back to the office and I told her that I had FEMA rental assistance, she went cold on me. I finally found an apartment about twenty-five minutes from my son's school, but the rent was very high and I would have had to pay out-of-pocket. Out of the blue, my phone rang and it was Reverend Nina telling me not to worry, the Pine Crest Presbyterian Church in Houston was going to help me with the rent. I went back to the office, and it was right before closing time. One of the nicest people that I've ever met, a Caucasian lady, was closing out and she told me that no matter what she would get me approved. I have shelving in my new apartment. I've placed the glass tea sets that I salvaged on the shelf, and I decorated around the angel.

Conclusion

After the storm, what happened was just far, far worse than the storm itself. I have heard a lot of stories of what happened since I started working at the Urban League. I think we live in a world where people will do what they do on a daily basis, and live side-by-side next to one another, but when the veil came off around some of the issues that were underlying in New Orleans, it was very reflective of how people felt. To move people around like chattel and drop them off essentially to die, without food and water, it is just unbelievable to me.

Now that we're five months out from Katrina, I worry most that hearts have grown cold. I remember going to talk to this furniture mogul in Houston who apparently has helped a lot of people. I began to explain my situation to him. He cut me off and instructed me to write a letter to one of his staff when Christmas comes around. This was October. I sat in the parking lot and cried. I never got the chance to tell him that my apartment was completely empty, I had been sleeping on the floor with a bad back, and I had no bed for my son to sleep in.

It's hard to get a footing here, to feel like I belong. I definitely feel disjointed. Right now I have no earthly idea what life is going to look like, just really from one day to the next, let alone six months or a year from now.

Age developmentally, I know that soon my son is going to be at that phase where his peers are his focus. I don't want to miss out on any of that time to teach, guide, and nurture him. He has made it very well known to me that he misses his mother. My son is up and down emotionally. Children can be very resilient, and he's doing as well as can be expected. My beautiful, vivacious, and witty daughter, Alana, was attending Xavier. This semester the school has made arrangements for them to live in the Hilton Hotel, without the frills of course, while the dormitories are being repaired. She still needs guidance and support.

What gives me optimism is that I've had a lot of harsh experiences in my lifetime, and to know that I've come out on the other side of them makes me very hopeful. My optimism comes with the energy that people are putting into trying to restore their lives and not giving up. My parents are still investing in rebuilding their home. I've feel like I've grown a lot in my relationship with Christ. I've learned that if God is trying to take me somewhere, go! I've learned not to be afraid, because if I get hurt, He will restore me, and if I fail, those lessons can become the keys to my success. I go back to my "church home" in New Orleans. The church is even bigger and it has even more energy, although the city is a smaller city now. You have more people coming from the surrounding areas, and I think people's level of awareness is heightened, so they perceive their relationship with God differently. I feel that's one of the few places, when I go back to New Orleans, that's so profoundly different in a positive sense.

These events make me think about Job from Scripture. He lost everything, all in one day. And the only sources of outside comfort that he seemingly had were his friends, who failed him too. Even in the state that he was in, Job was called on to pray for them! I'm praying for the people who did this to us every day. It is from a place of forgiveness that people can see God.

TWENTY-TWO

Yolanda Seals*

While growing up in the Lower Nine two blocks from Lawless High School, Yolanda Seals, the great-granddaughter of a Native American woman, longed for summer vacations in Natchez, Mississippi. Born in 1969, Yolanda grew up in a very different environment from that of Copelin, Roberts, Ferdinand, Salaam, or Duplessis. As the oldest sibling, she was deputized by her father, the manager of the A & P store, to make sure the other children came home safely. Before the storm, she was a homeowner in Violet, a hamlet in St. Bernard parish, and a social worker for Catholic Charities in New Orleans. On August 28, 2005, she left the city with her four children to go to Atlanta for what she thought would be an extended holiday weekend at Six Flags.

This chapter is based on two interviews recorded in cramped hotel rooms on the out- skirts of Atlanta on February 7, and March 31, 2006.[1] On both days, Yolanda arrived dressed professionally with her make-up still flawless at the end of a very long work day spent assisting elderly New Orleanians and overwhelmed single mothers transitioning out of hotels into apartments. Her narration deftly interwove her personal and professional observations.

The dream of a vibrant, socioeconomically integrated Lower Ninth Ward had died in the Tennessee Street neighborhood of Yolanda's youth, which was dominated by illicit drug trade. Yolanda represents members of a generation of New Orleanians who grew up without the love of neighborhood that makes exile painful for most New Orleanians. She therefore embraced the dislocation to Atlanta as an opportunity to start over. She enrolled her chil- dren in school, bought a new home, and landed two jobs within weeks of settling into her newly adopted city.

My father, Freddie Seals Sr., was from Natchez, Mississippi. He used to drive a truck. He went to make a delivery at a store, and it was raining, so he went to go through the front door because it was easier, and they told him, because he's a black, he had to use the back door. By the time he went to the back door, rain was pouring down. He was trying to get the stuff through the narrow door and he didn't see this lady. So they accused him of not speaking to a white woman. As a result, he had problems with the KKK (Ku Klux Klan) people burning crosses on his family's lawn because of that incident. He had fourteen sisters and brothers, so his mother made him leave. He had graduated from high school early and was very, very smart. He had two unfulfilled dreams: to be a lawyer and to major in business. So he always pushed me to go to law school.

When he was sixteen years old, he moved to New Orleans where he had relatives. He was living in Hollygrove. My mom, Iola, was born on Carnival day in Charity Hospital. She grew up on Jackson Avenue. My mom moved to the Ninth Ward when she was about ten. My mom and my dad, when they first got married, had an apartment on Tupelo. When he found out my mom was pregnant with me, he bought the house on Tennessee Street. I was born in 1969. I grew up there all my life. When they moved into the house where I grew up, it was a predominantly white neighborhood. The whites moved out to Chalmette, and the neighborhood became predominantly black when I was around seven or eight.

My father worked two jobs. He was a supervisor at an A & P (Atlantic and Pacific Tea Company) on St. Claude at Egania, until his health began to deteriorate because there was so much stress. He had hypertension and he was a diabetic, so he stepped down to a manager, but he started his own little business on the side cleaning up houses and buildings. My mom would work with him. My mom kept kids in the house, but there were five of us. I was the oldest.

I went to Lawless Elementary in the Lower Ninth Ward, and then I went to a brand new school on Caffin Avenue named after Martin Luther King Jr. I finished at McDonogh 35 High School.

I saw a lot of crime. It was tough growing up in that neighborhood. There was a crack house down the street, with a good twenty people living there. I did not feel safe. We had to fight every day to go to school, to do things, and we didn't have much ourselves. My father was a strict disciplinarian, and he was really, really strong. He had four girls and one boy, so he raised us pretty much like boys. My mom was very devout, so she was like, "Pray for them." But my dad was like, "Look if your sisters come home beat up, then you better come home beat up too." So me and my sisters always stuck up for each other.

I had a lot of my friends who went through some things, and didn't have no one to help them back then. I could name at least two people I know who've been molested or raped by family members. As they were coming up, they would reach out for help and tell their teachers. The teachers turned their heads or ignored them. As a result, a lot of my friends ended up on drugs and practicing alternative life styles. I understand why the anger and everything is there.

I remember we used to get in the street late at night, well not late at night, till the sun go down and play this game with a bat and ball. So there were some good memories, but there were a lot of bad ones. I remember policemen chasing criminals through the yard. I grew up across the street from this guy who was a police officer, Len Davis. We were the same age, but now he is on death row. Not only was he a police officer, but he was one of the biggest drug dealers in New Orleans. He was terrorizing people in the neighborhood. I just kept my distance.

Every summer we went to Natchez. I loved it! I always looked forward to the summer time to get away from the Lower Nine, oh my God, from those bad kids and the crime. It was extremely stressful, extremely. I remember having anxiety attacks as a young child.

It was so economically deprived in that area. People in Gentilly were considered rich coming up. If you moved to the East, it was just like moving to Beverly Hills, when

I was coming up. This is when the East first started growing. So whenever somebody in the neighborhood comes up, "We're about to move to the East," we were like, "ooh."

My dad got me my first job when I was fourteen. I worked as a cashier at the A & P, and moved my way up to head cashier, and did that up until the time I went off to college.

"Different Sceneries"

I left New Orleans in 1987 for Jackson State University in Mississippi. My father is from Natchez, so their family tradition is Jackson State or Alcorn State University. A drawing and a painting that I did, my art teacher put it into this contest, and I had won an art scholarship. So I met my first husband then, and he got called up into the military, and we went to the first place he was stationed. We traveled a lot. Wherever we moved, I just used my scholarship money there and went to college there. I got a degree in art, and someone tried to tell me about art therapy. So I went back and started taking classes in psychology, and then I started thinking about a lot of the things that I saw coming up, so that's how I got into social work and got a master's degree.

I liked Washington, DC best because it's a huge metropolitan area. I'm a big city girl. I love being able to get away into different sceneries. Everything about DC was close. If you got bored, you could drive to Baltimore, drive to Philly. I loved the art museums there.

I began to educate people, because there were a lot of people who never left New Orleans, a lot of people who never crossed the Mississippi river. I would come back and tell my sisters, "Guys, look, there's life outside of New Orleans." I would tell people, "Move away, get to experience other parts of the United States."

The Lower Ninth Ward in the Late 1990s

I came back to New Orleans in 1994 or 1995. My father had a massive heart attack and stroke, so I had to come back. I'm the oldest child. My mom had only gone as far as sixth grade. He was in a nursing home, and I ended up taking care of all that.

I lived on Tennessee Street with my mom in the same house. The crime was ten times worse. I didn't let my kids go outside to play. Whenever they wanted to play, I took them to the park. I had my car broken into three times. The house was broken into when they realized my dad was no longer there, and we were away at the hospital a lot. The burglary had happened on a Monday. The police didn't show up until Thursday. The police officer told me, "We only come out for important crimes, like murders, stabbings, and rapes. We don't come out for theft." I reported that to the chief of police, and he finally sent someone out.

There was a neighborhood store around the corner, and it was owned by my mom's cousin. She went down there to get some eggs. She came back and they had totally cleaned her house out in fifteen minutes. And I said to my mom, "It's time for you to move, because if they're crazy enough to do that in broad daylight in fifteen minutes, they're going to break in on you the next time." Everybody knew who was doing it because there was a crack house right down the street. It was drug motivated. The crack house was owned by the police officer that's now on death row.

St. Bernard Parish

I stayed in the Ninth Ward for a year, and then I met my second husband, who was from Violet. Violet/Chalmette, that's all St. Bernard's Parish. Chalmette was extremely racist. The bus going towards Chalmette from the Ninth Ward stopped right after work.

I got pulled over when I was about eight months pregnant. I had just bought an Acura with tinted windows. The police man said, "Get out of the car and stand behind the car." It took three police cars to deal with me. They were all white. I got out of the car and stood at the back. He said, "Turn around and stand like that." When the third police car pulled up, he said, "You got to let her go." And the cop said, "What you mean? She has tinted windows." They said, "That's Officer Wither's wife." Then they said, "Oh ma'am we're sorry."

But I loved Violet. Violet was mixed racially. I could sleep with my door open and no one bothered me. My kids were able to play outside. They left their bikes outside, and they'd wake up the next morning and the bike was still there. If you had a problem, the police were there in an instant. A lot of people were working for the refineries, and where I lived, the ferry was close by, and Belle Chasse Naval Base was only a ferry ride away.

The public schools there were truly prejudiced. They would hold kids back and not give black kids the education that they needed. My oldest son has a high IQ. When he was two years old, he put together a puzzle of one hundred pieces. When he was in kindergarten, he was explaining to his class about electoral votes. They were hesitant to put him in a gifted class, so that's when I made up my mind and took him back to Orleans Parish. I used my mom's address and got him into a magnet school in New Orleans. He went to Martin Luther King Jr. School for Science and Technology in the Lower Ninth Ward. After sixth grade, I put him in a private school. My daughter was at a magnet school uptown, Thurgood Marshall on Canal Street.

I drove my kids to school, went to work at Catholic Charities, and then picked them all up. I picked my kids up every day because I was so afraid for them to get on a public bus and catch the bus home. They could have been stabbed just for no reason, just to prove who's the baddest. Before the storm, I had four kids at four different schools.

Katrina

The Friday before the hurricane hit I was in the Lower Nine. My best friend, Kim, did hair in part of her house, so I used to go there to get my hair done. We were just sitting on her steps, talking and laughing. She didn't have reliable transportation, but she managed to get out.

We left early that Sunday morning. We decided to leave, after we saw the size of it. We had smooth sailing to Atlanta. My younger sister used to live here. I figured I'd take a second vacation with the kids. I thought it was going to be two days. I was like, "I'll take the kids to Six Flags," because if I'm stuck in a hotel with four kids all day, they're going to run me ragged. When I came back and was watching the news and what was going on, I was totally stunned. First thing they showed the Superdome blasted. A couple of hours later the levees are bursting.

So three or four days after the storm, that's when I went to Red Cross and started applying for benefits. I called FEMA. I was watching people stranded on the Interstate and started applying for benefits. The following week that's when I started looking for a job, because I found out that my house had fifteen feet of water in it.

I was in a hotel for about thirty to thirty-five days. Catholic Charities put me up at the hotel and they paid the down payment for the first house I was in. When I first got here, I went to them, and said, "I'm a hurricane evacuee. I worked for Catholic Charities in New Orleans."

This church named Destiny was helpful too. Destiny made sure we had clothes, food, and furniture. They would come and get my kids and take them places, and give me a break to help me try and find jobs. They helped with the house too, by paying for a couple of months rent.

I had to use my house insurance to pay off the house in Violet, so I wouldn't have a mortgage that I couldn't afford. Like a lot of other people, I was told that I didn't need flood insurance, but I was blessed that I didn't pay attention and got it anyway.

Social Work after the Storm

FEMA is horrible and I work for them.[2] They have re-traumatized the traumatized.

Post-traumatic stress is a major concern. I have a coworker right now who was left on the interstate. She can't even function. Like me for example, I left the Sunday before the storm. But hearing what people went through as a case worker still traumatizes me.

The elderly from New Orleans are used to jumping on a street car or bus, and going where they need to go. Here it's not the same because a lot of the buses shut down at 6:00. And the place here is so huge, and so they're finding it difficult to get around. A lot of them miss New Orleans too bad. One of the guys that I used to talk to kept talking about his church, sitting out on his porch, and doing things. Here the culture is totally different, whereas back in New Orleans everybody knew each other and if you were elderly, people in the neighborhood would try to help you. For example in the Lower Ninth Ward, when I went to the store, if I knew somebody elderly in the neighborhood, I would go ask them, "I'm going to the store, is there anything that you need?" When I would go to church on Sunday, I would ask the elderly, "Are there any errands you need for me to run this week?"

This one guy is eighty-three years old. We've been advocating for him for a month because they're about to put him out of the hotel. (People at the hotel are very unsympathetic.) A lot of the elderly don't know the area. People at the hotel were charging this man sixty dollars to take him a couple of blocks. We had to help him find an apartment and everything.

The cost of living here is much, much higher than it was in New Orleans, and if you don't have an education and you don't have the skills, you can't make it work here. There's a waiting list here for section 8 that's as long as the one they had in New Orleans. We have a lot of single mothers that I'm running into who are suicidal because they can't deal with the cost of living here. We've been able to go in there, take the kids, and put them in a foster home until the mothers get the help they need. Here in Atlanta, so far for the past couple of months, I've seen four murder-suicides that were Katrina related. Some are returning to New Orleans. I had a cousin here who was my mother's friend. She returned home to live in her abandoned house, because she couldn't find anything here. She said, "I've got to go back to a place that I know."

The ones who were working poor in the transition are worse off than those who were on welfare because they have to start all over here, and it's much harder here than it was there. So Atlanta's a paradise for middle-class, black families but it's not so great for lower-income working people.[3]

Conclusion

I went home one weekend in late January. The thing that really ripped my heart was where I grew up on Tennessee Street is where the first levee broke. Lawless High School is two streets over. Our family house is now in the back of Lawless in the yard. I went with a coworker to show them people with license plates that are not from Louisiana going through these people's houses. It's like a scavenger hunt. And there's no police presence. I saw someone coming out of my mom's house. One of my coworkers was like, "I didn't realize it was this bad."

My neighborhood in Violet is still vacant, no one's there. There's one house that was two stories on my block, and my neighbor is living on her second floor. She said, "It's a ghost town. There's no electricity." She's using candles.

I consider myself to be very, very blessed that I was able to bounce back and get on my feet. I think over the long run, I'll be much better off than I was back home. Here the standards of public education are much, much higher. My kids love it here. They say it feels so good to go to school without being shot at.

TWENTY-THREE

Aldon E. Cotton

Born in 1968, Aldon E. Cotton is the youngest of six boys raised in Back of Town by church-going parents who worked around the clock to provide for their family. As a child Aldon showed talent for music, determination, spirituality, reasoning ability, and sociability. These were already clearly visible in Aldon before the first of many "nobody-but-Jesus" moments at the age of fourteen, when a train accident changed his way of doing some things. At the age of twenty-two, he began pastoring at the only church he has ever known, Jerusalem Baptist Church (Jerusalem) in Central City. Although in ways he is an example of upward mobility, rising from his grandparents' plantation shack in Vacherie to a middle-class home on Lake Carmel with his wife and three children, he refuses to prioritize material values.

After doing everything possible to prepare his congregation of 160 members for the evacuation from the city, including Mapquest instructions and emergency phone numbers, he caravanned along with thirty church and family members to Greenville, Mississippi, on August 28, 2005.

This interview was conducted on February 14, 2008 in Luling, Louisiana, the site of Aldon's temporary home and gateway to New Orleans, the city of his calling. The small home was comfortable, pleasant, and unostentatious. Aldon was wearing a beige, silk turtleneck sweater. A spellbinding storyteller, Aldon's contagious smile, mellifluous voice, and hearty laugh animated his stories.

His narrative describes the experience of growing up in Back of Town, marked by the central landmarks of the Orleans Parish Prison (OPP) and the Orleans Parish courthouse at Tulane and Broad. Unswervingly upbeat, Aldon refuses to give depression a toehold in his psyche. Determined from day one of Katrina to return home, Aldon began working with coalitions of pastors and churches to facilitate the rebuilding process. Clearly articulated in his testimony is his vision of the role black churches can play in shaping the future of endangered neighborhoods.

I was born in '68. I'm the youngest. Growing up in a house with five older brothers, I always had noise. Silence disturbs me.

My mother, Marion, came from a town called Vacherie, Louisiana. It's about fifty to sixty minutes from New Orleans. I would spend my summers in Vacherie. My grandparents in the seventies were still living in a plantation house on the Laura plantation.

I didn't understand what a plantation was until I watched the movie "Roots." All I knew was I was going by Gram and Papa. As kids, we went in the sugarcane fields, but we played. I know what it is to take a bath in a number three tube. I appreciate the in-house because I know what the outhouse is. When I would stay out in the country in the summer, my grandmother, Olivia, was up at 4:30. She would say, "Son, know your people. Your car could break down right in front of your cousin's house. If you know your people, they'll help you." My grandparents didn't have a formal education, but you couldn't cheat my grandfather, John Joseph, out of a penny and my grandmother know how to feed thirteen children with one chicken.

My daddy, Eugene Cotton, III, had an uncle in Back of Town. My uncle owned a house, and so my parents started renting from him. At one point we were at 3232 Gravier Street. Where they first lived, they had the living room or the front room, a bedroom, another bedroom, a kitchen, and a bathroom. When I got on the scene, we were at 3234 Gravier. When my mama planted something, it would grow. She loved flowers in the front yard, along and up the fence, everywhere.

Mama was doing people's hair when I was little. I remember she got a job as a nurse's aide at Montelepre Memorial Hospital, a small, private, community hospital on Canal Street. She worked the seven-to-three, the three-to-eleven, or the eleven-to-seven shift. My father was working fulltime, trying to go to school in Baton Rouge, where he went to Southern for his master's degree. Sometimes he worked two jobs plus playing for three or four churches. He was working in Lockport, which from New Orleans at that time was like a two-hour drive. He was a vocal music teacher. That was the place that gave him a job. He would leave about 5:00 or 5:30 in the morning, drive, teach, come home, get about an hour's sleep, go to choir rehearsal, come home, sleep, and then get up. We'd have to wake him up for 10:15 p.m. He'd have to be in at 10:30 p.m., when he was working at Mercy Hospital, the gem of Mid-City, as an orderly. He was exhausted all the time. Sometime he would start fussing at us, or he'd ask us a question, and while we're answering, he'd fall asleep.

There was a lot that we could not understand as children from my father's perspective. My mother comes from a family of thirteen with both parents. We're a family of six boys with both parents. My father, a single child, had a single parent. There was a time that I didn't respect him, I was scared of him. I knew when around 4:30 was coming because I got sick in my stomach. It's time for my daddy to come home soon. There were times I went to the store with my mother not to be home with him. You never really knew his frustration level. Sometimes he was just frustrated with not being able to do for his family like he really wanted to, or from not fully understanding and dealing with a lot of the hurts and pains of his own childhood. Constantly, you're reliving your own pain. But he was adamant that he would not abandon his children and just always wanted better for us. And that's why, even though we grew up five blocks from Orleans Parish Prison, our parents never had to come five blocks to get none of us out.

All of us went to school with a great deal of knowledge. We would get our report cards, and all the teachers would say the same thing. "They are very good students, but they talk too much." My daddy would say, "Give them more work." In the summer time you had to do book reports for my dad. My daddy loved books. You had to pick some book out of his suitcase or go to the library, and there wasn't too much that we picked that he hadn't already read.

I spent a lot of time with my mama. We talked and we talked and developed a bond. My mother was always totally honest about the good and the bad. I would ask about her childhood, how was it growing up on a plantation. She taught me principles: "Treat people right, don't mess over people. You've got to learn how to cook, because if you ever get a wife and she get mad, you don't have to starve." I've been cooking meals since I was in the sixth grade.

I would ask my mama about things that I didn't understand from my daddy, because he grew up in kind of a strict environment, so it was, "You do as I say." And I'm like, "Why?" I would ask her stuff like, "Do God know where we really live? Because if so, why we here? We're doing the right stuff Mama. We love and worship God, and we read the Bible. We got family Bible class." There were these people who didn't go to church. And they're doing a whole lot better, even though we're in the same neighborhood. They've got a new car, new clothes, new name-brand shoes. We've got no-namers on. I just couldn't understand why God would do that. She'd say, "Maybe God has us here for right now so that we can show them His love."

I remember we used to cut grass or get an allowance. We tried to make birthdays special. We told my mama, "We're going to buy daddy a bike." Mama really paid for it, but I paid my twenty-five cents. Daddy just cried because he'd never had a bike. I remember days when he first started riding. He'd get up and he'd say, "You all feel like going riding?" We'd go from my house to City Park, a distance of two miles. He was able to enjoy a childhood with his children.

Back of Town

I grew up in a shotgun house in an area called Back of Town, a few blocks from an area called James Alley, where Louis Armstrong grew up at. It's adjacent to Central City. I grew up between Broad Street and Jefferson Davis Parkway. If we went on the other side of Broad Street, everybody lets you know you're on the wrong side. And when you went on the other side of the Jefferson Davis overpass, you were headed toward Gert Town. They would definitely let you know you're not a resident here. Once you got in Back of Town, you were pretty safe.

Growing up back there, you knew the families. You knew the guys who were messing with stuff, so you knew not to hang around them. We were known as the church-going family. We were always in the choir. We would have family rehearsals at home, and a lot of times we would be in there singing. Then we'd hear somebody say, "Sing another one!" We'd go to the door, and we may have had ten to fifteen people standing in front of the gate.

All of my brothers and I didn't have the faintest idea we were poor. I thought everybody had the silver can of peanut butter and the dried eggs and dried milk. We would go to the grocery store. Jiffy cornbread would be ten for a dollar. My mama would say, "Get thirty boxes." I would think, Oh Lord, we're going to have cornbread forever. I knew there were "rich, rich people," you know, when you'd go on St. Charles Avenue. We had to watch the local news. We would hear about the "lower class," but we didn't recognize ourselves in those depictions.

Saturday shopping for my mother was an experience. About 9:00 a.m., she'd go to Winn Dixie. Then she'd go to the A & P, then Schwegman's Grocery. She would come home, and we'd have to go over the receipts, and we had to pick out everything. Then she'd say, "Now put them five boxes in that bag. Bring that bag to Ms. Louise." We had a little neighborhood store around the corner. Sometimes it took us fifteen to twenty minutes to get there, because we had to go next door. "Ms. Ruth, my mama is sending me to the store. Is there anything you would like?" Then you'd have to go to the next house, Ms. Sister, and finally across the street to Ms. Gladys. You'd go to the store, get in line, bring everybody back their stuff, and when they offered you something, it was, "No ma'am, it was my pleasure." I don't care how bad you wanted that quarter, you better not take nothing. That's what you do because you're a neighbor.

"I Was Born Playing"

My father plays. He taught vocal music, and he has a master's degree from Southern. My daddy taught all of us how to play music. We all had to go through the John Thompson "Teaching Little Fingers How to Play." And my daddy said, "After you finish this, if you want to continue, fine. But I want to make sure that you all have a foundation." He treated us just like he treated the rest of his students. I played piano, but I kept quitting because I wanted to be outside playing football, riding bike, and all that. That practice stuff always got in the way.

I played the trumpet in elementary school. In sixth grade, I was in the band, and I wanted to march. The band teacher's rule was that only seventh and eighth graders could march. So the band director called my name, he said, "Cotton! Are you kin to the twins?" I said, "Yes, that's my brothers." So I asked him about marching. He said, "I normally don't let sixth graders march." I said, "Look, I'll do whatever I got to do. I'll challenge anybody." So we scheduled a challenge. When I got to the section leader, the whole band was there. I said, "no problem." I ran up and down the chromatic scales, just playing scales that he couldn't play. So then he slapped some music in front of us and I said, "no problem." And everybody was like, "oh man, oh man." The band teacher said, "I'm going to let you march, but he won the session." Then he pulled me on the side and said, "Cotton, really, you're better than him. But this is his last year. Let him go on out as section leader." I said, "I just wanted to march!" So when I got home with the marching band jacket, because Peter's had a little jacket with the emblem on the back, my brothers said, "What? We don't let no sixth grader march." I said, "This is where we're going to march, and you be out there watching."

Our parents said to us, "Always have a plan and have a backup plan just in case your first plan don't work, you've got something else." And music did that for us. Music gave us a platform where you could go to college. Churches would always need musicians.

"Nobody-But-Jesus Situation"

In 1982, August the 20th a friend of mine, John Buckley, and I were going to choir rehearsal at Jerusalem. Normally we would walk over the Broad overpass to church. The summer before, he had passed out from a heat stroke. When we got halfway up the bridge, and it was kind of hot that day, so I said to him, "Say Bro, let's get under this bridge, because if you have a stroke and fall over the side, I love you, but I ain't going over with you." Half way across the bridge, there's a ramp you can go down. When we got to the tracks, the Amtrak train was there.

The train was just sitting. They had maybe five different people standing there. I said, "Come on bro, we late." So we was about three cars closer to the end. So that's what we did. If the tracks are like this, you just hop in the middle, then on the other side. That's what my friend did. I was right there. The guy at the back of the train said, "No boy!" And when I looked up, I fell. The track was across my chest. I tried to get up and the rocks were slippery, and I couldn't get up. Then I heard them, "Start rolling." So I tried to roll and the train started moving and it dragged me. I was catty-corner to the train tracks. The first wheel went completely across both legs. I was pinned under the train by the second wheel. It was literally on top of my leg.

I said, "Jesus! Mama!" This calmness came immediately. My friend was losing it, because out of his peripheral vision, he saw the train backing up, and I think he saw it drag me. I think it was the cook or somebody on the back hit the emergency brakes and the conductor came out of the train, looked at me, and ran back on the train. The crowd got larger and larger.

The doctor who came to the scene said that he never traveled that way home, but the normal way he was going, there was traffic. He just got off from his shift, sees the crowd, and said, "I'm tired. I'm going to go home." He said, "Something just stopped the car." He came over and saw me. He asked the conductor, "Is there a train that has a jack?" The conductor said, "Yeah, but that train is an hour away." The doctor said, "This boy don't have an hour." He asked me where I was going. I said, "I was going to choir rehearsal." He said, "Well, we have a priest here." So they called for the priest, and he opened that black book. I just looked at him. "That ain't good." I'm hearing, "Lord receive his spirit." "That ain't the right prayer," I said.

Then they said, "Listen, in order to get you from underneath this train, I'm going to have to amputate. I can't give you nothing because we don't want you slipping off into unconsciousness. We need you talking." I said, "Go ahead, Doc. I'll be alright." He just kind of sat there and looked at me. I said, "You got something you want to go do?" They cut the leg off.

In the ambulance, I said, "I'm hungry. I want a Big Mac." I was so skinny, you could count my ribs. They got me to Charity. They had to take the other leg off. They told my family, "His heartbeat is low. Infection has gone throughout the body. We've done everything we could do. He's going to die." My family got on its knees and prayed right there in the hospital. They kept bringing me back and forth from surgery. When I woke up, I had IVs all over.

I never went through any type of depression. At one point this psychologist came in. "I want to ask you some questions. Were you trying to kill yourself?" I said, "I just walked over a bridge. If I was going to kill myself, wouldn't I have jumped off the bridge? And if I wanted to kill myself, why wouldn't I have put my neck on the track?"

Literally sacks of mail arrived for me. My mother and I read every piece of mail, and answered every piece of mail. My mother said, "You need to write them because they want to hear from you." But I was still me. Even with people writing me, we remained humble, and we were grateful, because people don't have to do that. My pastor would always say, "You be grateful because people don't have to be nice. They don't owe you nothing." My mama would say, "You still have your hands, you still have your mind. You just have to find a different way of doing these things." So that's the foundation of why I'm able to even handle Katrina.

I grew up poor and lower class, been called a whole bunch of stuff, but I never was called a handicapped. All of a sudden this accident happened to me and people wanted me to go to an amputee camp. I said, "for what?" If you really don't have a personal relationship with God, I think you're a handicapped. I was walking with prosthetic legs at one time. I gave them up. With them I'm dependent on people. Without them I'm independent. One day I was sitting in a class, me and this other guy. He's paralyzed. And the teacher said, "I know y'all two angry." I said, "Tell me something else you know about me that I don't know about myself." I'm too busy telling God thank you for saving my life to be mad with Him.

I was at the store minding my business, putting my groceries in the car. A man a little younger than me came up to me and said, "Is this your car?" I'm from Back of Town. I put the groceries in the car, and I got my hand on the crow bar. I say, "yeah. This my car." I'm thinking, I'm about ready to bust you in your head. He said, "You drive this car?" I said, "Yeah, I drive this car." "Nobody with you?" You really got my attention now. I said, "Do you see anybody with me?" So I'm about ready to just pull it out, because that's what my 'hood taught me. You pick it up, and you use it. He said, "And you make your own groceries." He's got tears swelling up in his eyes. "Yeah, I make my own groceries. I put my own clothes on. I'm going home, I'm going to cook this food, and I'm going to eat this food." He just looked at me and said, "Man, I'm not going to complain again." God gave him hope just by me being there.

Jerusalem

I grew up in a Christian home. I always had a belief in God. I may have been about eight years old when I told my mother I wanted to get baptized. She asked me why, and I told her because I believed in God, and I want to go to heaven. So what happened, as children do, we'd stay up for the singing, but then when the preacher get up and talk about stuff we don't understand, our eyes get drowsy. I was sitting next to my mother that Sunday and went to sleep. The next thing I know it was, "Stand up for the benediction." And I said, "Mama, why you didn't wake me up?" She said, "If you had really wanted it, you'd have stayed up." That next Sunday, Rev Oris Marshall Sr. got to preaching. I'm fighting, sitting on the edge, hitting and pinching myself, because I wanted to make sure I stayed up. You have to confess what you believe publically. That's the part about not being ashamed to openly declare your faith, and so I did that. And that was it.

Even in the hospital, God had told me, "I will call you to preach, but not now." We knew we were going to be in the church doing something with music. Choir director was as far as I wanted to go. So in '85, my brother Alex had to preach at the Fifth African Baptist Church on Robertson Street. I get there, get out the car, and there is absolutely nobody around. I got my chair and started pulling my chair up the stairs. I get half way up the stairs and I hear, "Aldon!" I hear it externally and I looked around. I'm waiting for the person to come. Nothing. I get ready to go in the church. "It's time for you to preach My word." Somebody's playing with me. I go in church, and the choir was singing. My brother gets up and opens the Bible. "It's time for you to preach My word." And I just said, "Yes, Lord."

I was a stickler for the Bible. I didn't worry about no books, none of that stuff. I wasn't burdened to go to the seminary. Matter of fact, God told me, "Don't go, because I'm the only one that's going to get glory out of your life." I get married in December. My pastor died that January. God called me to pastor Jerusalem at the age of twenty-two.

Jerusalem is in Central City. In our second year of pastoring, we drove around from Broad to Claiborne, from Washington Avenue to Martin Luther King Jr. Boulevard, and took down every church name. We invited all of those pastors to come to a fellowship breakfast and recognized that the largest congregation in that area may have had three hundred members. We said to them, "If we work together, we can change this community." We started having community revivals. We had some guys who were most likely selling drugs. When we would have church, they would sit across the street. And I would say, "Man, come on in." "Nah, Rev. We been listening." And they would tell me stuff I said two or three weeks prior. I would go out and play dominoes with these guys and just talk to them. Everybody would say the same thing: "Rev, I know I ain't right. But I can't come in there and be no hypocrite." I said, "It's not about being a hypocrite. All of us who come there, we know we're not right. That's why we come. Sick folk need a hospital." One guy told me, "Rev, I would come, but I don't have no church clothes." I said, "Look bro, this tic is yours. Now come on in." And so sometimes they'll see me just with a t-shirt going to church because I wanted them to understand, it ain't about this at all.

Katrina

What happened was about a year before Katrina, late '04, God had burdened me, and I told the church, "I don't know what's about to happen, but I feel this urgency from God to get you all ready for what is about to happen. Something is about to happen that's going to shake this city." The last three sermons I preached was: "God over us," "God with us," "God in us."

In '04, about four families, when they was telling us to leave, went to Rustin, Louisiana. One of our members was in Rustin, graduating from Grambling University and working with the Sheriff's Department. While up there, we were saying, "We really need to expand this and make sure we know where everybody is." So May or June they got together, had about four or five meetings, and came up with a plan. They shared it with the church and had assignments for everybody. If we had to go east, north, or west, who was to call, from what area you were to call, and what price range you were to call. We had a list of what everybody should pack for three days. We had all the members' cell numbers. We knew who were going to leave and where everybody's going. Those of us who were traveling had Mapquest. People said, "Pastor, I'm traveling with you." No one wanted to be by themselves that far a distance. We planned to meet at a central location, and we made sure we had walkie-talkies with a five mile radius.

At 2:00 a.m. Sunday morning, I got a call. "Pastor I found a hotel for everybody. They got rooms enough for thirty people. They're holding the rooms." I say, "Alright! Where we going?" She said, "Greenville, Mississippi." I said, "Do that show up on Mapquest?"

The city had called to find out if the church had a van and could people ride in it. I said, "No, we don't have a van, but let me ask you a question. What is the city plan?

Is there a phone number or something that I could call and give you the address of an elderly person, so that you all could go get them and know exactly where they are?" He said, "The city don't have a plan."

I wound up going to the church. I said, "I'm going to be gone for three days, still got to study." So I grabbed my computer. The Lord said, "get this, get that." I went into the safe and got the receipts from the previous Sunday and the checkbook. Then when I went back to the Sanctuary, the Lord told me, "This is going to be your last time seeing it this way." I went back to my childhood, seeing us crawling underneath the pews and reliving everything that went on in that church. Before we got ready to leave I said to my armor bearer, "Get my pastor's picture."

Pastor Sean Elder, my twin brother in the spirit, called me while I was in Mississippi. He said, "Doc, I'm standing in front of your church now. It's bad." I said, "really? Look across the street. Do you see that multipurpose building?" He said, "What?" I said, "Do you see a multipurpose building across the street? Do you see our new building?" He said, "Reverend, I see it." I said, "If you see it, I'm coming home!" So it's about what the pastor sees in that community.

"Pastor of the City"

In any issues that I'm dealing with and certainly something as traumatic as Katrina, I'm asking, "Where is God? What do He have to say?" I mean preachers will preach to one person. But to ask me to go and preach in a graveyard is a whole different story. None of us would want Elijah's ministry. But Elijah said, "God, you know whether or not anything can be transformed."

Before the storm I had about one hundred fifty, one hundred sixty members. Now I've got fifty people back. I already know what needs to be done. First thing I have to do is deal with this issue about people angry with God. Was Katrina something God did or man's neglect? And even if it was man's neglect, it still goes right back to God, because if he knew man was going to mess up like that, why didn't he do something to protect the city? I tell my members, "The question is now are you going to trust Him? Before Katrina, you said He was good. Can you still open your mouth and from your heart still say and acknowledge that God is good?"

I made myself available. I met with people one-on-one. You're not going to get a group of people to come together to say openly, "I'm mad at God. I've got to fend for myself right now." In the secret chambers of their hearts, they'll say that. People in the world are looking for Truth. The church is being what the church is supposed to be: a source of hope and inspiration. I have seen the transformation. People were coming with all of these questions. People were taking it personally. "How come this happened to me?" "If God just wanted to deal with you, do you think he would have put everybody out first?" I dealt with their pain and their questions. Rebuilding as a pastor has been exciting. I preached a sermon dealing with Peter. Peter's issue was he lost focus. What you have to do is remain focused. I can understand how somebody can be depressed and go through all of those emotional roller coasters. I'm just not going to do it.

It is not about my wants and my wishes. What the prosperity doctrine has done is give people this idea that God is an ATM machine. Say, "Hallelujah, praise the Lord," press the button, get a blessing. It don't work that way. Solomon told us, "The conclusion of the matter is this. Fear God, keep his commandments, this is the whole duty of man." That's my responsibility. I discovered He does what gives him the greatest glory. Me dying in 1982 was not going to give Him the greatest glory. So out of this, God has to get the glory and the church has to be the one to point people in the right direction.

It's not about trying to put no bricks and mortar back together. All of us need our houses cleaned up. So while we're cleaning each other's house up, we going to sing. While we clean each other house up, we going to encourage each other.

"I Cannot Let People Define Who I Am"

So we're at this trauma retreat for pastors after Katrina, with some group from New Jersey or New York. They say, "We want to help you all deal with trauma." I said, "What is trauma?" The response was, "I don't know." Right there you done lost me. I said, "Man, I don't even know if I'm traumatized, because this didn't affect me in the way...." He interrupted me, "You're in denial." I said, "Come on man. What I'm denying, that I've got a house that I can't live in? How do I deny that? How do I deny that I'm living in Greenville, Mississippi?" They said, "What do you do to handle this stuff?" I said, "Number one I find a preacher who is going to preach to me, because I'm a preacher and I need to be preached to. When you're preaching, you're pouring so much out of you, you need to be restored. Secondly, if I have ten dollars worth of quarters, I'll wear this Pac-Man out. I'm not trying to solve all of this in one day. I do what I can today, but I'm not about to lose my mind behind this. You going to die. It's still going to be here, huh?" That's the way my people are. You don't let stuff kill you. I said, "If I didn't go crazy lying under that train, you know I'm not going to go crazy over this."

I was at a meeting in the Lower Ninth Ward. We had a group come from out of town and they was trying to say that we could do this, we could do that. I said. "We don't need you all coming down here trying to tell us what to do. We know what needs to be done. We don't have the resources that we need to get it done." I say, "These pastors who the Lord has burdened to come back to the city, the first thing we needed to do is to get on the mountain with God. 'God, what do you want done?'" I said, "God is speaking to us and giving all of us our assignments."

Churches Supporting Churches (CSC) was a seed that has been deposited in the hearts of pastors that have flourished the hope. It never was, "I feel sorry for the people." It was, "We see the long-term, we see what it's going to take. The church has to come back." So the churches in New Orleans need to be strengthened by the churches outside of New Orleans. There is a part of the body that is wounded and those blood cells that rush to that area where the body's wounded, need to rush there and start helping heal that part of the body. What they did was find out where is the hurt for the pastors. "If you're burning to be in New Orleans, and God gave you an assignment there, how can we help you get back to that place? Once your needs are met, what are the needs of the community?" So you need to walk around, you need to look, and you need to talk to people. Tell us what you see. Once you see, the next thing you want to know as it relates to CSC, is what do you need in order to bring that vision to pass? We'll train you in those areas.

When you get help, then you start helping your brother next door, because you got ten churches across the nation who are helping you. Now if they have enough faith in you, if those ten can help you, then you should be able to help the other pastor around the corner. Now maybe you didn't know who he was before Katrina. It doesn't make a difference if he's Baptist, Catholic, whatever. Can you all work together for the betterment of this neighborhood? Do he see what you see? That's the church being the church.

Conclusion

God was in Greenville, Mississippi. Every preacher I've talked to, wherever they were, says, "God was with us. God provided." So what did we learn from this? There are certain sayings in the Baptist church among Black Baptists which you just grow up hearing. "He's a doctor for the sick, He's a lawyer in the courtroom, He's a mother for the motherless, He's a father for the fatherless, He's a bridge over troubled waters." We got a roll call. We used to make fun of that as little children, but I understand it in a clear way now, because He's all of that and more. All of these sayings has a purpose, a story behind them, and it is for my generation to understand, know, and embrace. This is how my people were able to deal with the slavery and segregation, because they knew some way, somehow, God was going to deliver them. But you don't really get it until you have an experience. Katrina is an experience. I lived in the Lake Carmel subdivision on the other side of I-10. Down the street is Eastover with the million-dollar houses. I drove through Eastover after Katrina. Katrina respected them the same way it respected our house. The lesson to be learned is not just for the city, this is for the world.

Katrina did not affect the older people the way it did the younger people. A man in the Ninth Ward was there by himself in his house. The man was eighty something years old. They asked him, "You ain't scared?" He said, "I was in World War II. What I need to be scared down here for?" It's because he knew how to survive. For our older people, it's the emotional thing that was hard. Far as their material stuff, they aren't tripping over that: "Oh man, I lost my flat-screened TV." They know how to survive. They know even how to recreate those memories. They know how to say, "Yeah, I lost that picture, but it was always in my heart anyway."

God challenged me one day. God said, "Who are you?"—"Aldon Cotton."— "That's your name. Who are you?" "I'm your child." "That's what you are. Who are you?" I finally said, "I am encouragement." That's what I am. I understand why God anointed me to come back to this city. Where else do people need encouragement? You send light to darkness.

I know what life is and what life is not. Jesus says it so well in Luke 12:15: "A man's life consists not in the abundance of things which he possesses." I don't need material things. It's convenient, but that's not my joy. When you see the lives of people change for the better, that's what I'm about. In the few years I've been on planet earth, I've learned what it takes people decades to learn because I've been in a nobody-but-Jesus situation under the train. You learn to enjoy life. God says, "The just shall walk by faith." I don't need physical legs to walk by faith.

COMING OF AGE

TWENTY-FOUR

Leslie Lawrence

Born in 1972, Leslie Lawrence was raised in a middle-class suburb of Denver, Colorado, where she lived until graduating from high school. Her summers were spent in New Orleans, where both of her parents, an engineer and a principal, had extensive family networks. From the age of fourteen, Leslie dedicated herself to the pursuit of medicine. She chose to attend Xavier University, where she was a pre-med student, majoring in biology. After graduating jointly from the Charles Drew University of Medicine and Science and the David Geffen School of Medicine at the University of California at Los Angeles, Leslie returned to New Orleans for her residency. At the time of Katrina, Leslie had just begun her tenure as Chief Resident of the Department of Psychiatry and Neurology at Tulane University's School of Medicine.

The interview was conducted on April 6, 2008, a Sunday, at a folding table in a makeshift office in a stately home in Mid-City.[1] A black-and-cream mud cloth dominated the wall opposite Leslie, who was wearing a magenta, loosely crocheted, cotton sweater with capped sleeves over a teal t-shirt with blue jeans and long, dangling, stained glass earrings that matched one of her four bracelets. Around Leslie's neck was a delicate necklace with two antique rings from her oldest surviving aunt, the only personal artifacts she salvaged after the storm. The tall granddaughter of B.W. Cooper has brown eyes, long, curly hair, and a captivating smile.

The weekend before Katrina, Leslie was in Atlanta, where she was visiting a friend. She adeptly used every form of technology available to her to gain safe passage for her mother and great-aunt out of the New Orleans area and to support her residents, at least two of whom were on duty at Charity Hospital. Leslie's actions are an illustration of the loyalty New Orleans can inspire, even among men and women who were raised elsewhere. Leslie is part of a group of African Americans who are reversing the trend of outward migration that began in the 1980s.

My maternal grandma, Noella Cooper Rogers, was born on Christmas Day. It is to her memory that I dedicate this chapter. She was the second to youngest of six children. They grew up in the Seventh Ward on Havana Street. Both of her parents were black but during that era they were described as "Creole." Her mother, Amanda Rogers, was of Mexican descent and Amanda's mother spoke Creole, which is broken French. Noella's father, Simon Rogers, is of French descent and actually changed their family name from "Rogèr" to "Rogers."

Growing up in the Seventh Ward, not only the family, but all of the people living there were close. The kids grew up together, graduated from school, and had successful professional careers. Most of the women, like my grandma's mother, were housewives. The men were bricklayers or worked on the railroad. They would walk to the Circle Foods Store.

My grandma was smart and witty. She even got accepted to Columbia University in New York to get her master's degree in teaching. She was ahead of her time. She didn't follow through with it because she wanted to marry. She did, however, pursue a career in teaching. To this day, many people tell me their fond memories of working with Mrs. Cooper.

Before my grandma began both her career and motherhood, she struggled with mental illness. I really believe that she is someone who persevered. She was a strong woman. My grandma was actually diagnosed at the time with schizophrenia. Because she was black, she wasn't accepted into any inpatient hospitals here in New Orleans, so she had to go all the way to Chicago to be treated. She received electric-convulsive therapy, and she was in an inpatient psychiatric facility for six years. My aunt Carmen (now Morial) was a big factor in getting my grandma well. They wrote like love letters to one another on a weekly basis when my grandma was in the hospital. Their brother Simon, my uncle, had a relationship with the nurse in Chicago so that she could get care. Once my grandma was finally discharged from the hospital, she was on no medications or follow up aftercare with a psychiatrist. The only "treatment" she did for herself was smoking cigarettes to "calm" herself from stress. We believe that she was bipolar more than anything else. When she was in her seventies, she had caught a small fire in her kitchen. She put it out with her hand, and then stopped smoking all at once. After that she started to have more mood swings and was more irritable, but other than that, she was fine.

She was great to be around, brilliant. All of my family members loved her. She was their favorite aunt or great-aunt. She had a gregarious laugh. I see myself in her all the time.

My great-grandmother on my mom's side, Edna Cooper, was Jewish, and I think she was some heir to Maison Blanche. My great grandfather on the Cooper side worked as a Pullman on the train and is of African descent.

My maternal grandfather is named Benjamin Walter Cooper. He was born on October 2. Unfortunately, I did not get to know my "papa"; I was two years old when he died of a heart attack on the interstate near the exit of Chef and the high rise. He was changing his tire. My papa fought in World War II. He was described as a very stoic, yet warm, gentle man once you got to know him. His nickname was "Gentle Ben." My papa by the way is Benjamin W. Cooper of the B.W. Cooper Projects. He was the manager, and to this day everyone that lived in the projects then tells me that they wanted to live there. When he would inspect the projects, they had their homes tidy. They were proud living there.

My grandparents had two children, my mother, Gaynell Cooper, and her sister, Penny Cooper. They're thirteen months apart. My aunt was born first, and when my mother was born, my grandma just found out that her older sister May had died, and that's when she had to go into the inpatient psychiatric hospital. So at that time, my mother was raised by the Cooper family on Conti Street. And then my aunt Penny stayed with my grandma's mother, Amanda, because my Aunt Carmen had to be the one traveling back and forth from Chicago. For six years, my mother and my aunt were separated from my grandma. They would see each other through the week, but they didn't live together. Today, no one would know that they were raised separately for six years because they act like twins. They didn't move to The Park until they were seven when my grandma was back. My mom went to Xavier Prep and then Xavier University of Louisiana.

My father's name is Arthur Leroy Lawrence. My dad's grandparents are from Labadieville. It's near Lafourche. He was born in New Orleans and raised in Algiers. He is the oldest of four siblings, and he lived for a while in the "Fischer Projects." It was when they were nice. My father has always been self-determined and intelligent. He graduated from Landry with honors, and then from SUNO with an undergraduate degree in engineering. He landed in Denver and had a good career in engineering at Martin Marietta in Denver. My mom was eager to go, because my "Nanny" (Aunt Penny) was already there teaching. Denver at that time was hiring African Americans because of the affirmative action requirements. My dad loves Denver. He doesn't like people being slow at the grocery store. New Orleans is not his pace.

"The Best of Two Worlds"

I was born on June 14, 1972, in Denver, Colorado. I lived in Denver for eighteen years, but every summer, beginning the first summer that I was born, I was in New Orleans. Back in Denver, my mom was a teacher. She taught second grade, and then fourth grade. Later she became vice-principal, and then principal. She would send me to be with my grandma, and then meet me and my brother later. It was exciting. Because my birthday fell in June, I always had two birthday parties. New Orleans was my second home, and I loved it. To have the cold climate and then come to the humid, warm climate in New Orleans was great.

My brother, David, is four years younger than I am. My nanny only has one son, Brandon, who is seven years younger than me, and like a brother to me. When I turned six, we moved to Heritage Village, a subdivision in Littleton, a suburb of Denver. The bedrooms were upstairs, the kitchen and den were on the ground level, and then we had a basement. The subdivision was definitely middle class. I was one of two or three blacks. I went to Littleton High School. I was the only black in all the honors classes.

My mom was light-skinned. She has straight, jet black hair and a radiant smile. I think she's beautiful. Being in Denver, her New Orleans accent prompted the question, "Where are you from?" She would proudly say, "New Orleans!" She drilled it in my head that she was black, and let me know that I was beautiful no matter what skin color or hair type I had, because she wanted me to have high self-esteem. She left nothing to chance.

I was always in some type of extracurricular activity. I did synchronized swimming with the Tarpons swim team for three years and dancing with the modern dance troupe Cleo Parker Robinson in Colorado. And then I was in Jack and Jill, Inc., which is a middle-class organization for African Americans. Because my parents provided me with recreational and social activities, I was able to have a balanced social upbringing. The summer months were the fun part, when I was able to connect culturally with my roots, be with my extended family, and be really comfortable with myself.

I had a very good upbringing in Denver, but I love New Orleans! I have an older cousin. She said, "I remember as a little girl, you said you were from New Orleans." I think I was drawn by the food, the culture, you know, the people coming in, and the kiss on the cheek. It's the nurturing part; everyone knowing everybody, just feeling safe. I didn't get that in Denver, even though we could leave our garage door open. There it was more of a superficial "hi." But here you might have gates on your door, and your door might be locked, but it's still, "What's going on?" It's just lovey-dovey. There wasn't anything for me as an adolescent to do in Denver. The only other place I can think of that is like New Orleans is New York, and I have traveled abroad.

I was raised Catholic. I have close ties to St. Leo the Great Catholic Church. It was my church before the storm.

Becoming a Doctor

I knew I wanted to be a doctor when I ended up getting this skin disease called *molluscum contagiosum*, when I was about fourteen. Basically it's a common skin disease similar to warts. The treatment for them was freezing them, and that was a very painful experience. I didn't want anyone to experience that pain. My mom had found a diary, however, that said I wanted to be a psychiatrist. I attended all of the summer schools at Xavier. They had BioStar and ChemStar. My mom got us into the programs.

I got accepted to Xavier. My grandma always told me that I should marry a doctor, not become a doctor. My major was pre-med biology. I ended up getting a partial scholarship to Xavier, and my grandmother paid for the rest of my schooling, so I don't have any debt from college. Once I got to Xavier, it was etched in stone that I definitely wanted to be a doctor.

For med school I chose Charles R. Drew University of Medicine and Science/ University of California at Los Angeles (UCLA) Medical School. It's a combined medical school. It's the only historically African American medical school west of the Mississippi River. Charles Drew Medical School only accepts twenty-four students a year. The first two years the medical students attended UCLA to do the academic work in Westwood, and the last two years, we did our medical rotations in Compton, California at Martin Luther King Jr. Hospital; hence I was able to see the best of both worlds, being exposed to Westwood, the academic world, and then being exposed to medicine/trauma and treating the underserved, both African Americans and Latinos. At that time, Martin Luther King Hospital was known for trauma surgery, and in Compton, the Crips and the Bloods (the two most prominent African American gangs in Los Angeles in the 1980s and '90s) were still fighting. I thought that psychiatry might be a job for me, but it wasn't until I did a rotation at UCLA that I confirmed it.

When it came down to choosing a residency, I knew I wanted to be home. I fell in love with the Tulane program, and the program director was great. When I did finally move to New Orleans to do my residency training at Tulane, Tulane was contracting with DePaul Hospital on Calhoun Street to allow their residents to do their inpatient and outpatient training at DePaul. My aunt Carmen told me that my grandma could not get into DePaul because of the color of her skin. The ironic thing was here I am a doctor at this hospital that didn't allow black people into their hospital facilities. My grandmother would have been really proud of that fact.

So when I moved down here and did my residency, I was living in my grandma's house on Mithra Street because she had died. It's right off of Press Drive near SUNO in Pontchartrain Park. I lived in my grandma's house just the way it was. I loved it. I entertained, and I used my grandma's dishes. I was a pack rat. If I went to weddings, I kept the wedding invitations. I also had pictures of everybody, including pictures of cousins at different ages.

My grandma's house was surrounded by a very warm community. I had two neighbors that watched me better than my alarm system. So whenever I was out of town and I forgot to tell them, I would hear about that! I lived across the street from Coghill Elementary School. My aunt Rose Jase was down the street within walking distance. My aunt Audrey Terry, B.W. Cooper's sister, was around the corner, and my godfather, Wilfred Duplessis, was down the street.

In the spring of 2005, I ran for chief resident. It was a peer election, and they elected me as chief. Right before the storm, I organized a retreat to the Children's Museum, so everyone could know one another. The fiscal year for residents is July 1, so I figured it needed to happen early so the interns and the second year residents could know the third and fourth year residents who never get a chance to see them. People exchanged phone numbers.

Katrina: "We Started Texting Each Other"

When Hurricane Katrina happened, I wasn't in New Orleans. I had gone to Atlanta to visit my friend for a weekend getaway. I went there for a trip two days before not knowing that Katrina was going to happen. I was trying to pack lightly, so I didn't pack my computer.

I was the chief resident of the Tulane University School of Medicine's Department of Psychiatry and Neurology residents. Whenever there's a code grey, like Hurricane Katrina, the chief resident is in charge of making sure that there's someone covering Charity Hospital's psychiatric ward. There was one adult faculty-staff psychiatrist and two interns covering seventy-two patients. Family members dropped off loved ones with a history of mental illness. The staff was in Charity Hospital for five days without electricity and food was getting scarce.

The Tulane Department of Psychiatry and Neurology had two chiefs. The other chief, Kris McCoy, and I were working together as fast as we could. Kris, who was located in Lafayette, was putting together e-mail access so we could all talk amongst each other on computers. The program director, Patrick O'Neill, received a text message from an intern, which he forwarded to me, and from then on, we started texting each other. The text messages went through. I had the log of all the cell phones of all the interns and residents. Once I found the interns, I called their parents to let them know that I had at least gotten in touch with them because the interns couldn't get in touch with their parents.

At the same time, I was trying to locate my family. My mother had moved down to New Orleans when my parents separated during Mardi Gras 2005, and she was living with my Aunt Carmen in Gentilly. At the time of the storm, they were still in the New Orleans area. I was terrified. I didn't talk to my mom for two days after the levee broke. It turned out that my mom and my Aunt Carmen went toward the storm to Covington in St. Tammany Parish, about forty-five minutes from New Orleans. They were in a mansion surrounded by gigantic trees. Through the night my mother could hear trees snapping like toothpicks. They just got wind damage. After the storm, she went out there with other men, trying to break the trees blocking the road down with their sawing equipment. My Aunt Carmen's best friend told her to go to Houston.

Reclaiming Existence

At the time though, I was saying that I was going to stay in Atlanta. My best friend, René Smith, was in Atlanta. I was thinking I would give up my chief residency and just complete my term and stay in Atlanta. Then I remember hearing Ray Nagin talk. He was really advocating for some help in New Orleans and at that moment, I was like, "I'm going to continue the fight. I'm not going to let down the residents. I have a job to do. I will not give up. I love my city too much. It's not about me, it's about other people." So I decided I was going to go back to Houston, where Tulane relocated, and continue my residency.

I have a lot of support systems. Friends that lived in Atlanta and that graduated with me from Xavier started contacting their friends. They were giving me clothes and money to get me to Houston. As I traveled to Houston to meet my mother, the engine blew out on my Audi. I had to take a tow truck all the way over to Houston, and it was a frightening experience. I prayed, "Please just let this tow truck driver be a good guy." And he was. But I remember seeing my mom at the Houston repair shop, and finally, I got to hug her. It was such a relief. She was safe. I don't remember the details because some of it I've blocked out because I remember that I didn't want to go into how traumatic it was. I remember having to do a job. That kept me focused.

I ended up living at an elderly retirement home in Houston. I continued my job as chief. Tulane University School of Medicine had negotiated with Baylor University School of Medicine so that the residents and medical students could continue with their training. I knew that the residents didn't have a place to live. My Aunt Carmen's friend had a daughter-in-law who was a faculty member at Baylor, and she gave me some numbers to contact to get some housing for the residents. I was able to do that for the residents, because all the official efforts were focused on the medical students.

Residents were getting in car accidents, they were having meltdowns, and the last thing I could do was have a meltdown. I had to be strong for them.

As people started getting into their rotations, I had the great experience of working with children from the Lower Ninth Ward. The children went to a school called New Orleans West KIPP (Knowledge Is Power Program) Middle School. Tulane was asked to help with behavioral problems that the teachers were seeing. I took on that responsibility as a community program. It was a mutually beneficial experience for the school, the kids, and the Tulane faculty that helped because we were all able to share our stories and provide healing in an innovative way by journaling through the workbooks. All of the kids loved the journaling because it was something no one could take away from them. They were separated from their parents and had been transitioned to a new city. I gave them pre- and postsurveys about their PTSD (Post Traumatic Stress Disorder) risk factors. The intervention with the workbook helped decrease their risk factors.

While I was doing that, in September 2005 I went back for the first time to see about my Aunt Carmen's house. It was very early, when there were no women around. It was very surreal to be there, to have the Red Cross meal and think that's the best thing ever. The flies and the odor you could smell coming in right through Kenner. We started gutting my Aunt Carmen's house first, because she was elderly. Before the storm, my Aunt Carmen's house was this beautiful house. She had furniture from Italy and people loved to come see her house because she had so many ornate things. My brother flew in from Colorado and we gutted the house. Then we went to see about my grandma's house. I was truly set on living there. My grandma's house got nine feet of water, and we weren't able to salvage much. We said a prayer.

When I came back, the neighborhood associations were meeting where people could voice their opinions on how they wanted to rebuild their communities. My mother's goal was to be active in Pontilly, the post-Katrina name of the Gentilly and Pontchartrain Park areas. She wants that neighborhood to be active again and the golf course to be manicured again.

I was going to look into buying some property in the historical Seventh Ward, which I did. I ended up living on one side of a duplex, and I rented out the other side. I knew that wasn't going to be my permanent place to live. It was just a place for me to lay my head down. I didn't decorate. It was a good place for me to be with neighbors. Everyone knew one another before the storm. When I moved on Havana Street, I knew Sandra, who lived across the street, and she ended up being my new "watch person." I needed that place for me to just be settled. I didn't have to rebuild my rental house, but I had to find another house. All of my other clothes were in storage. I was wearing the same stuff. I knew that I wanted a house. I was making choices.

Psychiatry and Katrina

Before the storm, my family would make jokes about me being a psychiatrist. Now it's a serious thing. Getting the funds from the federal and state government, we can now treat more people. We still don't have enough psychiatrists, but I think we are treating people differently.

Some people are still traumatized that they're not in their homes and that things are still not the way they used to be. I've asked people who visit the city, do they see a difference? And I like to ask tourists that a lot, because I think they are less biased than the ones who live here.

Violence has become an issue here, and I'm talking about the youth. Substance abuse has increased since Hurricane Katrina. People of all ages are drinking more now. There's more suicides. People are having sleep problems. A lot of the primary care doctors are treating people with other agents that may not be therapeutic for what's really going on. I work at a drop-in center that provides free mental health services for homeless youth between eighteen and twenty-four years old. Sometimes follow-up is hard, when you give them medications that need to be monitored with labs. It's hard to get homeless people to make appointments to come back. My mom asked me, "How do you get relief at work?" I said, "That's my job. I don't see myself in those people's places. I know how to separate myself from it. I work hard, but I also play hard."

Conclusion: "I'm No Longer a Gypsy"

We need more leaders in our African American community. Many of them are scattered because of the storm. Most of the middle class is gone. Many doctors and pharmacists left because their businesses were destroyed, or they got new jobs or promotions in other cities. It doesn't even have to be that far away, it could be Baton Rouge, but they're not here. They may be still hurt. They may feel they were forced out. But we have older people whose doctors have not returned, and they're not comfortable going to strange doctors. We have African American youth, graduates, who need mentors. I was at the beginning of my career. These people that left were middle-aged or nearing retirement, and they lost everything.

For me, I don't feel Katrina was a "woe is me." My Nanny said, "Leslie, I can't believe you've been through all of this, and you just keep going." But I keep hearing other people's stories. I mean, my story is nothing, nothing! No one in my family died. We all were able to get housing. And most of my family who evacuated are back in New Orleans. In December 2007, my Aunt Carmen was able to move in to her house. It's back the way it was. My Aunt Carmen thanks my mother for helping her rebuild. My mom went through a divorce at the same time.

I found a house, and I fell in love with it. It didn't get any water from the storm. It's in Lake Oaks, one of the wealthiest subdivisions in New Orleans, near Lake Pontchartrain and UNO. Before I moved into my house, I did some renovations to give the house an updated look. My cousin, Ryan Duplessis, was the contractor. He's a perfectionist. Now my house is perfect for entertaining and has a New Orleanian style for southern comfort. As you drive up, there's two French doors and lantern lights with a side, marble, gallery porch. When you enter the foyer, if you keep walking straight, there's the den, and the den is painted a deep orange. To your right is a brick fireplace and the mantle is white. If you keep going straight, you'll see the pool area. Then before the sliding doors to your right is my kitchen. It's a New Orleans-type kitchen with a brick tile counter. The dining room and living room have a formal look. My bedroom is a coral color with a teal-colored bathroom. The guest room is painted in gold. The following bedroom is painted rosebud and it's going to be like a TV entertainment center. My office is painted citronella and faces the pool. I had the house blessed yesterday, and I slept in it for the first time last night. I'm close to normalcy.

My long-term goals are that I want to get married and I want kids and a family. If my children were to read this, I would want them to live every day to the fullest, enjoy people's company, be good to people, listen to people's stories, and cherish family. I want to be in some type of dance class regularly for the exercise. I'd like to run my own private practice; possibly in my own home. I feel that working with the homeless people while I'm young and have energy is great. I don't want to burn out. I'd want to open something for working-class African Americans.

I'm so excited to be back in my city! I feel that it's an opportunity to be here. I think mental health would not have been looked at as strongly, if the New Orleans metro area wasn't devastated by Hurricane Katrina. I'm an Assistant Clinical Professor in the Department of Psychiatry and Neurology at Tulane University School of Medicine. I like the fact that we have a lot of young adult professionals that want to live here and make a difference. We are trying to network here and challenge the image of New Orleans as a place where people don't want to live. We want to show the world that there's opportunity here. I remember when Hurricane Katrina first happened, and people said, "It's going to take about eight to ten years for New Orleans to redevelop." It's been three years. I was in the French Quarter having beignets and café au lait, and it looks like a thriving city already. You have to think about how soon you think changes can happen when you have the worst tragedy of a city to happen in world history.

TWENTY-FIVE

Le Ella Lee

Born the second of three children in 1987, Le Ella Lee, or Brandy, as she prefers to be called, was raised by her mother, Velbert Stampley, who found expression through her cooking in a catering business she ran before a car accident left her in constant pain. Lee's maternal grandfather was an active participant in the freedom struggle in Natchez, Mississippi, for which he was imprisoned and tortured for six days in Parchman Prison.[1] Before he was taken away, he told his wife, Annie Stampley, he would gladly go to prison if it would lead to a better life for his children. He died shortly after he returned home. In the fall of 2005, Le Ella began her senior year of high school in Metairie, a predominantly white, middle-class suburb of New Orleans, and graduated in a suburb of Birmingham, Alabama, after a short stay in Cullman, Alabama, where Le Ella and many distant relatives ended up after the storm.

On December 15, 2005, Brandy came to the Birmingham Urban League with her mother and baby brother after school on an outing that had Wal-Mart as the final destination.[2] She was still dressed in school clothes with her hair pulled back in a ponytail. The interview was conducted in the Urban League's conference room with a buffet of food along one wall. Temporarily hanging on the inner wall of the room was a map of New Orleans, which drew tears when it was spotted. When it was Le Ella's turn to talk, Velbert and her brother Mikey left the room. Immediately, Brandy was transformed from a withdrawn listener into a self-confident performer.

Le Ella's trauma was primarily a product of her abrupt separation from her friends in her senior year. Several of her closest friends were never heard from again. On a trip back home to see what belongings could be salvaged, Le Ella discovered the body of an older neighbor from a neighboring apartment to whom she had been close. However, by December 2005, Brandy had begun to imagine an even brighter future than she had thought possible in New Orleans.

Before Katrina I spent my whole life, every minute, in New Orleans. I liked the French Quarter best. There's always a lot of action there. You can ride on a horse and carriage all the way through the French Quarter. I love the beignets. I like the Riverwalk. It's in the French Quarter too. It's kind of like a mall on the river. It's very beautiful, especially at night.

For grammar school I went to Hazel Park Elementary, and that was in Metairie. I liked it there. Very good teachers. I miss the schools and stuff in New Orleans. I really do. I think when you make learning fun for kids, they want to learn. And that's what our teachers did back there.

My dad's a character too. He used to tell me every year, if I'd be in the sixth grade, that year he'd say, "You need to go to school and get your education. I only made it to the fifth grade." Next year he'd say, "I only made it to the sixth grade." He was trying to motivate me.

I'm very good at artwork. I want to be an artist, but I like poetry. I love to write. I love to sing and dance. I like to talk in front of big audiences. I like to play volleyball.

"Dramatizing" Events

She flew through New Orleans / Something like a raven / Left the city flooded like a snow dome that's been shaken. / Many left, many stayed. / Others wept, but many prayed. / Lost families, separated lovers, crying children who can't find their mothers. / Diabetics without insulin, and / everybody coming together who never cared for one another.

We had just got off the interstate, and one of my in-laws was going into labor. We had at least eleven cars and trucks with us. My brother was driving the first car, and we noticed that everybody was slowing down. He gets on his two-way phone, "Why is everybody stopping?" They call him back, "My sister is going into labor." We're like, "Where's a hospital?" We didn't know where anything was. That was very dramatizing. We were thankful that she wasn't in New Orleans trapped in the water or in a house with no electricity.

We stopped in Cullman. At first, we were in the shelter. The shelter had to be the worst part. At 8:00 p.m. you had to be asleep. They'd cut the lights out on you. Grown people go to sleep when they want to. We only had cold water for our showers. You'd eat when they came and brought food in. We slept on air bags. We were all sick. Being in the shelter was very stressful. We felt like we were useless. We had nothing. Everybody was so depressed.

A lot of people in the area, church people and neighborhood people, brought us things: clothes, personal items, shampoo, and everything. The Red Cross in Cullman would not give us money, but they gave us the things that they wanted us to have. They went out and bought shampoo and hygiene things. We weren't of the same race as them, so we didn't use the same products as them, and we had to adjust to that because we didn't have anything. We get perms and stuff, straighten our hair out with relaxers. It wasn't working for us. But they were doing it out of the kindness of their hearts. Cooking for us three times a day and having church meetings and praying with us. It was mostly church people that were helping us.

There were a lot of problems with the shelters. The first shelter we was staying in, they booked a gun show there, so then we went to the hotels. The second shelter was like a miniature apartment complex, because it was like furnished with two bedrooms and a refrigerator and a TV and some stuff. And they promised us that we were going to stay there for two to three weeks, but they had booked a large amount of people to come stay there also. So we and our family members had to leave, and they promised us that we would come back.

At that point in time, we didn't have anywhere to go, so we were saying, "If we go to Texas, there might not be anywhere to go out there either." But after those three weeks were up, they never called us to move back into the shelter. And that was very traumatizing, because we were like, "We haven't heard from them, so what are we going to do? Where are we going to go?" It was very scary. We went ended up staying in another hotel. Moving from hotel to hotel was stressful.

In Cullman, I witnessed acts of intimidation.[3] But that didn't change me. I'm going to be who I am, no matter where I am.

Le Ella's mother, Velbert Stampley, described one incident. Brandy and two or three other girls were driving home from Wal-Mart. An apparently inebriated white man followed closely behind. Out of the window of his pickup truck he yelled, "You-all niggers get out of our town! We don't want no niggers in here!" Brandy called her mother, who made sure all of the forty-nine survivors were visible as the girls pulled up to the hotel. Eventually, the man drove off.[4]

Events in Cullman really made me want to better myself in life, so even if you want to discriminate against me or engage in violence, I'm going to be the better person about it. I tried not to let it get me down when I was up there. It motivated me. Get your education. Then you can be whatever you want, and nobody can look down on you, once you reach the top.

I had a lot of friends back in New Orleans. I talk to some of them. My school, East Jefferson High School, has opened back up. Some of my friends are back there, but others I haven't heard from. I don't know where they are. In a way I'd like to go back to East Jeff, so I can graduate with my friends and walk across the stage with them, but in a way, it's livable here.

I think if I could have changed anything, I would have convinced my mom to go with my big brother to Texas to be near family because it is lonely being out here with no family and no friends that I grew up with. I have a lot of friends and family in Texas. You can always call your family, "Come pick me up, or bring me by this place, or whatever."

Conclusion: "It's Like a Whole New Life"

At first, I just thought "Oh my gosh, my life is over. I have no friends, no family." I know most of my family's alright. There are still some family members I haven't heard from. I always wonder about them and friends and things like that. I was worried about our house in New Orleans for a while. But then I learned that materialistic things don't matter. It's life itself. You can always buy something to replace something else, but life itself is so beautiful to me now. Just knowing that I could have been one of the thousands of people who died is very touching to me.

Since I've been down here, it's been a whole new life for me. It's like I've been born again in Alabama. It's been fun. I've met a lot of people and made new friends. Sometimes you grow out of friends. You leave them and go off to college. It's best to do what's best for yourself. That's what I'm thinking right now. Now that we have a home, it feels a lot better.

Brandy lives in a three-bedroom house neatly furnished with dark green hedges lining the front porch. It is situated in a quiet, working-class suburb nine miles from Birmingham.

I got a camcorder as an early Christmas present. So I want to film a documentary. I just want it to be like, "What's Life Like Now: The Aftermath of Katrina." I really want to film how I'm doing and send it to my family and show them that I'm doing better.

I'm working on a book. I wrote a few pages in my mom's journal, but her book shows more of the spiritual, and how it's made her really turn to God for answers to her problems. She's very thankful that we weren't caught up or drowned in New Orleans. My book has that too, but my book is more like, "This is all the destruction and what happened in New Orleans, and this is what I'm living like now, and the people I've encountered in my school experiences." That's what I want my book to be: life. It is partly an adventure.

It's kind of boring sometimes because I don't know anywhere to go around here, but it makes me stay at home and just help my family. They're some characters. My baby brother, Michael, is the biggest character. He'll make you laugh just because he's laughing. He's a lesson too. He's teaching me how to take care of children. He makes me not want to have any children. I'm going to be thirty first. I'm sticking with my mom, and she's showing me how to handle my business and pay bills. I'm going to be out on my own after high school.

I will definitely go to college, so I can create a stable environment when I do decide to have children. I would like to be something like a marine biologist. I love water and animals. Southern Alabama University is right there on the beach. And if you take marine biology, you're going to spend a lot of time on the beach, and that'd be very beautiful.

TWENTY-SIX

Robert Willis Jr.

Born in 1983, Robert Willis Jr., better known to his acquaintances as BJ, was initially raised in a two-parent home on St. Maurice Street in the Lower Ninth Ward,[1] where he played baseball on his father, Robert Willis Sr.'s, little league team. Shortly after his parents separated, BJ and his mother, Paulette, moved into the Seventh Ward. Robert described his pre-Katrina life as a "hurricane." A rapper and an amateur barber, before the storm BJ had lived a life he felt lucky to have survived at the age of twenty-two. He was kicked out of several high schools for possession of marijuana and threatened with incarceration after a serious altercation, but his father's lawyer was able to keep BJ out of a home for juvenile delinquents. After graduating from John McDonogh Senior High, Robert left New Orleans twice in the fall of 2002, the first time to attend Jackson State University in Mississippi on a basketball scholarship, and then to audition in Manhattan for a film part as a rapper. Failing at both, he returned to New Orleans in December 2002, and lived with his widowed grandmother, a retired school teacher, in Pontchartrain Park. After his father overcame his own drug habit, BJ moved in with his father and they concluded that his only chance to escape the danger of street life was to get a fresh start elsewhere. His second chance began in Memphis four days before Katrina. BJ experienced Katrina vicariously via television and his cell phone.

BJ is short in stature. What he lacks in height, however, he makes up for in muscle tone and swagger. On the day of this interview, September 30, 2005, he was wearing name-brand tennis shoes and a loose, gray sweat suit. Several of his front teeth were gold-capped, and almost every inch of visible skin including his knuckles was tattooed with symbols, artwork, and words. The interview was conducted on the campus of the University of Memphis in the conference room of the Benjamin Hooks Institute. An ornately framed picture of a young Ben Hooks with Martin Luther King Jr. was on the brick wall directly opposite BJ. On the wall behind him was a street map of New Orleans. A blue, yellow, and red kente cloth adorned the otherwise austere conference table where the interview was conducted. Baderinwa Ain was also present.

BJ passionately and vividly describes the experiences, beliefs, and priorities of some of New Orleans' youngest generation. His narrative articulates the quandary of some New Orleans youth, who before Katrina did not see opportunities beyond the self-destructive third rail of crime, drugs, and violence. Shaken by the storm and the way he saw his fellow

New Orleanians treated, he renewed his spiritual faith and vowed once again to give up drugs and violence. BJ's experience provides a glimpse into the effect of the government's slow response on a subset of young men who are usually feared and rarely understood.[2]

I was born July 2, 1983 and raised in the Ninth Ward. We started off in the 'hood. We didn't have too much when I was little. That's where I stayed at most of my life, like up to my teenage years. That's when I moved out. I stayed in the Seventh Ward for a little while. I stayed in the Eighth Ward. Where I was from, we hardly ever seen white people.

My mama's been working my whole life. My mama ain't never got food stamps, ain't never been on Section 8, ain't never been on no government assistance. My mama had me, my little brother, and my three little sisters, man, and my mama worked every day. My mama had her own car, her own house, and nice things. She tried to provide for me my whole life. I felt like the best thing I could do is make my burden lighter on my mama, let her know that she ain't really got to take care of me as much as I got older. Once I got to maybe fifteen, I let her know, "Just do what you got to do for my little sisters and for my little brother. I'm straight right now." She ain't probably want me to be out there and was probably worried about what I was doing. By the life I was living, even with cutting hair and selling drugs, I always was getting money fast. Cutting hair without a barber's license paid ten dollars a haircut.

I grew up playing ball. I grew up going to school. I liked Eleanor McMain Secondary School on South Claiborne Avenue in Uptown. That's where I almost had a fight with L'il Wayne the rapper. He's one year older than me. I got expelled for smoking weed. I graduated from John McDonogh High School. I went to Jackson State University in Mississippi for one semester in the fall of 2002.

I rapped at the Apollo Theater in 2002. They wanted me to audition for a movie. They picked me up and brought me to 157th Street and Lenox Avenue. I stayed about three days in Manhattan and that was the only times I left New Orleans: to go to Mississippi and to go to New York. The movie must have falled through. They never called back.

Life on the Streets

My hurricane's been going on for years, for years I am talking about! I done seen everything already. Since I was fifteen, sixteen years old, I have been exposed to drugs, guns, and violence. I'm talking about my heart was cold for a long time. It was just like every nigger for themselves. If you got something better than the next person, they just want to take it from you. In New Orleans, you really raised to look at the bad in people before you look at the good in people. If you at the mall or somewhere in New Orleans, and somebody leave their trunk open, and you sitting in the car with somebody, the first thing they might be thinking, "Somebody going to steal them people stuff," or "Let's steal them people stuff." All the way from elementary to high school, you got to have the name-brand tennis shoes, the freshest things.

My mama was never involved in none of it, and she didn't raise me that way. But that's all I saw and that's all I knew from all my friends, my brothers, and my daddy. I have friends on crack or heroine—nineteen, twenty years old just feeling like they don't have nothing. My daddy was a drug addict for twenty-two years, you heard me? My daddy did his best with me, was always a part of my life but he couldn't do nothing for me because he was chasing the drugs.

BJ's father, Robert Willis Sr., was a video editor for a television station in New Orleans. The job allowed him to travel the world. On the road, he switched from using marijuana to cocaine. Willis Sr. recalled that he had BJ buy cocaine for him on the street.

The job I had was only paying minimum wage, like $5.15 an hour. So your check might only be like $200 or $250 every two weeks. If you do got a child or a girlfriend, what you going to do with $250 dollars every two weeks? So I started doing the things that I had to do. I cut hair, but I started doing a lot of things I had to do to keep money in my pocket. Even though I know that those things was wrong, I just felt like that's what everybody else is doing in New Orleans.

It wasn't older people. It was just getting younger and younger and younger. They made a video called *New Orleans Exposed* (2005) and went through the projects. They had little bitty kids five, six, seven years old smoking weed. They asked them what they want to be when they grow up. They saying they want to be killers or drug dealers. I got tired of living that life.

I feel blessed to see twenty-two years of age. In 2002 at my high school, John McDonogh Senior High, some dudes ran in there with three guns and shot a dude up on the gym floor. So you got students in the ninth, tenth, eleventh grades seeing murder at school. If they seeing murder at school, how they are thinking everybody else living out in the world?

I try to keep my eye on the future / but the second you slip be the second a nigger shoot ya / 'cause where I'm from you live by the gun / you die by the gun. Every day you walk out your door you know could be your last.

Like I had a friend, Jamar. He went to jail when he was seventeen, and he got out in May 2002. We were all living with my dad in my great-grandma's house in Gentilly Woods. On Christmas day 2002, he ran in my house, and they killed him.[3] He had changed. He was trying to go to college, and he had took his gold teeth out of his mouth. That was God showing me that regardless if you trying to change, if you still in that environment and you still doing some of those same things, ain't nothing going to change.

I can sit here and tell you all kinds of stories about things that I done saw. Right before the hurricane, my friend was riding a pedal bike up the street. They shoot him in the head one time and when he fall off the bike, they stand over him and shoot him fourteen more times. So what you shot somebody fifteen times for? All you doing is try to send a message to everybody else that you the baddest, that you don't want nobody to mess with you. We got murder after murder after murder. We got little babies getting killed. Right before the hurricane, they ran in a dude's house and shot his girlfriend and his seven-year-old baby in the face.

When events come around downtown, they'll take the whole police force and make sure nobody don't harm no tourists or make sure nobody don't harm all these people coming from out-of-town, when they got nothing but chaos going on in the rest of the city. So you could just do whatever you want in the rest of the city. We might have four or five murders a night. Nobody solving the murders. Instead they taking care of people who are only going to be here for three, four days, and I've been living here twenty-two years and can't get help.

I used to always be the one that try to let my partners know that even though we selling these drugs and this and that, we can do something better. Whether it be through music or through anything, we just can't feel like this is our last resort, our only alternative. That's what I used to tell my friends but I never knew how to change it because I'm doing the same things they doing. But I always felt like we could do something better with ourselves.

Before the hurricane, I lived with my grandmother on Providence Place in Pontchartrain Park. My grandfather passed away in 2001. Two weeks before the hurricane, my daddy finally got a chance to buy him a house. It was a nice house behind the Gentilly Woods Mall. He had just got it. It was around the corner from my grandmother house. Even though it got destroyed, I thank God he had that opportunity because at that time I had a lot of stuff going on with my life, and I had strayed so far away from my family. By him getting his house, I got to spend those two weeks with my daddy. So we got a chance to talk to each other and that's when we made a decision that I was going to leave, start fresh, and try to get myself together. Hurricane Katrina hit on the twenty-ninth. I had came out here to Memphis on that Thursday before the storm.

My daddy doing chemotherapy. He's a cancer patient. Him and his wife evacuated to Gonzales. When Katrina hit, my mama was already in Lafayette because of her job; she works at a check-cashing place in New Orleans and works in Lafayette sometimes. I had a cousin who was still in New Orleans who lived with my mama. It was my mama, her husband, my cousin, her two sons, my three little sisters, and my little brother. So my mama told my cousin to drive the car to Lafayette because that's where my cousin's father stayed at. So my cousin took her two sons by her father's house, and she drove my little sisters to my mama. My mama's husband stayed for the hurricane and videotaped it.

Watching Katrina on CNN

It's just really, really hard. The more I was watching TV and the more I was seeing how they making fun of us on TV, making fun of black people like we ain't nothing, you know? Like we ain't sh** to be honest with you.

So now CNN and the world really got a chance to see / what it was like living a life of a nigger like me. / We can't look back, / God did that. We wasn't in a war but every day was Iraq. / Made you ride through my city in a boat / like a third world country that ain't got no hope.

I just feel like we need to come together to get through these times. People got to start opening their hearts up. It ain't about money or anything. It's about people just being there for each other right now because they showing us that we don't have anybody. What it looking like to me on TV was if you come from where I come from, you ain't have sh** and you was in the poor class, you going to stay poor. That's what they trying to tell you. Don't get up.

I done seen everything already. I done seen dead bodies my whole life. It's just me seeing it being kids and me seeing it so, so lumped together like that, a bunch of dead bodies. I already knew about killing. But it's just me seeing the way we be treated. How they could have all those people in one area and not come and help them? You got people on the news saying "Help us, help us" and nobody coming just because of how they look. When we do get a hurricane, y'all send the National Guard in there and just tell them, "kill everybody. Kill all the young men, just kill them." How you go in a city in America and give them shoot-to-kill orders?

"Picking Up Our Lives"

I called my mama to find out what she was going to do. By a lot of people leaving New Orleans and trying to relocate all across Louisiana, my mama said the store in Lafayette had too many people trying to work there. So they gave her a job at the store in Lake Charles. She moved my little brother, my three little sisters, and her husband. In Lake Charles, my mama said that they wasn't telling her where the Red Cross is at. The Food Stamp office would close some days and some days they wouldn't open them up. She said that the lines would be long all the way out to the end of the door. She said that they wasn't talking about paying for any housing or any temporary housing. She said the places they did show her was like run-down, hole-in-the-wall places that she said she never lived before in her life, and she wouldn't have my little sisters and them living there. She just took the little money that she had, found somewhere to rent and tried to start her life over like that. She rented a little mobile home, paid for some furniture and everything, got my little sister and them in school.

Then they told her there was a mandatory evacuation for Rita. So she told me she was coming out here to Tennessee. She only thought she was going to be out here for two or three days, and she thought she was going to go back to Lake Charles. She finds out that Hurricane Rita was a direct hit on Lake Charles. So now everything is gone in Lake Charles. And now my mom is really out here with nothing. We got our car, we got each other, but she has to go through the same process all over again. My mama had a big screen TV, a computer, all those kinds of things. Now my mama ain't seen a computer in three or four days. She really don't know what her next move is. She walking around looking confused and bewildered and all I can do is just pray for her that God just give her some kind of guidance. My mama way out here and she just looking for help, and I ain't saying help from the government. She just want to be able to be settled and know what she is going to be able to do with herself. My mama don't like being confused and not knowing what's going on. My mama like being in control of her own life.

I lost a few little poems and stuff that I wrote that I got my mama to copyright for me when I flew out of town. I had a little motor bike I lost, a few clothes, and furniture and stuff like that, but I ain't really worrying about those things. In Memphis, I got the food stamps already out here. We just going to try to pick up our lives and start from where we at right now. We just going to be thankful for the things that we do have, that we are alive, and that we all together.

I told the man I talked to at the place for the housing what I was trying to do. I told him I was a good barber and I had real nice barber skills. I said, "Could you give me a list of all the different barber shops and I'll go on my own to the barber shops and let them know that I want to cut hair and see if they going to try to help me?" An older black man was like, "Here's the phone book. That's all you need. Go ahead." So I was like, "No problem, sir I appreciate it. I'll just do what I got to do." I guess because I dress how I be dressing, I guess he feel like I got him looking bad. I don't know what people be thinking no more. Everybody got their own problems.

I feel like regardless of what's going on and what everybody trying to do, we got to use this time to turn to God and just be thankful for the things that we do have. This is only temporary. Life ain't over. They got a lot of people who care out here. You got to be willing to open your mind up and your heart up to it, you got to really be willing to accept help and stop thinking negative about everything. By me coming out here to Memphis, just meeting all type of different people, seeing people opening their doors up, letting thirty people come live there, giving out furniture, baby clothes, and stuff like that. Everybody is not automatically bad.

Conclusion: "I Know I'm Somebody"

I have a friend, Chad Charles, who said they left him down there in that water around dead bodies and dead people. He's one of my close friends. He said they shooting people right in front of him. He said a lot of people thought the world was coming to an end. He seen dudes kill theyself. He said they came down through the Lower Ninth Ward with tanks, jumping out with their army suits on with guns, telling them to stop running or they going to shoot them and kill them. He was like, "Damn, you rescuing nobody?" He was like, "Man they ain't care about a nigger. They wasn't trying to save anybody." He said you had to find the best way you can to get out of there. A lot of dudes my age, by the things that's going on, they're feeling like f*** everybody. They feeling like it's only me, all I got is myself so I'm going to do what I got to do.

Even though in New Orleans, I saw all the murders, it took this hurricane and for me to watch all those babies on TV dying, all those old people dying, all these family members missing, that's what made me realize that my heart wasn't as cold as I thought. I just couldn't live that way no more, because the devil is just doing too much right now for us to not care and for us to want to harm people and kick people while they down.

I got a lot on my mind, but it's really been like God been my counselor. I just sit down and people might see me and they might think that I'm talking to myself sometimes if they see me, but I really be trying to establish my own personal relationship with the Lord because it's like I've been straying away from him for so long. I always knew there was a God. I got "Trust in God" tattooed on my neck right here. I got it when I was sixteen years old but I never fully knew the meaning of it till I see all this stuff going on. Since this has been going on, I have been getting on my knees and praying to the Lord every night hard.

Every day I wake up and ask God to guide me and direct me. I just ask Him to let me speak the word that He want me to speak, do the things that He want me to do, and be the man that He want me to be. I can't change the way I am, or what I've been through, or what I've done with myself. I made a lot of mistakes. But I've got to move from where I am at right now, I can't look back and I can't turn around. I feel like God made it a lot easier for me because it's like I've been the one that everybody looked down on for so long, so I can't look down on nobody. I'm the one that everybody feel like is a criminal and violent and a thug and all I do is steal and all I want to do is hurt people and mess up the world. That's how they portraying me on the TV.

I know I'm somebody regardless what they talking about on the news. I'm twenty-two years old. I got thirty to forty more years ahead of me to accomplish all kind of things. If Martin Luther King would have of died at twenty-five, he would have been a Deacon and if Malcolm X would have died at twenty-five, he would have still been in the street breaking into houses. I know I got a lot ahead of me, and I ain't going to let none of that discourage me.

We got a lot going for us. Even though they be trying to put us down, everybody getting a chance to start over no matter where they put you at and no matter what kind of housing they give you or no matter what your situation is—use this opportunity to just make the best of it man! Don't let this opportunity make you become racist. Don't let this change you just because they throwing roadblocks in the way man. Just move them out the way and keep moving.

I want to see my family situated and my mama back smiling and knowing that everything going to be alright. I really want to start enjoying life with my family.

Hurricane Katrina is really like a blessing to be honest with you. I felt there wasn't no opportunities down there in New Orleans. I felt like it was never going to change. It's like God made it to where now I don't have to worry about what they're doing in New Orleans or missing my friends or worrying about my family. It's making me see a different world.

There are a lot of people who would be thankful for the things that I have. That's why I'm so thankful and that's why I can sit here and really speak the way I'm speaking because a lot of people in one shelter, they family in another shelter. Some people ain't have cars. A lot of people have to walk certain places or are stuck on the highway.

All these parents got to do is to try to raise their own children. They can't let music and rappers raise they children, if they the ones who made them, even if they feeling like the government not giving them no assistance and they don't have no money. It don't take a whole lot of money to teach your child morals and values and responsibility. It don't take a whole lot of money to do those things, no matter where you stay at.

I just want to do my best to try to help change my generation, the young dudes around my age, to let them know that the way we going and the men y'all trying to become is not the right men to become. I'm saying, we running around here with our pants sagging, boxers showing, t-shirts not tucked in. I'm not saying there's nothing wrong with that, if that's how you want to be, but you can't have that mentality in your head that that's what's cool. Just because your pants sagging that make you a thug and that make you hard. There's more to life than that and that's just what I want to try to explain to a lot of people.

We need to start making music that's going to help the younger generation. If they going to let rap be the thing that teach us, we got to use rap for a better way than what we using it for, not to teach the children that all girls are whores and that life is about selling drugs and making money. We got to change it for each other.

Toussaint Webster

Born in 1985 to the renowned Baptist preacher, Dwight Webster, and his wife, Trudell, Toussaint Webster, the second of four boys, always felt a bit of an outsider in his own hometown because neither of his parents were New Orleanians. In junior high, he traveled with other students from Lusher Extension to Senegal and Gambia. He graduated from Ben Franklin High School, one of New Orleans' highest ranked public schools, and enrolled at Morehouse University, where he majored in international studies. In Atlanta, Toussaint witnessed Katrina as a junior, suddenly confronted with the need to comfort his displaced parents and brothers. In his senior year, he was chosen to go to Geneva, Switzerland, as part of a Model UN team.

On April 11, 2008, Toussaint arrived at 9:52 a.m. for a 10:00 a.m. interview. He wore a light tan, short-sleeved polo shirt with a collar and button-down neck over a longer white undershirt and neatly fitted, belted jeans. His shoes were ruggedly casual. Not towering in height, Toussaint's presence was nonetheless commanding, as he narrated his life story with the self-confident poise of a Morehouse man accustomed from a very early age to the competition of intense sibling rivalry in a religiously conservative, culturally African household.

Raised to consider himself a child of the African diaspora and a world citizen, Toussaint was named for the Caribbean military leader, Toussaint L'Ouverture,[1] who decisively defeated slavery in Saint-Domingue and helped to defend the most socially egalitarian of the eighteenth-century revolutions against Spanish, British, and French invading troops, leading ultimately to the creation of the Republic of Haiti. Toussaint Webster reminds us of his namesake for his decision to return to post-Katrina New Orleans before starting graduate school to donate two years of his life to creating a better environment for the youngest generation of New Orleanian public school children.[2]

I was born in Rochester, New York in 1985 and raised in New Orleans. I think we moved when I was about one year old. I'm two years younger than my older brother, three-and-a-half years older than the one directly below me, and seven years older than the youngest. Being one of four siblings, there was competition, always, especially between my older brother and I. We competed for everything from grades, at one point, to basketball in the driveway. We always had an ongoing joke about who's the favorite of our parents.

So after my first "D" on my report card, I got made fun of, like, severely. When I joined the Model UN team and found out I was going to Geneva last year, I was totally calling my brother. "I think I'm the favorite now."

My great-grandfather on my mother's side died when he was ninety-four. I remember a few stories, like one time he shot somebody during prohibition. He was selling alcohol. He was raised in Camilla, Texas which is maybe an hour outside of Houston. He owned two houses right next door to each other in Houston, and we'd always stay with him.

My mom's dad died when my mom was eleven or twelve, so at first my grandmother was a homemaker when her husband was doing insurance or real estate. After he died, she worked in a department store.

Both of my parents are not from here. I think it makes a difference in the way I see things, and the way people perceive me when they hear me speak. My dad was born in '51. He is from Philadelphia. My dad was adopted. He was at Howard University in the late '60s for undergraduate. By the time my dad was twenty-one, he was a minister, and when he was going out in college, he was going out with the gospel choir. When he finished Howard, then he went to Colgate Rochester Divinity School for the divinity degree, and after that, he went to GTU (Graduate Theological Union), so he was in the Bay Area. That's where he met my mom. My mom was from Berkeley and Oakland. My dad worked at Colgate Rochester for a while, and then Dillard University invited him to become the chaplain.

My mom went to San Francisco State University. She got a degree in broadcast journalism. When I was in high school, she decided to become a licensed massage therapist. Up until my mid-teenage years when she started going to night school, my mom was always there. She'd usually pick us up from school. We were cooking dinner for ourselves by thirteen years old. My mom was always on some new health food thing. Every now and then she'd try and make something happen, like wheat spaghetti, and we'd just have to rebel.

They were always hard on us, though. Nothing was ever good enough. I remember getting punished for getting to the breakfast table late. We were like, "Do you know what other kids are doing?" And we're getting punished for not brushing our teeth on time. Every time I got into trouble at school, it was like, "We're not going to even call your dad."

My dad is the Senior Pastor at Christian Unity, situated in the Fourth Ward near the Sixth Ward. Even though a lot of our members came from the neighborhood, I'd say just as many came from New Orleans East. Before the storm, we had like five hundred members, three hundred on a normal Sunday. My dad would usually not come home until 9:00. He'd stay up and watch the sci fi channel, read the newspaper, and do crossword puzzles all at the same time.

With my mom, she was really a nice person in public, so no one could imagine her being the authoritative figure that she was at home behind closed doors. When we were younger, she didn't understand noise, like bouncing the basketball in the kitchen or just giggling at night when we were supposed to be sleeping.

I remember when my maternal grandmother came to New Orleans. That meant we could have whatever we wanted: white bread, beef, pork, cheese, ice cream, and expensive tennis shoes. When we were kids, the first thing we would do when she came was go to the mall.

I felt like financially we had less than some of our friends, but opportunity-wise we had more. Just the access to someone so well versed and knowledgeable like my dad. And his influence: We were in this low-budget film called "Sho' Nuff." Definitely wouldn't have known about that. Every summer, the Progressive National Baptist Convention was in a different city, so we got to visit a lot of different cities that most people wouldn't just randomly visit like Kansas City, Missouri. People from New Orleans weren't going to St. Louis, New York, Buffalo, of course Atlanta, or Washington, DC. We traveled to Barbados.

I think anytime my dad wasn't open to some of the decisions or possibilities my life was going to have, it was because he wanted something bigger. On the lower end, if I had decided not to go to college, because I was really amazing at something else, even if I was culturally conscious, I really think there would have been a lot of pressure for me not to do it.

Neighborhoods

The first place we lived in New Orleans was on Dillard's campus in Gentilly. Then we moved to the Ninth Ward on Desire Street. Then we lived on Cartier Street near the lakefront by Robert E. Lee Boulevard and Paris Avenue after Jean Gordon Elementary School. It was a pretty quiet neighborhood. One of our neighbors was Emily, and that's who I hung out with the most. One day she said, "My mom said we can't play with you anymore." And I was thinking, "Why?" And she was like, "You know, because you're black." And I'm like, "That's pretty messed up." I knew then that there was something wrong with it, but I didn't know how wrong it was.

Then we moved to New Orleans East to the Lakewood East subdivision, where I lived the rest of the time. It was halfway between Crowder and Downman. There's town houses on one side and Lakewood East on the other side. It was middle class. By the time we got there, it was mainly black. There were two white families when we got there, and they left.

It was a one-story, four-bedroom house. We had a pool, so we had pool parties. The biggest ones were like church pool parties, and they were fun. We really enjoyed them. A good group of our friends came from church. So we'd have our own break-offs, and do what teens and pre-teenagers do, and talk. Up until high school, pastor's house was not a bad hang out place.

What I liked about New Orleans was the access to culture here in New Orleans. I don't remember a year of not going to Jazz Fest. They used to have these celebrations in Congo Square. We went to all those celebrations of the black child, family day at the park, gospel things. The music was always there, the architecture, the art. What I liked least about New Orleans was the weather, the humidity, roaches, and the closed-mindedness because people from New Orleans know how great of a place this is, so they're not open to anything else.

I never felt afraid for myself. A lot of the socializing in high school was like house parties, and my friends lived in safe neighborhoods too. When senior year came and we turned eighteen, from then on we'd go uptown near Tulane University. But I knew my dad was afraid for us. He'd always say, "The majority of violent crimes happen between these hours and these hours that you insist upon being out of the house."

Either every year or every other year, we'd be in the Bay Area. I have a lot of memories of my experiences there and hanging out with family members. My family members were always happy to see me, so it was always a treat. I have good memories of the green hills and Berkeley. People always did seem to be a little more laid back or easygoing in California. A lot of the people that we knew in New Orleans, and I can't say we knew a whole bunch of angry people, but the way they communicated was just a bit more harsh than the Bay Area communicator.

Schools

I went to Jean Gordon. It was on Robert E. Lee, a block away from Paris. I had a lot of friends there. My best friend in elementary school lived in the other cul-de-sac. After Jean Gordon, I ended up going to Robert Mills Lusher Extension Elementary School for middle school in Uptown. It's an old courthouse near La Madeleine and Louisiana Pizza Kitchen.

Lusher was a little more difficult to fit in, maybe it was just the age. When you're in middle school, there's some pressure to be cool. None of these schools had uniforms, so there was a pressure to dress a certain way and hang out with the right crowd, and get invited to the first teenage parties that end at 10:00 p.m. I had to work hard to fit in at Lusher. There were some neighborhood kids that didn't fit the Lusher mold. I tried to hang out with them. Some days I was cool enough for them, and other days they were like, "Go hang out with the white kids." This depended on which shoes I was wearing, or my outfit, or what music I could quote word for word. The middle class of black kids was too elite for me. Their parents were mainly doctors and lawyers. I took theater. And that's where like most of my white friends came from, because I was like the only black guy for most of the time. There's always the token black person. I didn't find it until like the end of eighth grade, but the group I was most comfortable with included a white guy from Uptown and a black guy from Gentilly. The white guy was a white person that a lot of black girls liked, so it worked. That was our crew. We shared in common WWF (World Wrestling Federation) wrestling. We would always talk about that, and for some reason, it was cool enough for the girls that we knew. In '97, my older brother and I were part of Kids to Africa. We went to Senegal and Gambia. We were the first group to go. I was in middle school, so everybody knew that going to Africa was out of the ordinary, and the school made a big deal about it. I was twelve. Just being out there and figuring out how similar we were to kids on the other side of the world was huge. We did the Door of No Return from the slaves' dungeon. We had time to think about how far we've come and having come back voluntarily to see it, we've come even further. Just being exposed to things that aren't the norm for kids in New Orleans, even privileged kids, and being able to talk about it gave me a sense of confidence. I felt like I could stand in a circle of adults and have a conversation about some of my experiences, and I would bring something to the table. For high school, I went to Benjamin Franklin High School. I started at middle linebacker my junior year. I don't remember struggling educationally until I got to Franklin. Franklin is actually on the UNO campus on the lakefront. At Franklin, we had access to the UNO library. There were a lot of resources, but there was so much pressure academically. I just remember being stressed out for four years. I used to hate the days when my parents did make us go to church after school. I was like, "I really do have three hours of homework." By my dad being pastor of the church and always wearing African clothing, people at school would ask me if we were Africans, and the answer would be weird, because they taught us that we were, but not just directly from Africa. We were like, "kind of." My mom had dreadlocks and wore a lot of African clothing too. My dad played the drums. Black kids made their assumptions. And the name that they gave me made life difficult, because I always had to explain it. But by the time I got to high school, everything became cool all of a sudden. Overall, I didn't have any traumatizing racial experiences growing up in New Orleans. I don't know if it's because people knew me already, or knew who my dad was and knew not to offend him, or maybe Franklin, Lusher, and Jean Gordon had liberal enough people that I didn't have to worry about it. Part of it may have been my middle-class background. Let's say I had come from the Iberville and gone to Franklin. I think I would have been perceived in a different way no matter how smart I was. I did know one person from the Calliope projects, and he went to Franklin. I don't think anybody was blatantly racist with him, but they definitely talked down to him. That was what he dealt with every day for four years. And he was really, really smart.

I got into some program at Howard for international affairs the summer between my junior and senior years of high school. It was interesting, and it changed my mind about some things. When we were in DC, they took us to the State Department and they sold us on it. They explained what category it was, and what an opportunity it was to live in different countries.

When I left for college, I knew I wasn't coming back to New Orleans. I'd gotten accepted into Howard and Morehouse and was trying to make a decision. My dad was really pushing Morehouse. His good friend Walter E. Fluker was over the Leadership Center at Morehouse and could watch out for me. Some friends of mine, white and black, asked me, "Why are you going to Morehouse? Not only is it all black, it's all male!"

At first I wanted to do International Business. When I found out that Morehouse didn't have it, I stayed in business my first year before I even took any business courses, and then decided that I was more interested in the international side of it, so I switched to International Studies, which was like political science with an international edge.

Overall, Morehouse was a great experience. It made me appreciate the diversity in the black community. A lot of guys were like upper, upper-middle class and when sophomore year came, they came back with their Mercedes. I appreciate that a rich black person from L.A. is so different from a poor black person from Georgia. I mean worlds different. Sometimes, I felt like I had more in common with my white friends at Franklin than with some of the black people I met at Morehouse, especially the extremely wealthy ones. I knew a lot of people who had golfed in like Europe and things that just totally weren't on my radar, and I had friends who had been shot at. And they were all at this same school, supposedly working towards the same thing, which is uplifting the black community one way or another. So for the first time, I definitely fit in. I didn't have to pretend to do anything. I didn't have to prove myself at all about anything, because there was somebody there just like me. There were two people named Toussaint other than me there.

At Morehouse, in the dorms at night, especially freshman year, but since I was an RA in a freshman dorm for two extra years, the amount of debating that went on at 2:00 in the morning was unreal, and could be about anything. Maybe about who's the best rapper, or it could be about religion and politics, or it could be about a football team, or about gay marriage and abortion.

I went to Johannesburg for four days with Dr. Fluker. At Boston University, there's the African Presidential Research and Archives Center, and every year they bring in a former African president, and he lectures at the school for a semester, and then every year, there's a roundtable discussing African development. That year's roundtable was in Johannesburg. So I got to go and observe. Having that experience made it easier for me to make the Model UN team.

Katrina

September 2005 was the beginning of my junior year. My long-distance girlfriend at the time lived in New Orleans. She and her mother took a weekend to visit me right before. I think maybe we had already started classes, and she was one weekend or two weekends away from starting class. That weekend trip became a semester stay, which worked out in my favor. I think I was in denial a lot. Everything still had to keep going. It was my first year as a lead RA too. I watched it on television the day of and the day after, and after that, once I started to see the way it was handled and the way we were portrayed, I was totally disgusted. When someone would come up to me, "Do you see how they are portraying you guys? Aren't you mad?" I didn't want to hear that either. I felt like I didn't have to prove it to anyone.

It would come up in class, and I'd be the authority on it. They'd all look at me, and say, "What do you think?" I didn't want to hear everyone asking me five times a day, "Is everything ok?" My cell phone was being pounded with text messages from people that I knew. My mom's college friend's son went to Morehouse, and we became friends. He was like, "Man, I can't believe it." He kept going, and he was starting to depress me. I didn't really have the time to get depressed. I had things I had to do. None of my closest friends were stuck there.

Partially, I was still privileged because my family wasn't stuck there, and I had all of my belongings. I took everything with me to college. At that point I was able to live simple enough to have everything I owned in one dorm room. I had never planned on moving back into that house in the East anyway. That was supposed to be my last summer in New Orleans. For my parents, they were in their house one week, and then not the next. There was no way I could empathize the way people wanted me to, even the way my parents wanted me too.

When my younger brothers ended up in California, Amir went to Head-Royce School, a private school, and Kwame went to Berkeley High School. At both schools people were always amazed at how articulate they were. They had expectations of kids from New Orleans, even though they didn't take into consideration that they had made it all the way to California.

Returning to New Orleans: Teach for America

My first semester senior year at Morehouse, I had the worst case of senioritis. This was 2006 and on to 2007. I got remotivated when I got on the Model UN team, but I still didn't want to work on my senior thesis. I knew that I had some growing up to do. I decided to take a break, and teach for a couple of years. Teach for America just made everything happen. They are a well-oiled machine. Everything I needed to know was on the Web site. Once I got in Teach for America, I didn't have to find my own school or anything like that. It just made life a lot easier. My number one choice was in New Orleans. A big part of it was, you're doing Teach for America, you're giving back and everything, why not help the region that may need you the most, and that you're already familiar with because you've grown up here?

I'm at McDonogh 15, which is now a KIPP (Knowledge Is Power Program) school. It's in the French Quarter. KIPP is a high performing, high poverty school. The way McDonogh 15 started was as a college prep school in Houston under Gary Robisheaux, a couple of veteran teachers, and a lot of new Teach for America teachers. They went to the shelters and got New Orleans kids. And they did really well. They had amazing test scores last year, when they were back home. The kids come from all over the city. Everybody who was at NOW (New Orleans West) College Prep in Houston was automatically in. I think there's some preference to the students who were McDonogh 15 students originally, and after that, first come, first served, then lottery. It's like 99 percent black, 1 percent white French Quarter students, mostly free and reduced lunch. We've had maybe four or five fights all year. The entire K-8. That's amazing by New Orleans standards. All my colleagues have two to three fights to break up a day. A day! We have unlimited resources, like copies, money to get programs, field trips.

The amount of hours that I put it in, it's unreal. We have students from 7:30 in the morning until 5:00, and on Thursdays I do after school until 6:15. Saturday school every other week from 9 to noon. Just the time I spend on preparation! I feel like I have more to prove as a black man from New Orleans. I am the only black teacher there without a New Orleans accent. The PE coach went to St. Aug and Southern. A lot of the white teachers are young and single and aren't from the South. About 40 percent of the school is Teach for America alumni. The black teachers tend to be a bit older than the white teachers, so I look like a black teacher, but a much younger version, and I sound like a white teacher. This is nothing like a break! They have to pay teachers more because I'm thinking about the black males that I know that would make much better teachers than I. They're all like in Ph.D. programs or on Wall Street. When I told people that I'm teaching, they were like, "Really? That's nice." I'm like, "No, you really have to be smart for this, and you really have to work hard."

Teaching is fun. The kids make me laugh every day. When you see them reproduce something that you know you taught them, it feels really good. I know some of the boys actually look up to me. Some of them had never heard of Morehouse before this year. They always come in and tell you random facts, like, "Did you know Martin Luther King went to Morehouse?" Even if it's just an exposure impact, maybe I'm not doing what I need to do with the test scores, but I am having a positive impact. We're learning more than the students in most cases.

I've learned to deal with white people who are different than from before. At Franklin, they were a certain type. They were from New Orleans, they were way more open-minded. They were elitist in a different way. Franklin was like a UC Berkeley, and the rest of the white middle-class world is like Stanford. I had a level of comfort that I don't have now. Teach for America has a lot of social events where I meet a different type of person who claims to be liberal. I'm learning how to really be myself with them there.

I am the only person from New Orleans this year of 130, and there's one girl from last year, and she's also black. I'm totally under a microscope at all times. So I always try to put my best foot and my most honest foot forward at the same time. It does get tiresome. Teach for America flew me up to Morehouse for two days to recruit, and I was driving through campus. I'm like, "Look at all these black people!" In New Orleans, most days, the only black people I see are little black people, little twelve-year-olds.

Conclusion

When I left New Orleans, I didn't want to come back, and now I like New Orleans even less, because a lot of the people who made it home aren't here. A lot of people from church are in Houston. All my high school friends went away to college anyway. Some of them came back, but they were like Uptown, privileged people that I really don't want to associate myself with. I could move back to Atlanta and have a better social base there, than I have in the city that I grew up in. I think I'm more looking forward to leaving New Orleans than leaving the teaching.

This year for the first time I'd even heard gun shots. Part of that is living in Uptown and having like such sharp contrasts and being so close. When the weather changed like last week, I found the first roach in my house. I was like, "not again! Why am I here?"

Just last night, I was with one of my black friends from Teach for America, and he had a friend in town down from Brown University. We went out to dinner last night, and we parked at the school in the French Quarter. The gate was open. So I was unlocking the lock, because the gate was locked open, and I was trying to unlock the lock to close the gate, and lock it. This older white guy walks up and says, "What are you doing?" I looked up, and I said, "I work here." He stood there for maybe ten seconds trying to think of what to say, and then he just walked off. And we were like "Racist!" And they were like, "You handled that well."

I'm worried about the pace at which things are happening. Uptown and the French Quarter, that's my life right now. I'm in this five-mile radius like 90 percent of the time. I live Uptown on General Taylor Street. I went to the East last week, and in comparison to the way things should be happening, it's really discouraging. Driving on my block and seeing my own house and our neighbors', it makes me know it's never going to be even some semblance of the same five to seven years down the line when I plan on being long gone anyway. The rebuilding thing is like stealth gentrification. When certain people move in the neighborhood, and paint the houses a certain way, all of a sudden property values go up and other people get pushed out. I think that's going to change the cultural dynamics of the city, and what made it most valuable may not be present. The richness, the culture, is kind of being diluted.

Every now and then, I'll like purposefully drive through Central City. My favorite thing to do is take Washington Avenue from Central City past St. Charles, past Commander's Palace to get the contrasts. Central City is like a different country almost. Still I'm lacking a connection. My New Orleans with the Uptown/French Quarter perspective is so different compared to Central City. When I'm hungry, I only know to go to Magazine Street. And there are restaurants all over. When I do go out,

Kasey likes spoken word poetry. She wants to get the black experience, so we'll go to open mic night. I'm that black guy who's there with the white girl.

What makes me optimistic is the grit that I'm seeing in the middle classes. How they're not leaving, they're staying here: "That's the way it is, and we're going to stick it out." But usually those people have their entire family in the New Orleans area, so they kind of have to do it. That makes me optimistic. Just like after 9-11 there were more American flags out. After Katrina, everyone has their fleur-de-lis tattoo or bumper sticker.

I totally don't approve of the direction we're going in as a country. I'm definitely not even thinking about joining the military, but I do have faith that with our resources and the world perception of Americans that I would be a valuable person and that I could bring about some changes. I know I want to be in Africa. I wouldn't mind living in Europe. I've only been to Geneva, Switzerland. It was like being in a boring New Orleans. It was more beautiful, and it was old, like New Orleans, but it was just really stiff. I didn't get the same exciting feeling that I felt when I was in Johannesburg, Senegal, and Gambia.

I want my future kids to always have a plan in case of emergency. Like there were so many people that should have plans here and nationally that just didn't, and it showed so much. Don't forget that racism and classism are alive and ugly, and just know that's going to be something you're going to battle at least the next generation for sure. I think for my kids, I want to give them more of a sense of home because of this. When I'm talking to other Teach for America people, they always say, "I'm going home this summer," and I'm like, "Am I home already? Or is California home now because my parents are there most of the time?" When it comes down to just financial planning, just being more frugal. I read this book, *The Millionaire Next Door*. I think after learning what happened to people with Katrina, like how little we can trust those institutions that are in place when those things happen, I would probably be a lot more frugal and just plan for the future. So in case of emergency, I'm definitely self-sufficient, and if there's no emergency, I have a leg up. I'll have that story to tell my kids.

Conclusion

We were people who said, "I support my troops. Go USA!"

—Kevin Owens

Collectively, these narratives remind us of the major contributions African Americans have made to New Orleans and the United States since their arrival on the continent. Indeed, their contributions go back much further than even the oldest narrator can remember. This crescent-shaped bend in the Mississippi River was ideally situated, or so it seemed to Jean-Baptiste Le Moyne, sieur de Bienville, in 1763, to grow into a major commercial port that would bridge the old and new worlds. "Though mortality rates were high for the slaves, those who survived built levees, dug drainage ditches, cleared forests, and prepared timber for building boats and houses," writes Elizabeth Fussell. "In the city's antiquated architecture," writes Ned Sublette, "slaves' handiwork is everywhere—in the grillwork, the tile work, the mortuary work, and the carpentry." The first Senegambian slaves, for example, included goldsmiths, silversmiths, and ironworkers.[1]

During the first period of French colonial rule over New Orleans, rural slaveholders were required to build levees that would help to safeguard the city.[2] In the early American period of Louisianan history, African American slaves performed the taxing, dangerous labor that kept New Orleans' levees safe for commercial traffic and compensated for the city's poor drainage system.[3] Their efforts were so crucial that one of the most pressing concerns troubling New Orleans' leadership after the Civil War was who would maintain the levees now that slavery had been abolished. Only in 1882 did the United States Army Corps of Engineers finally step in to fill the gap and increase the reliance on technology.[4]

Fighting under the Spanish flag, slaves and a free black militia supported the American Revolutionary War by participating in the Louisiana militias that stopped British ambitions in the Gulf region and shored up the Continental Army's southern flank.[5] One way Creole and black leaders fought for increasing political and civic rights in the years after the institutionalization of Jim Crow was by volunteering for military duty at every opportunity. Pierre Carmouche, a blacksmith from Donaldsonville, Louisiana, personally inspired 250 newly disenfranchised African Americans to fight with him in Cuba on behalf of the United States. His goal, according to Rebecca Scott, was to "prove his community's patriotism and valor to a broad North American public in the face of crushing defeats for equal rights at home in Louisiana."[6] The wave of first-time black homeowners that greatly expanded settlement in the Lower Ninth Ward and pioneered Pontchartrain Park was, in part, financed by blacks' willingness to risk their lives during World War II and the Korean War. Even in the era of defiant Black Power, two of the most radical of the twenty-seven narrators served during the Vietnam War. The seventeen-year-old daughter of one of the narrators who endured the convention center was in the ROTC in New Orleans before the storm. The grandfather of the youngest narrator lost his life as a result of his involvement in the civil rights struggle.

Katrina's Barometer: Race Relations in the South in the Aftermath of the Storm

When I got to the convention center, I saw one white lady and maybe fourteen Chinamen. If you make an effort to bring twenty thousand of us in one area, and don't bring no whites, then something looks wrong.

—Dwayne Chapman[7]

Even this prejudiced German lady—I kept feeding her. And I'm looking at the fright in her face and not just the fright that she in that situation, but I'm around all these black people inside the convention center. And the few Asian people that was in there didn't say nothing but got in a corner. I went back and gave them water and food too. That's the time to be selfless. Blacks will reach out to our enemies because we already know the struggle.

—Eleanor Thornton

Peggy McGaughy came to the hotel that we were staying at in San Antonio, and she said, "Denise, how can I help you? My heart's broke after what you and your family have gone through." I said, "Peggy, all I want to do is get my kids together, get us all in one place." Peggy started looking into leasing houses and apartments, so she leased us a house in a gated community, kind of like what we're accustomed to, in Cypress, a well-to-do suburb of Houston. The Episcopal Church and Peggy got my house furnished. Our neighbors egged and spray painted "nigger" on our cars, and flattened our tires. They said, "We paid two hundred and fifty thousand dollars (a fraction of what my house in Eastover cost) so we are particular about who our neighbors are." The six-year-old son of a neighbor came up to me, "My daddy said all black people are poor and on welfare." He said, "Someone else bought you your car because my daddy says you couldn't afford a car like that." At Christmas time, I gave my neighbors a present of store-bought cookies in a sealed container; they placed them unopened at the top of their garbage cans with the lids off.

—Denise Johnson

Catastrophes are like social barometers: They allow us to gauge the intensity of racial attitudes when lives and property are at stake. Hurricane Katrina flooded 80 percent of New Orleans for at least two weeks after landfall and dispersed almost half a million people across the country, especially to other places in the South. These narratives of survivors give us abundant insights into race relations in the South in the wake of the storm some thirty-seven years after Martin Luther King Jr.'s assassination in Memphis, Tennessee. The results are mixed.

Media coverage of the aftermath, as we noted in the introduction, exaggerated the amount of chaos, theft, and violence, especially by blacks. Narrators did indeed describe the presence of some armed men seeking or selling narcotics after the storm in at least three settings: in and around the convention center, around University Hospital, and on the Fort Chaffy military base. However, it is not obvious from these testimonies whether the men were black. Some of the trauma experienced by working- and middle-class people stranded in New Orleans after the storm could be attributed to the threatening environment produced by potential violence as the result of disrupted supply-and-demand patterns for narcotics. In addition, narrators volunteered possible explanations for the well-publicized incident of a flood victim shooting at a military helicopter attempting to evacuate medically critical patients from the Superdome: After days of seeing helicopters all around them, even the educated middle-class narrators began to despair that they would ever be rescued. Shots fired near the helicopters, many suggested, were aimed to command the attention of the pilots.

All but one of these twenty-seven narrators, who come from a wide range of class, generational, political, and religious backgrounds, strongly favor safer schools and neighborhoods. Indeed, many African American church and community leaders have worked for years to address the needs and problems created by drug addiction in the city's neighborhoods. Aldon E. Cotton, for example, envisions his future multiuse church building in Central City playing a role in reducing crime. A possible launching point for collaboration between groups of all colors could center on how to make post-Katrina New Orleans a safer place for everyone. In the post-Katrina environment, where the murder rate remains the highest in the United States on a per capita basis despite a marked decrease in total population, the need for coalition building is acute.

At the same time, examples of interracial cooperation and harmony abound. Some African Americans with the means, experience, and knowledge to leave after the mandatory evacuation was declared chose instead to stay. Some did so in search of adventure; others fulfilled a sense of duty to safeguard an entire extended family. Survivors reported being fed, cared for, and taken to dry land by black (and white) men of all ages and social backgrounds. African American men recalled rescuing not only other blacks, but also whites, Asians, and Hispanics. The narrators witnessed the effective intervention of local black men, from church deacons to young, unemployed, streetwise men, who stayed behind and organized house-to-house rescues and food and water distribution for the children, the disabled, and the elderly in their self-appointed zones of responsibility. Without them the final death toll from the aftermath of Katrina would have been significantly higher. In turn-of-the-century urban America this outcome cannot be taken for granted, as Eric Klinenberg shows us in *Heat Wave*, a study of an unusually hot summer in Chicago in 1995. The extensive social networks in Chicago, so similar a few decades ago to those in black New Orleans, had become unraveled, thereby compounding the effects of the natural disaster that led to the deaths of more than seven hundred people in one week.[8] We wonder how displaced African Americans would have been received in Mississippi and Texas if these stories of black selflessness and heroism had received equal media attention in the aftermath of Katrina.[9]

Some narrators credited white owners of establishments with opening their doors to poor, even homeless African Americans, during the immediate period following the storm. In these situations, survivors came together as fellow New Orleanians. Months and even years after the event, narrators reported a sense of spiritual connectedness to these men and women they had not seen since the storm. Some men in the "Cajun Army" mentioned how their wives feared for their safety in New Orleans. What is significant is that they overcame their fear to fulfill the honor code of boaters who live in hurricane-prone areas. Rescued eyewitnesses remarked on their compassion and delighted in their homemade sandwiches. In one case, these efforts were so appreciated that one narrator dedicated her Katrina memoir to the Cajun fisherman, whom she credited for saving her family's life, even though she ultimately ended up outside the convention center for four, long, traumatizing days. There are countless reports of white strangers all across the South plying New Orleanians with food, clothes, and even cash, as FEMA buses passed through their cities. During the interviews, these acts of kindness were recalled with great emotion and, in many cases, were identified as the first compassion from outsiders they had experienced in over a week. These gestures helped to counter the feelings of isolation that descended upon them after the lengthy delays in relief.

These moving accounts of interracial collaboration are offset by the descriptions of racist responses to the displaced. One young woman who endured the Superdome recounted being greeted at a local Arkansas rest stop by a handful of locals brandishing shotguns and nooses that they threatened to use if the bus did not whisk the New Orleanians out of their county immediately.[10] Such intimidation tactics may have been in part a response to the media's decision to make the allegedly rampant looting their main story. However, as Duplessis' account notes, owners of gas stations in three southern cities were engaged in racist behavior well before Katrina's landfall. The customers they refused to serve were culturally conservative Baptists. Pockets of the South still remain inhospitable to African Americans of any class, even after tragedy has struck.

Although independent studies by engineering specialists like the team headed by Professors Ray Seed and Bob Bea of the University of California, Berkeley, and others concluded that the levees were not dynamited, but rather were poorly constructed and very negligently maintained,[11] several narrators continue to believe that the flooding of the black sections of New Orleans was caused by the use of explosives. This is sometimes regarded as being on par with the belief that the U.S. government introduced the AIDS epidemic into African American communities.[12] The narrators who believe that the levees were blown found the most convincing evidence to be the loud booming noises heard when the first floodwaters began pouring into the city. This rumor then spread by word of mouth and text messaging, even to those New Orleanians watching the disaster on television in places across the South. Whether these reports are believed by any particular survivor depends heavily on the credibility of the person making the claim and the depth of the survivor's relationship to them.

Sections 1 and 2 of the book are strewn with memories of threats, perceived threats, and unequal treatment by whites in a variety of contexts. A survey of the history of racial relations in present-day Louisiana, from the introduction of the first sizable influx of African slaves in 1719 through to the former Klansman David Duke's popularity as a political candidate in the 1980s and '90s, reminds us that the narrators' memories fit within a much larger context of unequal treatment for almost three centuries. This helps us to understand the continued psychological trauma experienced by evacuees who believe they survived a terrorist attack.

People with a history of feeling deceived tend to find the assertions of people they trust more credible than those of outsiders, even if they are from respected institutions like UC Berkeley. The memory of the dynamiting of the levees protecting St. Bernard Parish in 1927 and the official evacuation of whites from Tennessee Street during Hurricane Betsy in 1965 were enough to spark belief during Betsy that the levees had been blown again. The dramatic breaking of the levee protecting the Lower Ninth Ward during Katrina was viewed in this context. The belief that whites tried to kill blacks by blowing up the levees contributed to the decision on the part of some not to return to the home of their ancestors. This, in turn, inadvertently assisted those who have unabashedly seized upon Katrina as an opportunity to remake New Orleans as a smaller, less black city.[13]

The more serious question is the perception of racial privilege and differential treatment during the aftermath of Katrina in New Orleans. A rigorous analysis of documentation and a variety of sources would be necessary before definitive conclusions could be drawn. Some whites with cars also stayed despite the mandatory evacuation. According to some estimates, 20 percent of the residents remaining in New Orleans after Katrina's landfall were nonblack.[14] The death toll among whites in the immediate aftermath, though still not final, is proportionally high enough when compared to the deaths of blacks to indicate that significant numbers of whites in Lakeview and the racially mixed lakeside neighborhoods of Gentilly and Mid-City must have attempted to ride out the storm in their homes.[15] The Jefferson Parish Sheriff's Office reported that on August 30, 2005, alone, approximately five hundred people from Lakeview were rescued by helicopter. Subsequent efforts were concentrated in Mid-City and Gentilly, where Demetrius White heard helicopters for days.[16] Although whites were rescued from their rooftops and attics in Lakeview, New Orleans East, Mid-City, and Gentilly, on September 3, 2005, CNN reported that "as more and more eyewitness accounts of conditions in the convention center and Superdome surface, it becomes plain that most if not all of those who survived unspeakable days and nights under inhuman conditions were black."[17]

It strains the imagination to envision three thousand whites of any U.S. city moved to an interstate highway and forced to spend the night on the roadway with armed military personnel preventing their escape. If similar occurrences to what happened to blacks in New Orleans happened to whites in the aftermath of Katrina, publicizing such events would likely relieve some of the sting of the demeaning treatment.

Hyperbole is a common tactic used when other remarks have failed to draw attention to an underlying concern. The fundamental issue at the root of the beliefs about the levees is distrust. Local experts on the history of race relations in New Orleans assert that African Americans as a whole distrust whites in New Orleans more now than at any time in recent memory.[18] The most effective short term way to quell these rumors would be for African American leaders with long-standing credibility in the community to offer other explanations for what might have caused the sounds interpreted as explosions.

Collaboration between different generations, cultures, and classes might create new memories of interracial cooperation and assistance to add to the history of the

South. In the heroic efforts to save lives undertaken by the narrators during the after-math of Katrina, and in the simple kindliness of strangers enjoyed by survivors pouring off dilapidated buses across the country, we begin to see the outlines of a different kind of legacy that we might bequeath to the next generation.

Katrina's Long Shadow: Death, Trauma, and Depression

I have a ninety-four-year-old lady that was like my mom, I call her Mother Baker, who was removed from her granddaughter's apartment and taken to a landfill and left for three days by the National Guard. After the storm, her son found my number, and we connected by phone. She's West Indian, so she said, "Oh my child! I've prayed so much and asked God to make you be safe, and to let me talk to you." They took her grandson and granddaughter, because she had two young kids. And they said, "We're going to have to come back for you." When they came back for her a day later, they took her to a landing field. As time went on, they brought more and more people. They were outside on the black tar road. And she said to me, "Child, it's but God that kept me on that hot, black tar three days. I had nowhere to go to the bathroom." I mean just listening to her!

—Cynthia Banks

The guards at the Birmingham Interim Shelter, all they do is laugh in your face and throw darts at your character when you turn your back. Like one night I passed by. They were talking about Hurricane Katrina people here. So I went back and I said, "Let me tell you all something. I'm here, but I own my own home. My mother done left me a home, the house my dad built. And I worked just like you worked for years. It's not like we want to be here."

—Carl Singleton, retired hopper from the Lower Ninth Ward

A lot of our old people are dying from broken hearts. One of my mama's best friends off the West Bank, Ms. Frances, never lived outside New Orleans. She went to California and she died. It ain't about no money. It's about how you killing our people by separating them from their surroundings, everything they know and love, from people who took care of them. Those people never even crossed the Mississippi River Bridge before. How you going to send somebody from New Orleans to Boston, Utah, or Alaska? We are a community of people, and we took care of each other.

—Eleanor Thornton

The men, women, and children swept out to the bayous will never be counted. Those unable to scramble up ladders to their attics drowned and were discovered, often by loved ones, weeks and months later. Many medically fragile people who obeyed the mandatory evacuation order died far removed from the petroleum-and-waste-filled floodwaters engulfing New Orleans for more than three weeks. Even a more effective rescue effort would not have prevented most of these deaths.

Between Friday morning and Sunday afternoon when the military took charge of the evacuation effort, more than sixty thousand people were evacuated from the Superdome, the convention center, the Causeway Boulevard staging area, and the

Louis Armstrong International Airport.[19] In the late summer heat and humidity of New Orleans, the four-day delay in providing a reliable food and water supply to survivors had serious, long-term health repercussions. Individuals who narrowly escaped the water's mad rush by taking refuge in their attics died from dehydration, if they were not saved in time by rescuers such as Willie Pitford, Demetrius White, a volunteer in the Cajun Navy, a New Orleans firefighter, or a National Guardsman.

Public health officials and demographers have been unable to precisely detail the extent of death caused by Katrina, but it has been estimated that including those who died shortly after evacuation, 1,577 deaths could be related to the storm, thereby making it the largest natural disaster ever experienced in the United States. In the chaos during the immediate aftermath, it cannot be ruled out that some deaths went unnoticed and unaccounted for. Experts in the field of public health recognize that the exact mortality due to the aftermath of Katrina, or the so-called second wave of Katrina deaths, has not yet been fully documented. One novel attempt, spearheaded by Kevin U. Stephens, Director of the New Orleans Health Department, to document this surge in the death rate utilizes death notices from the *Times Picayune*, the New Orleans daily newspaper.[20] The study found that during the first six months of 2006, an average 1,317 deaths per 100,000 people a month were reported in comparison with an average of 924 deaths per 100,000 people a month between 2002 and 2004. The study definitively concludes that the stress of losing everything, being abandoned, and fearing for one's life caused a massive wave of strokes and heart attacks in the weeks and months after Katrina. To give just one example from the narrations, Mother Baker died shortly after she spoke to Cynthia Banks.

Many of those who endured the aftermath of Katrina in the city continue to fight against depression, helplessness, rage, and posttraumatic stress disorder.[21] Longing for the comfort of an ordinary existence, now elusive, is the fate of tens of thousands of Americans.

The survivor's trauma is due in part to the experience of being treated like a criminal during moments of great vulnerability. This injury is partially healed when survivors recount their experiences to family and friends, who are usually of the survivor's ethnicity. Listening to detailed accounts of the indignities suffered by loved ones, however, increases the emotional burden on the listener and may also produce secondary trauma in the listener. Physicians carry with them their worries about their patients, teachers about their students, pastors about their flocks, and social workers about their clients.

Anger and despair followed in the wake of home damage caused first by the hurricane, then by the flood, and finally by the black mold that took over the houses. In some cases, individuals were prevented from returning to their homes for months. The emotional impact of this loss does not appear to have been related to the income of the narrators or the value of the houses. Before the storm, most houses in the Lower Ninth Ward were appraised at between twenty and forty thousand dollars. Even a fragile, wood-frame, shotgun-style house may reflect a lifetime's investment of resources and heart. It may also symbolically represent the ties that bind one generation to the next, and be the only tangible link to parents or grandparents no longer living. Those who tenaciously believe that they were deliberately flooded out of New Orleans carry an additional trauma beyond the loss of their possessions and home.

A home (or business) built against tremendous societal odds, in the face of segregation, institutional racism, and blacklisting, reflects a substantial emotional investment. The destruction of the heart of black middle-class New Orleans—Pontchartrain Park and New Orleans East—also meant the splintering of dreams. The floodwaters of Katrina washed away the fruits of a lifetime of sacrifice and hard work.

Another source of the emotional toll springs from the disruption of the extended network of family and friends. Black New Orleanians are now scattered throughout the United States, weakening their sense of community and interconnectedness. Social networks allowed determined African Americans to overcome the obstacles of poor education, low wages, inadequate health care, and crime to live passionate, gracious, and fulfilling lives. These support networks have been broken artificially. As survivors make decisions about where and how to rebuild, it is the loss of this community and the awareness of its collective mistreatment that deepens the mourning. Is New Orleans still home if your neighbors, cousins, and lifetime friends no longer live there?

Recent data from the *Archives of General Psychiatry* documents that the adults who experienced anxiety and stress after the attacks of September 11, 2001, were at higher risk for heart disease, high blood pressure, and strokes. Those who developed depression were three times more likely to have heart-related illnesses one to two years later.[22] Studies from previous catastrophes, including wars, have demonstrated that with the collapse of the health infrastructure, such as New Orleans has undergone, preventable deaths continue for many years after the event.

For those of us not from New Orleans, who live eternally positioned at metaphorical crossroads, eager and ready to trade up not only jobs and partners but also cities and continents, these narratives provide detailed maps into now effaced communities with different rhythms and social values. Collectively, these histories may assist the concerned outsider, the guidance counselor, the community health worker, or the pastor to gain a better grasp on the stress and trauma still faced by survivors.

Epilogue: The Future of New Orleans

The thing that gives me the most optimism for the people of my city is rebuilding. In spite of all they've been through, they're willing to go back again because they love it so much. That's where they feel they're at home.

—Cynthia Banks

Three years after Katrina made landfall in Louisiana, local businesses and city planners are trying to recapture the glitter that New Orleans was once known for. The Superdome has been refurbished and continues to host national and international events. Uptown homeowners, flush with hurricane insurance payoffs, have visibly improved their homes. If you only visited the tourist areas, you would conclude a revival is sweeping New Orleans. Drive five minutes beyond these tourist destinations, however, and you will see littered wastelands of closed businesses and abandoned houses cluttering the horizon as far as you can see. As "salvageable homes continue to sit empty and further deteriorate," as some commentators have noted,[1] it is increasingly likely that entire neighborhoods may never return. The Lower Ninth Ward remains a motley mixture of damaged properties and vacant lots, where only concrete steps remain as evidence of previous family life. Many residents from heavily damaged areas have now permanently resettled in other southwestern or southern cities. Some clearly harbor distrust of local officials, the federal government, and the Army Corps of Engineers, whose job it is to maintain the levees, as it was before the storm.

From Aline St. Julien to Le Ella Lee, the narrators have faced challenging, if diverse, obstacles in their lives, of which Katrina is only the latest and not necessarily the most daunting. Those born before World War II faced an economic and social landscape created by Jim Crow laws. Those born between 1945 and 1955 grew up during the civil rights movement and could participate in social movements designed to challenge race-based discrimination. The generations that came of age in the 1980s and '90s faced obstacles created by worsening public education, a culture of drugs that affected even middle-class neighborhoods, and street violence.

For African-born slaves, surviving the ocean voyage from African shores to their port of embarkation in the Western Hemisphere, a voyage known as the middle passage, was enormously difficult. A conservative estimate of the number of Africans who died at sea is two million. To stay alive long enough to bear children in the New World was an accomplishment itself. Slaves working in the sugarcane concentration of southern Louisiana risked their lives and health while the plantations contributed to the young state's prosperity.[2] Until emancipation, the slaves of southern Louisiana who worked in the sugarcane fields experienced negative birthrates.[3] For enslaved sugarcane workers, one way to escape slavery was to live in the "mosquito-infested lowlands, crawling with alligators."[4] "We are the children of the ones they could not kill," the Poetic Panther reminds us, in the first line of his poem by the same name.[5]

This history of overcoming a wide variety of obstacles provides a shared reservoir of resilience that displaced black New Orleanians take with them wherever they choose to settle permanently. Those narrators who saw active duty in one or more of the twentieth century's major conflicts were not threatened by Katrina's challenges. Narrators got ahead in the segregated South through the common qualities of neighborliness, the insistence on quality education for themselves and their children, and the stamina to perform two or three jobs. The practice in the Lower Ninth Ward before Hurricane Betsy of picking up litter and mowing your neighbors' grass to foster community pride was reinvented after Katrina as the practice of keeping vigil for those still working their way back home. This was a metaphor for hope in the Havelian sense, meaning the poorer the odds of victory, the greater the hope.[6]

Taken together, these narratives challenge the widespread portrayal of the poor, lazy, and loud black New Orleanian looking for a handout, with stories of flesh-and-blood, hard-working, religious, family-oriented men and women with diverse peccadilloes like everyone else. The overwhelming losses after the storm: material, familial, and psychological, required tremendous fortitude to overcome, a project that all of the middle- and most of the working-class narrators went about quietly accomplishing for themselves and those around them.

For some, the rhythm of manual labor helped to fill the hours of what would otherwise have been long days of disorientation and mourning. Other individuals who had jobs before the storm appeared unfocused after the storm, but this should be understood in the context of trauma, depression, and culture shock. To lose your social network, job, home, church, and community all at once is a staggering blow. By the spring of 2006, the working energies of most had been redirected. This was the result of healing, the passage of time, and of necessity.

One of the most passionate currents running through these interviews is the love African Americans had for their neighborhoods and their city. New Orleans has a longer African-centered history than any other city in the United States.[7] It was one of the first ports of entry into the country from the Caribbean and Africa. Sundays in New Orleans, even during slavery, were a time for blacks to celebrate a few hours of freedom, and commune with other recent African arrivals.[8] This celebration and mixing of cultures was, in part, responsible for the birth of jazz. Before Katrina, many New Orleanians who did not eventually become professional musicians still had some contact with music, in school, in a marching band, a church choir, or a second line group.

Throughout the years that Jim Crow laws limited economic and political opportunities, and during the height of vigilante violence in Louisiana and the rest of the deep South, New Orleans stood as a beacon, an oasis of safety from the intimidating reach of groups such as the Ku Klux Klan.[9] Endangered after a cross was burnt on his family's lawn, Freddie Seals Sr. escaped to the Crescent City. Former slaves who had purchased their freedom headed to New Orleans to congregate and find work as craftsmen or tradesmen.[10] At the age of nine, Keith Ferdinand's father walked for days from Napoleonville to reach New Orleans. However low were the wages available to service-class blacks in New Orleans, they were still higher than any available in rural Mississippi, Alabama, or Louisiana.[11]

Arguably the most important factor in the romance between New Orleanians and their city was the way family histories became intertwined with the history of the city. Seventy-seven percent of people from New Orleans were born in the city.[12] This was a community where you could show up to an event and be sure to find a friend or relative in the crowd. People lived from event to event. On any given Sunday, there might be a second line parade in the middle of the neighborhood. Before Katrina, the city possessed a rich network of extended families.

Resonating through these pages is a contagious spirit of generosity, magnanimity, and tolerance. Black New Orleanians took pride in their lives before the storm. Waitresses served with care despite contrary wishes from management. Nurses helped the homeless on their days off. Maintenance men and security guards barbecued for neighborhood kids and offered red beans and rice to strangers passing by. Black New Orleans with its high crime rate in August of 2005 may not have been paradise, but its loss leaves us all impoverished.

The contributions of New Orleanians to the history and culture of the United States before Hurricane Katrina's landfall are irreplaceable. The struggling city now needs all of their creativity, genius, resilience, heart, and skills if it is to make a full recovery. It needs all of the people who contributed to this book and the thousands more who stand behind them.

Notes

Introduction

1. Sewerage and Water Board of New Orleans, "Report on Hurricane Betsy, September 9–10, 1965," (10/8/65), 23 and "State Death Toll Mounting," *Times-Picayune* (*T-P*) (9/13/65), 3.
2. Personal letter from Inola Ferdinand to Narvalee Copelin, 9/21/65.
3. On Lower Ninth Ward beliefs, see Kalamu ya Salaam, *What Is Life? Reclaiming the Black Blues Self* (Chicago, 1994), 85–86. On the televised blowing of the levees in 1927, see John Barry, *Rising Tide: The Great Mississippi Flood of 1927 and How It Changed America* (NY, 1998).
4. Juliette Landphair, "'The Forgotten People of New Orleans': Community, Vulnerability, and the Lower Ninth Ward," *Journal of American History* (*JAH*) 94 (12/07).
5. Kent B. Germany, *New Orleans after the Promises: Poverty, Citizenship, and the Search for the Great Society* (Athens and London, 2007), 69–75.
6. This section relies heavily on Shana Agid, "Locked and Loaded: The Prison Industrial Complex and the Response to Hurricane Katrina," in *Through the Eye of Katrina: Social Justice in the United States*, ed. by Kristin A. Bates and Richelle S. Swan (Durham, NC, 2007), 56; Felicity Barringer and Maria Newman, "Troops Bring Food, Water and Promise of Order to New Orleans," *New York Times* (*NYT*) (9/2/05); Douglas Brinkley, *The Great Deluge: Hurricane Katrina, New Orleans, and the Mississippi Gulf Coast* (NY, 2006); *CNN Reports: Katrina—State of Emergency,* with an introduction by Ivor van Heerden (Kansas City, 2005); Richard Campanella, *Geographies of New Orleans: Urban Fabrics Before the Storm* (Lafayette, LA, 2006), 385–405; Gardiner Harris, "Police in Suburbs Blocked Evacuees, Witnesses Report," *New York Times* (*NYT*) (9/10/05); and R.B. Seed et al., University of California at Berkeley and American Society of Civil Engineers, Report No. UCB/CITRIS-05/01, "Report on the Performance of the New Orleans Levee System in Hurricane Katrina on August 29, 2005," at 1–4 (7/31/06).
7. malik rahim, "this is criminal" in *What Lies Beneath: Katrina, Race, and the State of the Nation*, ed. by the South End Press Collective (Cambridge, MA, 2007), 65–68.
8. For pictures of this display of military might, see *CNN Reports*, 81 and 85; Robert Caldwell, "New Orleans: The Making of an Urban Catastrophe," *Monthly Review* (12/9/05); and Spike Lee's use of media footage in *When the Levees Broke*.
9. Rebecca S. Chopp, *The Praxis of Suffering: An Interpretation of Liberation and Political Theologies* (Maryknoll, NY, 1986), 2.
10. See for example, Congressional Hearing on Katrina at Katrina.house.gov, Lee, and Brinkley. "Through Hell and High Water" is an oral history project of first responders from the NOPD and the New Orleans Fire Department, spearheaded by the Historic New Orleans Collection.
11. See for example, Sally Pfister and Melody Golding, *Katrina: Mississippi Women Remember* (Oxford, MS, 2007); The University of Southern Mississippi Center for Oral History and Cultural Heritage's "Hurricane Katrina Oral History Project" includes "emergency management officers, local officials, residents, volunteer relief workers, and those displaced by the storm"; Alan H. Stein and Gene B. Preuss, "Oral History, Folklore, and Katrina," in *There Is No Such Thing as a Natural Disaster: Race, Class, and Hurricane Katrina*, ed. by Chester W. Hartman and Gregory D. Squires (NY, 2006), 44–45; and lakeviewcivic.org.
12. Russell McCulley, "Healing Katrina's Racial Wounds," *Time* (8/27/07).

13. On individuals, see Sheila Crowley, "Where Is Home? Housing for Low-Income People After the 2005 Hurricanes" in *There Is No Such Thing*, 123; on households, see David Dante Troutt, "Many Thousands Gone Again" in *After the Storm: Black Intellectuals Explore the Meaning of Hurricane Katrina*, ed. by David Dante Troutt; with a foreword by Derrick Bell and an introduction by Charles J. Ogletree Jr. (NY, 2006), 10.

14. McCulley, "Healing Katrina's Racial Wounds."

15. Tom Piazza, *Why New Orleans Matters* (NY, 2005).

16. Rob Walker, *Letters from New Orleans* (New Orleans, 2005).

17. Penner field notes (PFN), 11/16/07.

18. Kalamu ya Salaam, "All Hands on Deck: Katrina Communiqué #3," (9/6/05) at nathanielturner.com.

19. *After the Storm*, Michael Eric Dyson, *Come Hell or High Water: Hurricane Katrina and the Color of Disaster* (NY, 2005), and *What Lies Beneath*.

20. "Scumbags" was used by Chris Beck, Clear Channel radio host, as quoted in *CNN Reports*, 29; Shana Agid, "Locked and Loaded," 56; suheir hammad, "on refuge and language" in *What Lies Beneath*, 167–69.

21. As quoted in *CNN Reports*, 36.

22. Brian Thevenot and Gordon Russell, "Reports of Anarchy at the Superdome Overstated," *Seattle Times* (9/26/05) and Andrew Gumbel, "After the Storm, US Media Held to Account for Exaggerated Tales of Katrina Chaos," *Los Angeles Times* (9/28/05).

23. Darrel Creacy and Carlito Vicencio, *The Real Guardians* (Chicago, 2007).

24. See quotes by Brian Williams, Pat Buchanan, and Neal Boortz, WSB-AM radio talk show host, and Robert Tracinski, editor of TIADaily.com, as quoted in *CNN Reports*, 91, 117, and 155.

25. Letter from a nurse (9/05) at aliveintruth.org.

26. "GW," Worker and Shelter Volunteer, Alive in Truth Interviews, 9/16/05.

27. "More Than 87% of New Orleans' Inner City Residents Were Employed Prior to Katrina, ICIC Study Reveals," press release, icic.org, 9/19/05, as cited in John Valery White, "The Persistence of Race Politics and the Restraint of Recovery in Katrina's Wake" in *After the Storm*, 47.

28. Barbara Ehrenreich, *Nickel and Dimed: On (Not) Getting by in America* (NY, 2002).

29. U.S. Census, 2000, SF3 HCT25.

30. Landphair, "The Forgotten People," 837–45.

31. Troutt, "Many Thousands Gone Again," 12.

32. See *The Sociology of Katrina: Perspectives on a Modern Catastrophe*, ed. by David L. Brunsma, David Overfelt, and J. Steven Picou (Lanham, MD, 2007).

33. Mitchell Duneier, *Slim's Table: Race, Respectability, and Masculinity* (Chicago and London, 1992), 121–23, 128, 142, and 164.

34. Mary Pattillo-McCoy, *Black Picket Fences: Privilege and Peril among the Black Middle Class* (Chicago and London, 1999), 1–2.

35. Andrew Stuart Bergerson, *Ordinary Germans in Extraordinary Times: The Nazi Revolution in Hildesheim* (Bloomington and Indianapolis, 2004), 240–46, and Dori Laub, "Bearing Witness or the Vicissitudes of Listening" in *Testimony: Crises of Witnessing in Literature, Psychoanalysis, and History*, ed. by Shoshana Felman and Dori Laub (NY and London, 1992), 57–74.

36. Richard Campanella, "An Ethnic Geography of New Orleans," *JAH* 94 (12/07), 704–15 and Campanella, *Geographies of New Orleans*, 369–80. The next two paragraphs are based on Campanella's work.

37. Craig E. Colten, *An Unnatural Metropolis: Wresting New Orleans from Nature* (Baton Rouge, 2005), 81 and 99; Troutt, "Many Thousands Gone Again," 8.

38. For an excellent essay on Uptown and Downtown through the decades, see Campanella, *Geographies of New Orleans*, 157–68.

39. Robert K. Whelan, "An Old Economy for the 'New' New Orleans? Post-Hurricane Katrina Economic Development Efforts" in *There Is No Such Thing*, 217; White, "The Persistence of Race Politics," 46.

40. Michael Casserly, "Double Jeopardy: Public Education in New Orleans Before and After the Storm" in *There Is No Such Thing*, 197–204; Landphair, "The Forgotten People," 837–45; Troutt, "Many Thousands Gone Again," 8.

41. Leah Hodges, "Testimony before the Select Bipartisan Committee to Investigate the Preparation for and Response to Hurricane Katrina" (12/6/05) at katrina.house.gov.
42. Kevin M. Kruse, *White Flight: Atlanta and the Making of Modern Conservatism* (Princeton and Oxford, 2005), 3–15.

SECTION ONE: RETIREES
One Aline St. Julien

1. See Aline St. Julien, *Colored Creoles: Color Conflict and Confusion in New Orleans* (New Orleans, 1974); Campanella, *Geographies of New Orleans*, 205–8; and Ned Sublette, *The World That Made New Orleans: From Spanish Silver to Congo Square* (Chicago, 2008), 79–80.
2. For a summary of both the ordeals of enslaved women and the system of plaçage among free women of color, see Sublette, *The World That Made New Orleans*, 111, 222–33, 284.
3. See William Loren Katz, *Black Indians: A Hidden Heritage* (New York, 2005).
4. The Lafitte Housing Project was completed in 1941.

Two Narvalee Audrey Copelin

*Narvalee dedicates her chapter to the memory of Noah Copelin.

1. Juliette Landphair, "Sewerage, Sidewalks, and Schools: The New Orleans Ninth Ward and Public School Desegregation," *Louisiana History*, 40 (Winter 1999), 35–62.
2. This chapter also draws from letters written by Copelin to Penner, 7/11/06 and 2/18/08.
3. In the thirty years following World War II, the proportion of blacks in the Lower Ninth Ward rose from 31 to 73 percent; Landphair, "The Forgotten People," 5.
4. On de facto segregation in leisure spaces as late as 1964, see Germany, *New Orleans after the Promises*, 31–32.

Three Leatrice Joy Reed Roberts

*Leatrice dedicates this chapter to the memory of her grandparents and mother, and to her husband, son, and grandson.

1. The text also draws from PFN, 8/16/07, 11/14/07, and 2/10/08.
2. John Salvaggio, *New Orleans' Charity Hospital: A Study of Physicians, Politics, and Poverty* (Baton Rouge, 1992).
3. For more on Albert Dent's insurance plan, see "Negro Health," *Time* (4/8/40).
4. See "NOPL Centennial Exhibit: Extension," at nutrias.org.
5. On the bullets shot at two African American men by white LSU students in 1955, see "2 Students Attacked: Negroes at L.S.U. are Hit by Buckshot on Campus," *NYT* (7/20/55).
6. Howard Witt, "To Some in Paris, Sinister Past Is Back," *Chicago Tribune* (3/12/07).
7. Robert O. Zdenek et al., "Reclaiming New Orleans' Working-Class Communities" in *There Is No Such Thing*, 170.

Four Irvin Porter

1. blackamericaweb.com.
2. Scott, *Degrees of Freedom*, 154, 62, 93.
3. Edward Humes, *Over Here: How the G.I. Bill Transformed the American Dream* (Orlando, 2006).
4. ehistory.osu.edu.
5. PFN after conversation with J. Herbert Nelson II, Memphis, TN, 9/2/05.
6. Crowley, "Where Is Home?" 129.

7. Zdenek et al., "Reclaiming New Orleans," 167–71.
8. For less pleasant experiences in Baton Rouge, see jordan flaherty, "corporate reconstruction and grassroots resistance" in *What Lies Beneath*, 107, and White, "The Persistence of Race Politics," 41.
9. Zdenek et al., "Reclaiming New Orleans," 169.

Five   Leonard Smith

1. kreweofzulu.com.
2. This chapter also relies on PFN, 4/21/08 and 4/30/08.
3. "The Hurricane Pam Exercise," from PBS's *Frontline* (11/22/05) at pbs.org.

Six   Pete Stevenson

1. See also Ashley Nelson, *The Combination* (New Orleans, 2005), and Amistad Research Center (ARC), Sharmaine Shelton-Penner interview, Houston, TX, 1/05/06.
2. Bob Dart, "Tommy Lee Hines and the Cullman Saga," *Southern Changes* 1 (1978) at beck.library.emory.edu.
3. This chapter is also based on PFN, 4/15/08, 4/21/08, and 6/4/08.
4. Germany, *New Orleans after the Promises*, 65–70.
5. On Angola, see Deborah Willis, "Angola Bound," *Aperture* (Spring 2006); ARC, Mitchell Casmere-Penner interview, Birmingham, AL, 2/1/06 and ARC, Carl Singleton-Penner interview, Birmingham, AL, 12/27/05.
6. ARC, Velbert Stampley-Penner interview, Birmingham, AL, 12/15/07.
7. *CNN Reports*, 82; Office of the Governor, "Alabama Marking National Day of Prayer and Remembrance for Hurricane Victims," (9/12/05) at governorpress.alabama.gov.
8. See V. Stampley-Penner interview and ARC, Angie Ratliff-Penner interview, Fairfield, AL, 12/26/05.
9. For other accounts of survivors' distrust of local, state, and federal officials, see Ceci Connolly and Manuel Roig-Franzia, "Grim Map Details Toll in 9th Ward and Beyond: Katrina Proved Deadly in Every Section of New Orleans, *Washington Post* (10/23/05), A14.
10. For similar expressions from other distraught evacuees, see "New Orleans Tragedy: The 17th Street Levee Was Bombed" (9/9/05) at bellaciao.org.
11. V. Stampley-Penner interview.
12. hano.org/FAQ.pdf.
13. See also Jay Arena, "The War at Home," *ZNet* (11/12/05).
14. Charles Babington, "Some GOP Legislators Hit Jarring Notes in Addressing Katrina," *Washington Post* (9/10/05), A04.

Seven   Parnell Herbert

1. This chapter is also based on an e-mail from Herbert to Penner, 2/24/08.
2. Campanella, *Geographies of New Orleans*, 17.
3. On Haiti, see Paul Farmer, *The Uses of Haiti*, 3rd ed. (Monroe, ME, 2005).
4. Kate Tuesday, "Chief 'Tootie' Montana Dies of a Heart Attack at a City Council Meeting," (6/25/05) at neworleans.indymedia.org.
5. Joan Treadwell, "Groups Tackle Recent Racial Tensions," *T-P* (6/27/05).
6. For more information, see Casserly, "Double Jeopardy," 204–13.

SECTION TWO: AT THE HEIGHT OF THEIR CAREERS

Eight   Harold Toussaint

1. This chapter is also based on the second interview, 2/3/06, and on PFN, 10/27/07.
2. Sublette, *The World That Made New Orleans*, 37.

3. Greg Simon, "Al Gore Leads Charity Hospital Airlift," at tpmcafe.com.
4. Crowley, "Where Is Home?" 130–31.

Nine   Cynthia Delores Banks

*Cynthia dedicates this chapter to her children.

1. This chapter also relies on PFN, 2/14/06, 4/6/06, 6/4/06, and 6/11/06.
2. Peirce F. Lewis, *New Orleans: The Making of an Urban Landscape* (Santa Fe, 2003), 77.

Ten   Denise Roubion-Johnson

*Denise dedicates this chapter to Peggy McGaughy.

1. This chapter is also based on Johnson-Penner e-mail, 02/4/08 and PFN, 8/25/07 and 2/4/08. It benefits from a Mary Abell-Penner phone interview, PFN, 12/28/07.
2. The Road Home was intended to be the federal government's program for assisting low-income homeowners; Elizabeth Fussell, "Constructing New Orleans, Constructing Race: A Population History of New Orleans," *JAH* 94 (12/07), 846–55.
3. On the homelessness in New Orleans, see Bill Sasser, "Surge in Homeless Hits New Orleans," *The Christian Science Monitor* (3/28/07).

Eleven   Kalamu ya Salaam

1. This chapter also incorporates lines from Kalamu ya Salaam, "what Langston did." *Art for Life: My Story, My Song*, nathanielturner.com; and Salaam, *What Is Life?*, ii, 34, 85; Kalamu ya Salaam, *My Story, My Song* (New Orleans, 1995).
2. Kalamu ya Salaam, *Our Women Keep Our Skies from Falling: Six Essays in Support of the Struggle to Smash Sexism, Develop Women* (New Orleans, 1980).
3. Germany, *New Orleans after the Promises*, 34.
4. Guy Coates, "Tear Gas Used at Disturbance," *T-P* (5/13/69), 5.
5. On McKeithen as an overtly segregationist governor, see Germany, *New Orleans after the Promises*, 49.
6. "Award-Winning Year!," *The Black Collegian* (10–11/81), 15.
7. Eugene Genovese, *Roll Jordan Roll* (London, 1975), 411.

Twelve   Keith C. Ferdinand

1. This chapter also draws on seven taped interviews conducted between 1/4/06 and 7/11/06.
2. Evangeline (Vangy) Franklin, "A New Kind of Medical Disaster in the United States" in *There Is No Such Thing*, 188–90.

Thirteen   Charles W. Duplessis

*This chapter is dedicated to all the family members Thirawer and Charles lost as a result of Hurricane Katrina and its aftermath; it is also a tribute to those family members who continue to soldier on.

1. This chapter also draws from Thirawer Duplessis, "Hurricane Katrina 2005: My Story," (working paper, 2007), a Duplessis-Penner email, 3/20/08, and PFN, 5/15/08.
2. Anne Rochell Konigsmark, "Barge a Contentious Symbol of Hurricane Katrina," *USA Today* (2/17/06) at usatoday.com.
3. Email from Mtumishi St. Julien to Penner, 6/11/08.

Fourteen   Willie Pitford

1. Bill Smith, "Faces of Katrina: The Strong, the Resourceful and the Eternally Hopeful," *St. Louis Post-Dispatch* (9/18/05), B1.
2. This chapter is also based on PFN, 5/29/06, 2/10/08 and the Katrina video footage shot by Joe, Nakia, and Willie, including the exchange with the boatload of people who attempted to "rescue" them later in the week.

Fifteen   Mack Slan Jr.

* This chapter is dedicated to Bernadine and Amy Slan.

1. This chapter also draws on PFN, 10/9/05, 2/15/06, 8/6/07, and 4/19/08.
2. On the frustration experienced by the Casey Foundation in New Orleans, where progress resulting from their jobs initiative lagged, see Whelan, "An Old Economy," 217–18.
3. Gary Rivlin, "Wealthy Blacks Oppose Plans for Their Property," *NYT* (12/10/05).

SECTION THREE: THIRTY SOMETHING

Sixteen   Rochelle Smith

* Rochelle dedicates this chapter to the fishermen of northern Louisiana.

1. This chapter also relies on PFN, 2/8/08 and 3/13/08, and Rochelle D. Smith, "The After Path: A Katrina Story" (working paper, 1/06).
2. For more eyewitness accounts of Katrina's aftermath in and around the convention center, see chapter 17 and ARC, Huey P. Collins-Penner interview, Birmingham, AL, 12/29/05; ARC, Carl Singleton-Penner interview, Birmingham, AL, 12/29/05; ARC, Anika Pugh-Penner interview, Birmingham, AL, 12/15/05; and ARC, Sharmaine Shelton-Penner interview, Houston, TX, 1/05/06.
3. "Martial Law Clarified," *T-P* (8/30/05).
4. Renae Stephens was in charge of the outreach programs for survivors of Katrina and Rita at the Houston branch of the Urban League.
5. femaanswers.org.
6. Whelan, "An Old Economy," 219.

Seventeen   Eleanor Thornton

1. This chapter also relies on PFN, 12/29/05, 1/4/06, 2/4/06, 5/21/06, 6/5/06, and 3/15/08.
2. On health insurance rates among Gulf coast women, see Avis A. Jones-DeWeever and Heidi Hartmann, "Abandoned Before the Storms: The Glaring Disaster of Gender, Race, Class Disparities in the Gulf" in *There Is No Such Thing*, 85–102, here at 85.
3. See also Kim Severson, "Austin Leslie, 71, Dies; Famed for Fried Chicken," *NYT* (9/30/2005).
4. Crowley, "Where Is Home?" 137.

Eighteen   Kevin Owens

1. This chapter also relies on PFN, 12/12/05, 12/14/05, 12/24/05, 4/3/06, 2/10/08, and 3/13/08.
2. On the Superdome, see also ARC, Shriff Hasan-Penner interview, Houston, TX, 10/05.
3. See also Erik Brady and Andy Gardiner, "Upstarting Five Takes Court," *USA Today* (3/15/06).
4. "UAB Guard Finds Blessing in Katrina," *USA Today* (1/16/06) at usatoday.com.
5. Crowley, "Where Is Home?" 128.

Nineteen   Jermol Stinson

1. This chapter also relies on PFN, 1/15/06, 2/11/08, and PFN from a Banks-Penner phone interview, 1/18/08.

Twenty   Demetrius White

*Demetrius dedicates this chapter to the memory of his grandmother Gussie White.

1. This chapter also includes material from PFN, 2/08/08.
2. On the Causeway experience, see also ARC, Rogers Branche-Penner interview, Birmingham, AL, 12/05, and flaherty, "corporate reconstruction and grassroots resistance," 104–5.

Twenty-One   Senta Eastern

*Senta dedicates this chapter to those who were unable to endure the aftermath of the hurricane as well as to those who continue to struggle.

1. This chapter also relies on PFN, 1/9/06, 1/10/06, 1/11/06, 6/1/06, and 4/21/08.

Twenty-Two   Yolanda Seals

*This chapter is dedicated to the memory of Freddie Seals Sr.

1. An email from Yolanda dated 3/10/08 augments this chapter's source base.
2. Crowley, "Where Is Home?" 139.
3. On Atlanta's appeal, see also Larry Copeland, "Katrina Exodus Leads Thousands Home, to Atlanta," *USA TODAY* (4/07) and Jason deParle, "Storm and Crisis Housing: Lack of Section 8 Vouchers for Storm Evacuees Highlights Rift over Housing Program," *NYT* (11/8/05).

SECTION FOUR: COMING OF AGE
Twenty-Four   Leslie Lawrence

1. The chapter also draws on PFN, 5/18/08.

Twenty-Five   Le Ella Lee

1. ARC, Annie Stampley-Penner interview, Fairfield, AL, 12/26/05. See for historical context on her grandfather's Parchman ordeal, see David M. Oshinsky, *"Worse Than Slavery": Parchman Farm and the Ordeal of Jim Crow Justice* (NY, 1996), 237–38.
2. This chapter also relies on PFN, 12/27/05 and 4/2/08.
3. For a flavor of written expressions of hostilities to the New Orleanians in Cullman, see "Listeners' Comments: Sundown Shows," *Uprising* (12/28/05) at uprisingradio.org.
4. V. Stampley-Penner interview.

Twenty-Six   Robert Willis Jr.

1. This chapter also draws on PFN, 2/16/08 and PFN of an untaped interview with Robert Willis Sr. in Gonzales, LA, 2/08/08; See also ARC, Chad Charles-Penner Interview, Memphis, TN, 10/05.

2. See Angela Davis's comment in Gary Younge, "We Used to Think There was a Black Community," *The Guardian* (11/8/07).
3. "3 Gunmen Kill Man in His Home in Gentilly," *T-P* (12/27/02).

Twenty-Seven Toussaint Webster

1. For more on L'Ouverture, see C.L.R. James, *The Black Jacobins: Toussaint L'Ouverture and the San Domingo Revolution*, second ed. (NY, 1989).
2. The continued challenge of public education in New Orleans is well summarized by Casserly, "Double Jeopardy," 197–214.

Conclusion

1. Sublette, *The World That Made New Orleans*, 7 and 59.
2. Colten, *An Unnatural Metropolis*, 20.
3. Ari Kelman, "Boundary Issues: Clarifying New Orleans's Murky Edges," *JAH* 94 (12/07), 695–703.
4. Fussell, "Constructing New Orleans, Constructing Race.
5. Sublette, *The World That Made New Orleans*, 11, 100.
6. Scott, *Degrees of Freedom*, 3–4, 161–62.
7. ARC, Dwayne Chapman-Penner interview, Smyrna, GA, 1/4/06.
8. Eric Klinenberg, *Heat Wave: A Social Autopsy of Disaster in Chicago* (Chicago and London, 2002).
9. For the need to create more positive "social frames" that include law-abiding black men during less traumatic times as well, see Cheryl I. Harris and Devon W. Carbado, "Loot or Find: Fact or Frame?"in *After the Storm*, 102. See also Nancy Scheper-Hughes, "Specificities: Peace-Time Crimes," *Social Identities* 3 (1997), 471–98.
10. PFN, 9/23/05.
11. Seed et al., "Report on the Performance of the New Orleans Levee System."
12. Jeffrey Weiss, "Obama Pastor Jeremiah Wright's Incendiary Quotes Illuminate the Chasm between Races," *Dallas Morning News* (4/8/08).
13. Peter Marcuse, "Rebuilding a Tortured Past or Creating a Model Future: The Limits and Potentials of Planning," in *There Is No Such Thing*, 271–72; Crowley, "Where Is Home?" 141 and 151.
14. McCulley, "Healing Katrina's Racial wounds."
15. Connolly and Roig-Franzia, A14, and Laura Maggi, "Official List of Disaster Victims Still Untabulated," (8/29/07) at blog.nola.com.
16. "Hurricane Katrina Rescue Efforts" at jpso.com.
17. *CNN Reports*, 100.
18. McCulley, "Healing Katrina's Racial Wounds."
19. *CNN Reports*, 95.
20. Kevin U. Stephens et al., "Excess Mortality in the Aftermath of Hurricane Katrina: A Preliminary Report," *Disaster Medicine and Public Health Preparedness*, 1 (2007), 15–20. Columbia University's Earth Institute, "Accounting for Katrina's Dead," at earth.columbia.edu.
21. Jose Calderon-Abbo, "The Long Road Home: Rebuilding Public Inpatient Psychiatric Services in Post-Katrina New Orleans," *Psychiatric Services*, 59, No. 3 (2008), 304–309; David Abramson, Tasha Stehling-Ariza, Richard Garfield, and Irwin Redlener, "Prevalence and Predictors of Mental Health Distress Post-Katrina: Findings from the Gulf Coast Child and Family Health Study, *Disaster Medicine and Public Health Preparedness*, 2 (2008), 77–86.
22. E. Alison Holman et al., "Terrorism, Acute Stress, and Cardiovascular Health: A 3-Year National Study Following the September 11th Attacks," *Archives of General Psychiatry* 65 (01/08), 73–80.

Epilogue: The Future of New Orleans

1. Zdenek et al., "Reclaiming New Orleans," 189.
2. Scott, *Degrees of Freedom*, 24.

3. Sublette, *The World That Made New Orleans,* 135–36, and 225.
4. Scott, *Degrees of Freedom*, 14.
5. ARC, Herbert-Penner Interview, Houston, 1/6/06.
6. Václav Havel, *Disturbing the Peace* (NY, 1990), ch. 5.
7. Scott, *Degrees of Freedom*, 15.
8. Sublette, *The World That Made New Orleans,* 61–62, 97, 104–5, 114, 119–21, 175, 238–39, 274, 282, and 286.
9. Campanella, *Geographies of New Orleans*, 201.
10. Scott, *Degrees of Freedom*, 26.
11. Germany, *New Orleans after the Promises,* 38; Campanella, *Geographies of New Orleans*, 200.
12. US Census SF3 P21.

Index

Note: The Photo pages are identified with a "P".